D0161493

Mechatronics

Dan Necsulescu

University of Ottawa

Prentice Hall
Upper Saddle River, New Jersey

Library of Congress Cataloging-in-Publication Data
Necsulescu, D. S. (Dan S.)
Mechatronics / Dan Necsulescu.-- 1st ed.
p. cm.
Includes bibliographical references and index.
ISBN 0-201-44491-7
1. Mechatronics I. Title.

TJ163.12 .N43 2001
621--dc21 2001036584

Vice President and Editorial Director, ECS: Marcia J. Horton
Acquisitions Editor: Laura Fischer
Vice President and Director of Production and Manufacturing, ESM: David W. Riccardi
Executive Managing Editor: Vince O'Brien
Managing Editor: David A. George
Production Editor: Scott Disanno
Director of Creative Services: Paul Belfanti
Creative Director: Carole Anson
Art Director: Jayne Conte
Art Editor: Greg Dulles
Manufacturing Manager: Trudy Pisciotti
Manufacturing Buyer: Lisa McDowell
Marketing Manager: Holly Stark

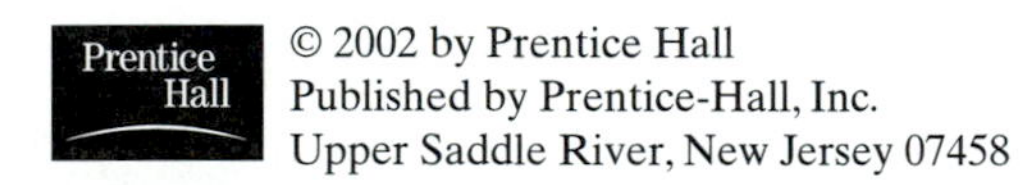

Published by Prentice-Hall, Inc.
Upper Saddle River, New Jersey 07458

Printed in the United States of America

10 9 8 7 6 5 4 3 2

ISBN 0-201-44491-7

Pearson Education Ltd., *London*
Pearson Education Australia Pty. Ltd, *Sydney*
Pearson Education Singapore, Pte. Ltd.
Pearson Education North Asia Ltd., Hong Kong
Pearson Education Canada Inc., *Toronto*
Pearson Educaíon de Mexico, S.A. de C.V.
Pearson Education—Japan, *Tokyo*
Pearson Education Malaysia, Pte. Ltd.

Contents

Preface

FEATURES OF THE BOOK

Mechatronics is an engineering field which, together with intelligent structures, robotics, micro-, and nanoelectromechanical systems, etc., appears as part of "high-tech" fields of mechanical and electrical engineering. Mechatronic systems represent integrated mixed systems. Besides a mechanical structure and mechanisms for motion transmission, such systems contain electrical and electronic components, as well as sensors and actuators, under computer monitoring and control.

Mechatronics is the engineering field of the design of mixed computer integrated electromechanical systems. Modeling mixed systems, virtual prototyping, and hardware in the loop experimentation are some of the methods used in mechatronic system design.

This book was written with a senior undergraduate-level mechatronics course in the mechanical, electrical, or computer engineering departments in mind. The prerequisites for using the book are a basic course in computers and an electrical engineering fundamentals course. A course in control systems is useful, but not a prerequisite. This text provides a self-contained, modern treatment of the computer-based mixed systems integration. The book covers fundamental topics starting with the physical properties, continuing with mathematical modeling and computer simulation and ending with applications illustrated by numerically and experimentally generated results. It explores all major topics in mechatronics and offers a detailed presentation of selected devices as well as numerous MATLAB®, Simulink™, and LabVIEW™ examples of applications. MATLAB™ and Simulink™ are registered trademarks of The MathWorks, Inc. LabVIEW™ is a product of National Instruments™.

This book contains a unified presentation of sensors, actuators, signal conditioning, and control parts (as signal transmission and conversion components) as well as mechanical power transmission, and electrical power transmission parts (as power transmission components). Power amplifiers and servovalves are typical signal-to-power conversion components, while sensors convert variables of power transmission components into signals.

Power transmission components are represented in a generic framework using cuts and two-port elements for both electrical and mechanical components. Two port elements are known as quadripoles, and are used for electric circuit representation. Free body diagrams are used for solid body mechanics calculations. This generic approach is a basis for modular modeling and encapsulation in the transition toward object-oriented modeling. Cuts and two port element representations also facilitate the introduction of virtual prototyping and hardware in the loop experimentation for mechatronic systems.

Simulation examples are developed using a popular graphical programming simulation language, Simulink. Subsystem encapsulation procedure from Simulink is employed to illustrate modular modeling and interchangeability of mechatronic system components.

System monitoring and control examples are presented using another popular graphical programming language, LabVIEW, as well as MATLAB Data Acquisition Toolbox. It is important to mention that LabVIEW and MATLAB/Simulink are available in Student Editions.

The book also presents various solutions for real-time monitoring and control implementation. PC, DSP, target PC, and embedded computers are discussed as alternative solutions for computer integration of mechatronic systems. Examples of applications with a 16-bit microcontroller and a simple 8-bit embedded computer are presented as examples of solutions for controlling DC motors.

Examples of laboratory experiments for mechatronics are described in the last chapter. These examples start with simple experimental set-ups for interfacing readily available sensors and actuators with a PC using LabVIEW. Other examples use MATLAB Data Acquisition Toolbox and Windows Sound Card for sound acquisition. Finally, robotics examples illustrate more complex data acquisition and control using dSPACE development system.

In conclusion, the main features of the book are as follows:

1. Ample coverage of all important topics in mechatronics for teachers and students.
2. Innovative organization and approach: It begins with fundamentals and physical principles of instruments, actuators, devices, hardware, and software; continues with a systematic treatment of signal and power conversion, data acquisition, and analysis topics; and concludes with an examination of mechatronic systems design issues.
3. Teaches basic concepts from numerical examples and computer simulations using student editions of popular engineering software (MATLAB, Simulink, and LabVIEW).
4. Balanced treatment of various mechanical, electrical, and computer engineering topics of mechatronics.
5. Focus on a more detailed presentation of selected devices and methods, rather than an incomplete overview of the large variety of industrial solutions.
6. State-of-the-art approaches in mechatronic systems design and integration:

   —object-oriented approach for modular modeling, virtual prototyping, and hardware-in-the-loop experimentation.

   —virtual instrumentation for system monitoring and automation.

   —real-time and embedded solutions for mechatronic systems integration.

7. Detailed presentation of MATLAB, Simulink, and LabVIEW examples such that students can easily practice and modify these examples.
8. Description of laboratory experiments for a mechatronics course. Experiments are described in sufficient detail such that the teacher can reproduce and adapt them for a particular laboratory equipment available.

9. Numerous homework problems and computer programming exercises to reinforce concepts presented in each chapter. Various numerical examples give structured solutions to be followed by the students when completing homework.

Acknowledgments

In developing the content of this book, I benefited from the research work with my former graduate students: Jean de Carufel, Rahim Jassemi-Zargani, Victor Lonmo, Robert DeAbreu, Mohammad Eghtesad, Kazuo Kiguchi, and Bumsoo Kim, as well as with my current graduate students, Nisan Rowhani and Marcel Ceru. In particular, they contributed in the detailed design and execution of the experimental set-ups: dual-arm robot, mobile robot, piezoelectric experiment, thermocouple experiment, strain gauge experiment, and the interfacing of DC servomotor with a PC.

Also, several graduate students enrolled in my mechatronics course—M. Panait, C. Giurgea, S. E. Zandi, W. Saksawangcan, S. Ahmed, A. Baig, T. Boutros, A. Patel, and T. McCullogh—contributed to the finalization of the design and in the execution of experimental set-ups for interfacing a PC using LabVIEW with a thermistor, piezoelectric actuator and sensor, light emitting diode and photosensor, and DC motor velocity control and Hall sensor measurement. M. Makasare, chief technician in my department, contributed with improvements in the realization of these experimental set-ups.

CHAPTER 1

Computer Integration of Electro-Mechanical Systems

1.1 MIXED SYSTEMS INTEGRATION

Traditional mechanical systems (mechanisms or machines) can be built with mechanical components only. In the case of James Watt's steam engine, steam energy is converted into kinetic energy of the rotational motion of a shaft. The incoming flow rate of steam is controlled by a fly ball governor. This is an example of a mechanical feedback controller for velocity [1].

Electric energy and electric signals were made available for industrial applications in the twentieth century and proved to be more efficient for the conversion into mechanical energy and easier to process as signals for measurement and control. In the last few decades, digital computers (for example personal computers and embedded microcontrollers) replaced most of the analog devices used previously for signal processing. As a result, most systems that provide motion and force (for example vehicles, manipulators, etc.) contain a mixture of mechanical, electrical, electronic, and digital components. In fact, today most systems are mixed systems. The design of these mixed systems requires knowledge from all these fields. To be able to communicate, engineers specialized in any of these fields need working knowledge from other fields. The purpose of mechatronics is to provide knowledge regarding the mechanical, electrical, electronics, and digital components needed for choosing the components of the system; for analyzing the interface requirements for signals, power transmission, and conversion; and for investigating the performance of the integrated system. Mechanical engineers have to deal with issues regarding electrical and computer engineering design, whereas electrical and computer engineers have to understand mechanical structure as well as mechanism machine design issues. Mixed systems design requires a Mechatronics specialization [2-4].

Electrical, mechanical, structural, and electronic components of a mechatronic system operate as an integrated system under real-time computer monitoring and control with interface for operator supervision and direction. Mixed system integration requires energy transmission and conversion (for example electrical in mechanical energy) such that power output is modulated to be in agreement with operator direction and computer control algorithms [3]. Using a terminology in fashion today, this is the case of "intelligent" energy that uses a raw energy source and modulates its output rate for end-use requirements. In the example of a car, the raw energy of the gas stored in the tank is modulated in the carburetor to follow the driver's control of the acceleration pedal. Car velocity is modified by changing the gas flow rate. Given the constant specific energy content of the gas, the change of the flow rate changes the rate of energy transfer from the gas, after combustion in the engine, to the motor shaft. Because it follows the driver's commands, this can be seen as an "intelligent" power output (or, less than intelligent in the case of careless driving).

In mechatronic systems, not only the energy is subject to conversion, but also the signals. The integration of the system requires sensors to convert physical variables in analog or digital signals that carry information. The energy consumption for sensing is irrelevant to the information content transmitted. Analog-to-digital and digital-to-analog conversion of signals is needed for interfacing digital devices (often, computers) with sensors and actuators. Monitoring and control algorithms are executed mostly by digital computers [5]. The commands to actuators, in the form of pulses or analog signals, require a conversion from signal to power form. In the case of a car, the velocity commands of the driver are converted from mental form into a position signal of the acceleration pedal, then in a flow rate command for the gas flow rate, and finally in the modulated power output of the engine. The conversion from signal command to modulated power form uses the control of the flow rate of gas. The specific energy content of the gas being constant, its multiplication with the time-varying flow rate of the gas gives the desired modulated power output.

In mechatronic systems, power transmission, conversion and modulation, signal transmission, digital-to-analog and analog-to-digital conversions as well as signal conditioning and processing are aspects of mixed system integration. Mechatronics issues appear in system design, fabrication, and operation. In the design stage, system modeling, virtual prototyping, and hardware-in-the-loop experimentation are used to achieve system integration. Similarly, modeling and simulation tools might be used in parallel with actual testing for justifying production changes and operations planning. Development of operating strategies involves modeling and simulation as well as testing.

The integration of signal and power transmission issues in mechatronic systems appear also in hardware-in-the-loop experimentation. In particular, signal conversion in power form is a common issue and, in fact, mechatronics provides a theoretical background and practical solutions to hardware-in-the-loop experimentation.

1.2 MECHANICAL STRUCTURE, SENSORS AND ACTUATORS, COMPUTER MONITORING, AND CONTROL

1.2.1 Components

An obvious mechanical part of a mixed system is represented by the mechanical structure, which has to be able to sustain static and slow-varying forces and show an acceptable vibration response. Dynamical parts of the systems refer to transmission and conversion of signals and energy. Actuators transform chemical energy of fuels (e.g., combustion engines) into rotational or translational kinetic energy or, more often today, transform electric energy into rotational kinetic energy (direct current motors, stepping motors, etc.). In most cases, sensors convert (transform) measurement signals into electric signals, which are easy to transmit and process. For example, motion sensors transform velocity and acceleration into voltages, whereas force and torque sensors convert these variables also into voltages.

Electrical part refers to analog signal transmission (wiring) and can be modeled by electric circuits containing resistances, capacitances, and inductances. The electric part of sensors, which give their voltage output, is also modeled by electric circuit components.

Electronic part refers to thyristors, triacs, power transistors, and operational amplifiers, used in analog signal conditioning devices and power amplifiers.

Digital part refers to analog-to-digital converters (denoted ADC or A/D), digital-to-analog converters (denoted DAC or D/A), microprocessors, and other devices associated with computer monitoring and control. Hardware and software used in mixed systems permit computer integration of system monitoring and control and are the main reasons for the development of mechatronics engineering. The level of system integration with digital computers is much higher than with mechanical, electrical, and electronic components only. Until 1970, cars did not require the level of mechatronic system integration needed today. Today, however, the performance of the engine and the maneuverability of the car are highly dependent on the use of digital control.

1.2.2 Computers for System Integration

Products and production systems share most of the concepts of computer integration, but often differ in actual implementation. Most production systems (plants or factories) are fixed and do not have significant weight and volume constraints. This distinguishes production systems from products for which weight and volume constraints are common due to mobility requirements. For production systems, mainframe and minicomputers can be used, but lately more and more powerful PC networks have replaced them. For products, in most cases the integration is achieved with single chip embedded computers. In this book, the focus will be on microprocessor-based computers (i.e., PCs, microcontrollers, and embedded computers) for system integration of production systems as well as products.

PC Configuration Since the 1970s, arithmetic logic units (ALUs), registers, and other digital components were combined in a single-chip microprocessor which, over years, changed the word length from 8-bit to 32-bit. In a PC, besides the microprocessor chip, there are volatile read-only memory (ROM), nonvolatile hard drive memory, and input–output (I/O) devices (keyboard, monitor, disk drives, mouse, etc.). A PC is normally used in a operator-in-the-loop configuration, such that I/O devices provide the interface to the operator (Fig. 1.1).

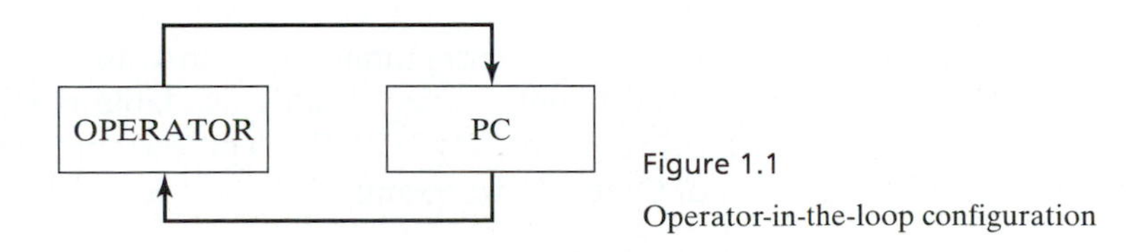

Figure 1.1
Operator-in-the-loop configuration

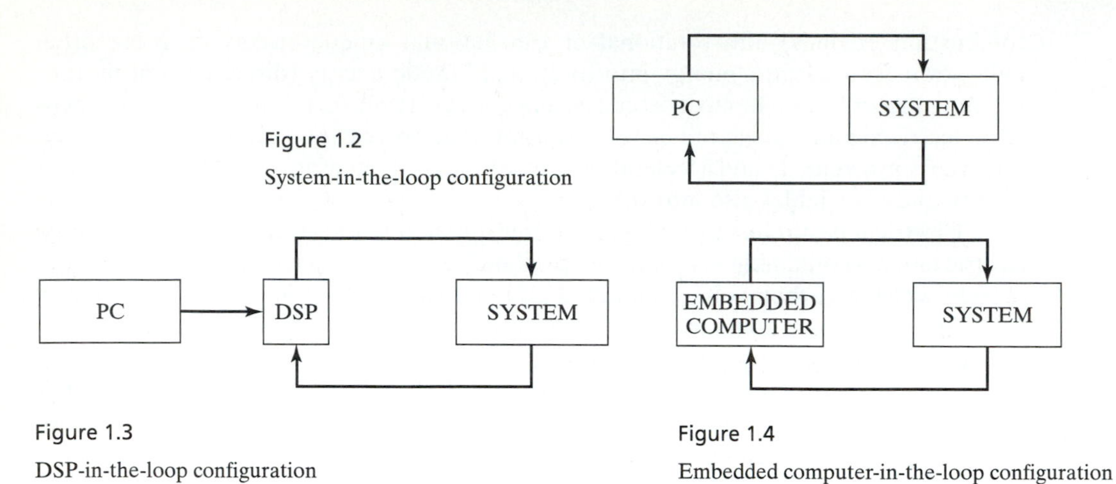

Figure 1.2
System-in-the-loop configuration

Figure 1.3
DSP-in-the-loop configuration

Figure 1.4
Embedded computer-in-the-loop configuration

A PC can be used for computer system integration (system monitoring and/or control) when equipped with devices to transmit measurement signals from sensors to the computer and command signals from the computer to the actuators. This requires analog-to-digital converters (ADCs) and digital-to-analog converters (DACs).

In this case, a system-in-the-loop configuration, shown in Fig. 1.2, is used.

Some real-time control applications require very fast operation that can not be guaranteed, given the interrupt priorities in a PC. This problem can be alleviated by introducing a digital signal processor (DSP) between the PC and the system. The DSP would be dedicated to real-time control computations and permit fast and uninterrupted operation [13] (Fig. 1.3).

Embedded Computers Embedded computers include the microprocessor, memory (ROM, random-access memory (RAM), and I/O devices (parallel port, serial port, digital-to-analog converter, analog-to-digital converter, pulse-width modulation (PWM), etc.) on the same chip. Embedded computers do not normally connect to a keyboard, monitor, or mouse, but to sensors and actuators using a limited number of chip pins. For the same number of word bits, embedded computers tend to appear on the market a few years after the microprocessors. Embedded computers are used in a system-in-the-loop configuration (Fig. 1.4).

Given the absence of operator I/O interface, an embedded computer is programmed differently from a PC (for example, using a specialized development system).

1.2.3 Virtual Instrumentation

PC-based integration of systems benefits from various software packages that often use graphical programming for creating virtual instrumentation. In this case, displays, controls, and wire connections of the physical instrument are replaced by a monitor-based instrument display and controls. Graphical programming for instrument design is based on click-and-drag method on the monitor using readily available components. LabVIEW, a product of National Instruments (NI); HP VEE, a product of Hewlett Packard; and DASYLAB™, a product of Omega™ are examples of such software [12, 14, 15].

The presentation in this book uses LabVIEW, available also in a student edition. All virtual instrumentation packages share a similar conceptual design.

NI Hardware A basic National Instruments data acquisition hardware is the plug-in DAQ board.

A portable data acquisition NI hardware for PC integration of measurement and control using LabVIEW is DAQPad-MIO-16XE-50, available as an external box that communicates with a desktop or laptop PC through the parallel port [10, 12]. For connections to sensors and actuators, the DAQPad has

- 16 single-ended (in case of one wire with common ground return for each sensor) or 8 differential (two wires for each sensor) analog inputs that go to a multiplexor, a 16-bit ADC, a buffer, and a central DAQ-STC, which is a system timing controller
- Eight bidirectional digital I/O lines connected to DAQ-STC
- Two analog outputs from two ADC receiving digital outputs from DAQ-STC

Actual connections to sensors and actuators are made in a detachable Terminal Block DAQPad-TB-52, with screw terminals.

LabVIEW Software LabVIEW software is a program development application using the graphical programming language G. LabVIEW programs are not textual programs (like C or Fortran programs), but block diagram programs. Programming with LabVIEW consists in graphically clicking and dragging given icons of components and wiring them together with a mouse.

1.2.4 Computer Monitoring and Control

Figure 1.5 shows the schematic diagram of computer-based monitoring and control of a process, where the thin lines correspond to signal flow and thick lines correspond to

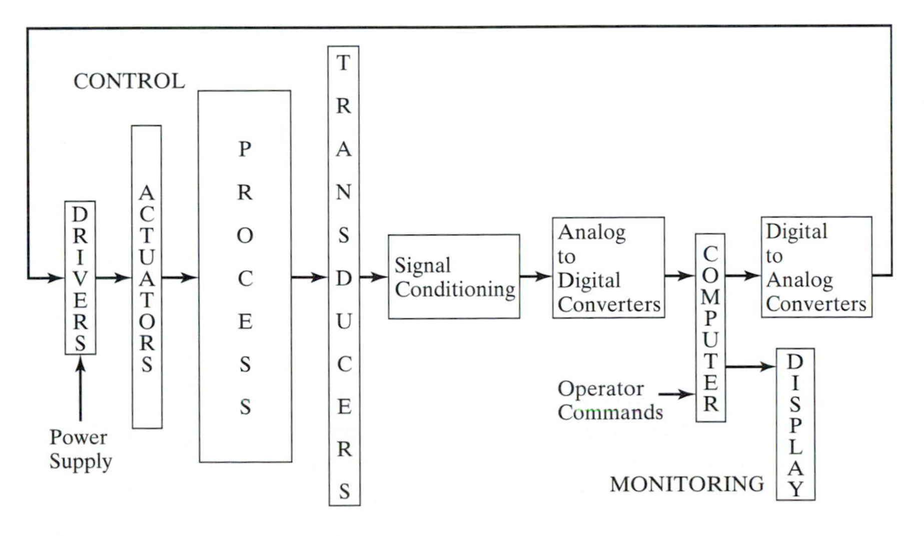

Figure 1.5

Schematic diagram of computer-based monitoring and control of a process power flow. In Fig. 1.5, the process variables, measured by transducers, are signal conditioned and converted from analog to digital form and then transmitted to the computer. Transducers with digital output (for example, optical encoders) can be directly compatible with computer digital input. The computer performs real-time monitoring and control as well as signal analysis and has two types of outputs. One type of output is for actuator commands and the other type of output goes only to the display for system monitoring.

Computer output for control sends commands to actuators. The signals are first converted from digital to analog form. The commands are either operator commands, which can be applied in open loop control configuration, or computed commands (based on processed signals from transducers), which are applied in a closed loop control configuration.

Both open-loop and closed-loop control commands are signals sent to drivers that modulate the power from an external Power Supply for supplying the actuators.

CHAPTER 2

Sensor Modeling

2.1 SENSORS AND TRANSDUCERS

Although there are a large variety of sensors, issues in sensor modeling as well as interfacing with the process and with a computer are, to a large extent, common. In this chapter, modeling and interfacing of temperature sensors, strain gauges, piezoelectric sensors, and position and velocity measuring devices will be presented as examples in order to clarify general issues in modeling and interfacing sensors and in preparation for the signal conditioning and analog-to-digital conversion topics presented in Chapter 4. For detailed presentation of the large variety of sensors and transducers, specialized books can be consulted [1, 3, 7, 16–19].

2.2 TEMPERATURE-SENSING THERMOCOUPLES

2.2.1 History

One of the inventions of Galileo (1564–1642)—the "air thermoscope"—used air expansion as thermometric phenomenon. This invention was enabled by the scientific discovery in Italy of the expansion owing to heat.

Currently, three scales are used:

Fahrenheit (owing to the German–Dutch physicist Gabriel Fahrenheit, 1686–1736, also the inventor of mercury thermometer);

Celsius (owing to the Swedish physicist Anders Celsius, 1701–1744);

Kelvin (owing to the British physicist William Thomson, Lord Kelvin, 1824–1907).

The conversions are

(degrees Celsius) = 5 (degrees Fahrenheit − 32) / 9;

(degrees Fahrenheit) = 9 (degrees Celsius) / 5 + 32; and

(degrees Kelvin) = (degrees Celsius) + 273.15.

Common temperature transducers for industrial measurements are [16–19]

thermocouples;
resistance temperature detectors (RTDs);
thermistors; and
integrated circuit sensors.

The thermocouple is a frequent choice the following presentation will focus on.

2.2.2 Thermoelectric Phenomenon

The physical thermoelectric phenomenon used for thermocouples is the Seebeck effect, which was discovered in 1821 by the German physicist Thomas J. Seebeck (1770–1831). Two dissimilar wires of the junction, shown in Fig. 2.1(a), generate an open-circuit voltage V that changes with the junction temperature T according to the formula

$$V = \alpha T, \text{ zz}$$

where

$$\alpha = \text{Seebeck coefficient.}$$

Figure 2.1(b) shows a thermocouple with metallic sheath, Omega [16].

Thermocouple sensors use various pairs of dissimilar metals:

J Iron/Constantan for less than 760°C (Seebeck coefficient at 20°C) = 51 μV/°C
K Chromel/Alumel for less than 1372°C (Seebeck coefficient at 20°C) = 40 μV/°C

The National Institute of Standards and Technology (NIST) published tables for various types of thermocouple sensors giving precise values for the input temperature T given the thermocouple output voltage V [17].

The typical values of the voltage V are in the millivolts range. (For example, the thermocouple type J at 21°C produces $21 \cdot 51 = 1071$ μV = 1.071 mV.) Ideally, this voltage has to be measured consuming no current (i.e., by a voltage measuring device with very high input impedance—for example, a digital voltmeter).

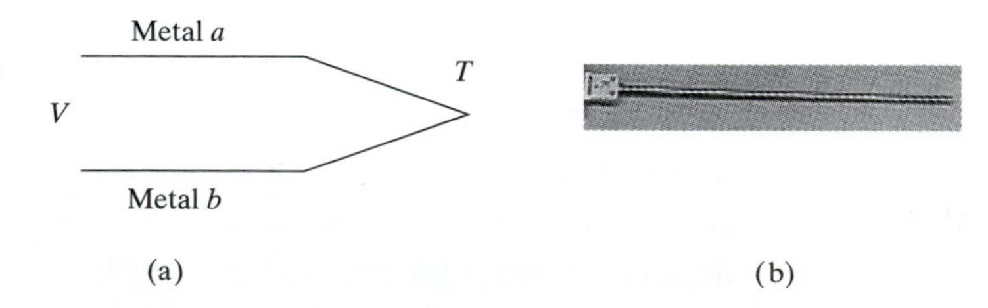

Figure 2.1
Two-metal junction (a) and an Omega thermocouple (b). (*Source*: OMEGA [16].)

2.2.3 Thermocouple Voltage Measurement Using a Potentiometer

A potentiometer-based scheme for thermocouple voltage measurement is shown in Fig. 2.2, where "Cu" denotes copper.

The moving junction ((Cu)–(potentiometer wiper wire)) and the fixed junction ((potentiometer wire)–(Cu)) normally have the same temperature and produce equal value and opposed sign voltages. This permits us to ignore the effect of these junctions on the measurement of thermocouple voltage.

The potentiometer is supplied with DC voltage U. The galvanometer G is used to detect current flow in the wiper. The wiper is moved until G indicates no current flow (when $V_p = V$) and the voltage V value can be read on the potentiometer scale. The input resistance of the potentiometer is assumed extremely high to satisfy the requirement of extremely low current flow in the wiper–thermocouple sensor circuit.

Connecting a voltage-measuring device (in the foregoing case, a potentiometer) to the thermocouple sensor is made normally with copper (Cu) wires, and new junctions result ($J_{Cu/a}$ and $J_{Cu/b}$). The measurement of the voltage produced by the junction J is affected by the voltages produced by junctions $J_{Cu/a}$ and $J_{Cu/b}$; a signal conditioning is needed to eliminate the parasite voltages produced by these new junctions.

Signal conditioning requires the measurement of the temperatures of the two new junctions, $J_{Cu/a}$ and $J_{Cu/b}$. One practical solution for computer-based monitoring is to enclose the two new junctions $J_{Cu/a}$ and $J_{Cu/b}$ into an external reference junction isothermal block that is equipped with a thermistor to measure the common temperature of the junctions (Fig. 2.3).

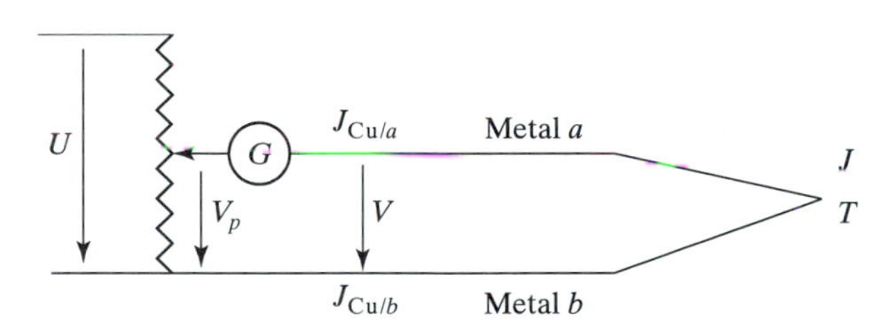

Figure 2.2

Potentiometer-based scheme for thermocouple voltage measurement

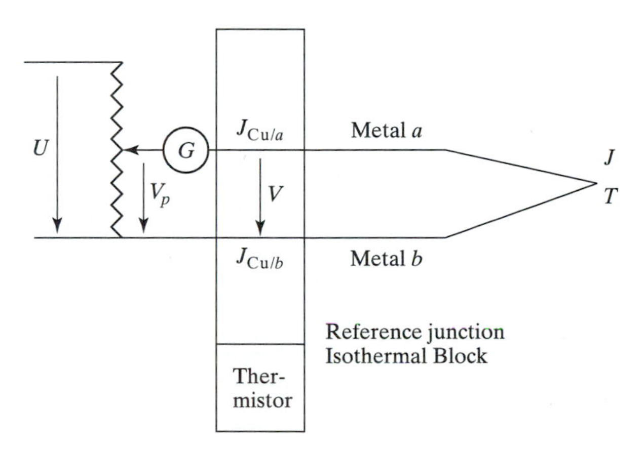

Figure 2.3

Thermocouple measuring device equipped with a reference junction isothermal block

For the known common temperature of the two junctions $J_{Cu/a}$ and $J_{Cu/b}$, the corresponding junction's voltage is determined using NIST conversions of temperature into voltages. Junction J voltage is determined from the measured voltage by subtracting the $J_{Cu/a}$ and $J_{Cu/b}$ voltages.

Example: *J*-Type Thermocouple

A J-type thermocouple (shown in Fig. 2.3) produces a voltage measured by the potentiometer as 9.78 mV. Determine T when the reference junction isothermal block temperature is indicated by the thermistor as 0°C.

Solution: At 0°C, the two junctions $J_{Cu/a}$ and $J_{Cu/b}$ each produce voltages of 0 V. For the (Seebeck coefficient at 20°C) = 51 μV/°C, the temperature at junction J is

$$T = (9780\,\mu\text{V})/(51\,\mu\text{V}/°\text{C}) = 191.76°\text{C}.$$

This temperature differs substantially from 20°C at which the Seebeck coefficient is defined. The value from the NIST table gives 182°C, which shows that there is significant nonlinearity and that the Seebeck coefficient is not a constant, but rather temperature dependent [17]. The NIST table is convenient for applications that are not computer based. For computer conversion of voltage V in temperature T, polynomials $V = p(V)$, with coefficients determined experimentally for each thermocouple type, are more suitable. In computer-based instrumentation, the voltage output of a thermocouple has to be compensated (for the new junctions resulting from the measuring device) and linearized (to account for the temperature-dependent Seebeck coefficient). This is achieved by software- or hardware-based signal conditioning [16].

2.2.4 Software Compensation

A software compensation scheme is shown in Fig. 2.4. It consists of a thermocouple connected to a reference junction isothermal block (shown also in Fig. 2.3). V denotes the analog voltage produced by the thermocouple, V_M denotes the analog output voltage measured by the DAQ board, and v is the voltage output of a temperature trans-

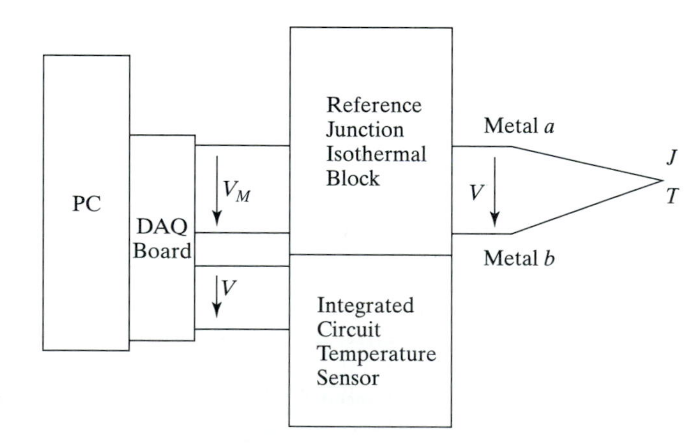

Figure 2.4
Software compensation scheme

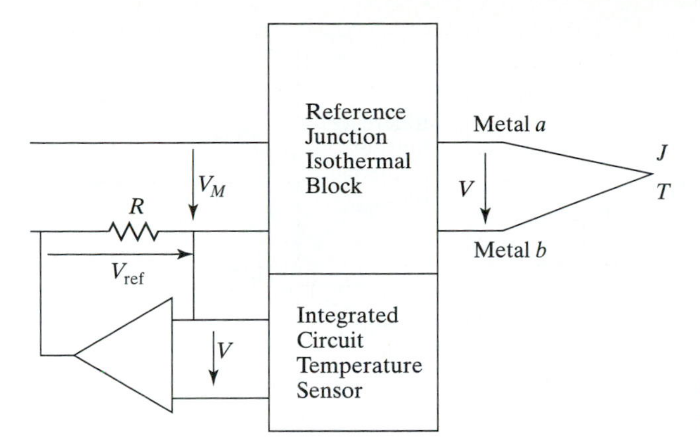

Figure 2.5
Hardware compensation scheme

ducer for the junction block (in this case, an integrated-circuit temperature sensor). The PC has the required data acquisition software and is programmed to calculate the temperature of T_{ref} of the reference junction isothermal block from the voltage v and then to calculate the voltage V_{ref} corresponding to the voltages produced by the junctions enclosed in the reference junction isothermal block. The equation is

$$V_{ref} = \alpha_{ref} T_{ref}.$$

By design, the reference junction isothermal block is maintained at a temperature not far from 20°C and average Seebeck coefficient α_{ref} can be assumed to give accurate enough results. The voltage V_{ref}, corresponding to the junctions enclosed in the reference junction box is then subtracted from V_M to determine the voltage V produced by the junction J. Finally, given the known polynomial $p(V)$, the actual temperature T of the junction J is calculated as

$$T = p(V).$$

2.2.5 Hardware Compensation

An operational amplifier with feedback permits us to use voltage output v of the integrated circuit temperature sensor, or another temperature transducer, to generate a voltage drop V_{ref} on the resistance R such that $V_M - V_{ref} = V$ is obtained (Fig. 2.5). Operational amplifiers are presented in detail in Section 4.2.

Example: PC-Based System for a *K*-Type Thermocouple

A *K*-type thermocouple is connected through a hardware compensation scheme to the PC-based data acquisition system as shown in Fig. 2.6. The DAQ board is plugged into the PC and has analog input terminals to which the thermocouple measurement system is connected. The analog voltage from the thermocouple is converted in digital form by the ADC from the DAQ board. The DAQ board is presented in detail in Section 4.4.

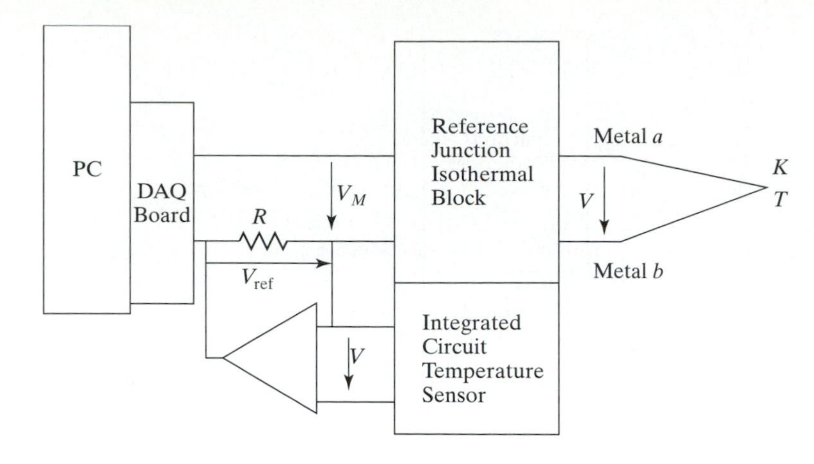

Figure 2.6

DAQ board-based thermocouple measurement system using a hardware compensation scheme

Two limit designs will have to be evaluated:

Design A—a cheap and less accurate design with

a DAQ board with an eight-bit ADC (analog-to-digital converter);

a standard electronic reference junction isothermal block with a temperature uncertainty of ±0.5°C at 0°C [17];

a standard thermocouple with ±2.2°C error limit [16].

Design B—an expensive and more accurate design with

a DAQ board with a 12-bit ADC converter;

a special electronic reference junction isothermal block with a temperature uncertainty of ±0.1°C [17];

a special thermocouple with ±1.1°C error limit [16].

Compare the two designs and

1. determine the scale range for the analog input voltage in ADC;
2. calculate the resolution; and
3. calculate the combined uncertainty (of ADC, reference junction and thermocouple) regarding the temperature measurement.

1. A typical ADC has various ranges for the analog input voltages (for example 0–0.1 V and 0–10 V). For all these ranges, the input voltage is eventually rescaled, with an appropriate amplification gain, to a standard 10-V scale. For a unipolar voltage input in the range of 0–10 V, a gain of 1 is needed (i.e., no rescaling). For a range of 0–0.1 V (0–100 mV) the rescaling gain is 100. When the latter range is chosen, the thermocouple output, with unipolar voltages in the domain 0 to tens of mV, covers most of the 0–100-mV range.

2. For an eight-bit converter, $2^8 = 256$ distinct digital readings are possible (i.e., a resolution of 100 [mV]/256 = 0.39 [mV] or 0.39 [mV] · 100/100 [mV] = 0.39% of the scale range).

For a 12-bit converter, $2^{12} = 4096$ distinct digital readings are possible (i.e., a resolution of 100 [mV]/4096 = 0.0244 [mV] or 0.0244 [mV] · 100/100 [mV] = 0.0244% of the scale range). This is an improvement of $2^4 = 16$ times.

3. The combined uncertainty u of the uncertainties of the ADC uA/D, of the reference junction u_{RJ} and of the thermocouple u_{TC} can be evaluated with a square root of sum of squares (RSS) method defined in this case by

$$u = \pm \sqrt{(u_{\mathrm{A/D}}^2 + u_{\mathrm{RJ}}^2 + u_{\mathrm{TC}}^2)}\,.$$

The RSS method accounts for the effects of both positive and negative values of the component uncertainties and gives more weight to uncertainties having larger magnitudes [17, 30].

For the case of no calibration error, the (resolution)/2 in mV of the A/D converter can be converted into temperature (±) uncertainty $u_{\mathrm{A/D}}$ using the Seebeck coefficient for 20°C for a K-type thermocouple of 40 μV/°C or 0.04 mV/°C, such that

for Design A,

for an eight-bit converter,

$$u_{\mathrm{A/D}} = \pm(0.39\ [\mathrm{mV}]/0.04\ [\mathrm{mV/^\circ C}])/2 = \pm 4.88^\circ\mathrm{C},$$

and the combined uncertainty is

$$u = \pm \sqrt{(u_{\mathrm{A/D}}^2 + u_{\mathrm{RJ}}^2 + u_{\mathrm{TC}}^2)} = \pm \sqrt{(4.88^2 + 0.5^2 + 2.2^2)} = \pm 5.38^\circ\mathrm{C}.$$

for Design B,

for a 12-bit converter,

$$u_{\mathrm{A/D}} = \pm(0.0244\ [\mathrm{mV}]/0.04\ [\mathrm{mV/^\circ C}])/2 = \pm 0.305^\circ\mathrm{C},$$

and the combined uncertainty is

$$u = \pm \sqrt{(u_{\mathrm{A/D}}^2 + u_{\mathrm{RJ}}^2 + u_{\mathrm{TC}}^2)} = \pm \sqrt{(0.305^2 + 0.1^2 + 1.1^2)} = \pm 1.15^\circ\mathrm{C}.$$

The large temperature error ±5.38°C of the Design A is mainly due to $u_{\mathrm{A/D}} = \pm 4.88^\circ$C, while the temperature error ±1.15°C is determined mainly by $u_{\mathrm{TC}} = \pm 1.1^\circ$C error limit of a standard thermocouple.

The reduction of the values of the dominant component uncertainty is the most effective overall uncertainty u. For Design A, an analog-to-digital converter with higher than eight-bit resolution, and for Design B, a thermocouple with lower error limit can reduce significantly the overall uncertainty u.

2.3 STRAIN, STRESS, AND FORCE MEASUREMENT USING STRAIN GAUGES

2.3.1 History

Lord Kelvin noticed in 1856 the physical phenomenon of the change of the resistance of a copper wire due to the change in the strain applied on the wire. This physical phenomenon was later used to make resistive strain gauges.

2.3.2 Gauge Factor Definition

A wire of diameter D, length L, cross section A, and resistivity ρ has the resistance

$$R = \rho L/A,$$

where $A = \pi D^2/4 = kD^2$, for $k = \pi/4$. That is,

$$R = \rho L/(kD^2).$$

When this wire is subject to a strain ε [m/m], L, D, and ρ change. Infinitesimal changes $\mathrm{d}L$, $\mathrm{d}D$, and $\mathrm{d}\rho$ are first considered: The differential of $R(L, D, \rho)$ is

$$\mathrm{d}R = \frac{\partial R}{\partial L}\mathrm{d}L + \frac{\partial R}{\partial D}\mathrm{d}D + \frac{\partial R}{\partial \rho}\mathrm{d}\rho,$$

where

$$\frac{\partial R}{\partial L} = \frac{\rho}{kD^2},$$

and

$$\frac{\partial R}{\partial D} = \frac{-2L\rho}{kD^3},$$

$$\frac{\partial R}{\partial \rho} = \frac{L}{kD^2},$$

so that

$$\mathrm{d}R = \frac{\rho}{kD^2}\mathrm{d}L + \frac{-2L\rho}{kD^3}\mathrm{d}D + \frac{L}{kD^2}\mathrm{d}\rho.$$

Or, using $R = \rho\, L/A$ and dividing by $\mathrm{d}L/L$, we get

$$\frac{\mathrm{d}R/R}{\mathrm{d}L/L} = 1 - 2\frac{\mathrm{d}D/D}{\mathrm{d}L/L} + \frac{\mathrm{d}\rho/\rho}{\mathrm{d}L/L}.$$

Denoting

$$\varepsilon_x = \mathrm{d}L/L \text{ average axial strain [m/m]},$$
$$\varepsilon_y = \mathrm{d}D/D \text{ average lateral strain [m/m]},$$

and

$$\mu = -\varepsilon_y/\varepsilon_x \text{ Poisson ratio},$$

the following result, defining the gauge factor G, is obtained:

$$G = \frac{\mathrm{d}R/R}{\mathrm{d}L/L} = 1 + 2\mu + \frac{\mathrm{d}\rho/\rho}{\varepsilon_x}.$$

For finite changes ΔR,

$$G = \frac{\Delta R/R}{\varepsilon_x}.$$

This is a linear relationship of relative change of resistance $\delta = \Delta R/R$, due to an applied average strain $\varepsilon = \varepsilon_x$. Another form, after dropping the subscript x, is

$$\varepsilon = (1/G)\,\delta.$$

The resistance of the strain gauge changes from R to $R + \Delta R$, as a result of the strain ε, yielding

$$R + \Delta R = (1 + G\varepsilon)\,R.$$

For a Poisson ratio $\mu = 0.5$ and an insignificantly low $d\rho/\rho \cong 0$,

$$G \approx 1 + 2 \cdot 0.5 = 2.$$

For a nominal resistance of 350 Ω and a small strain of 1 μm/m $= 10^{-6}$ m/m,

$$\Delta R = GR\varepsilon = 2 \cdot 350 \cdot 10^{-6} = 700 \cdot 10^{-6}\ \Omega,$$

a very small resistance change, which, without signal conditioning, cannot be directly measured by an ohmmeter or transmitted accurately to a computer-based measuring system. Strain gauges are normally connected to a voltage-measuring instrument (voltmeter or computer-based data acquisition system) using a Wheatstone bridge [1, 6, 17].

2.3.3 Wheatstone Bridge

Figure 2.7 shows a resistive strain gauge, OMEGA [16] (a), and a Wheatstone bridge (b) used to measure strain gauge resistance variation.

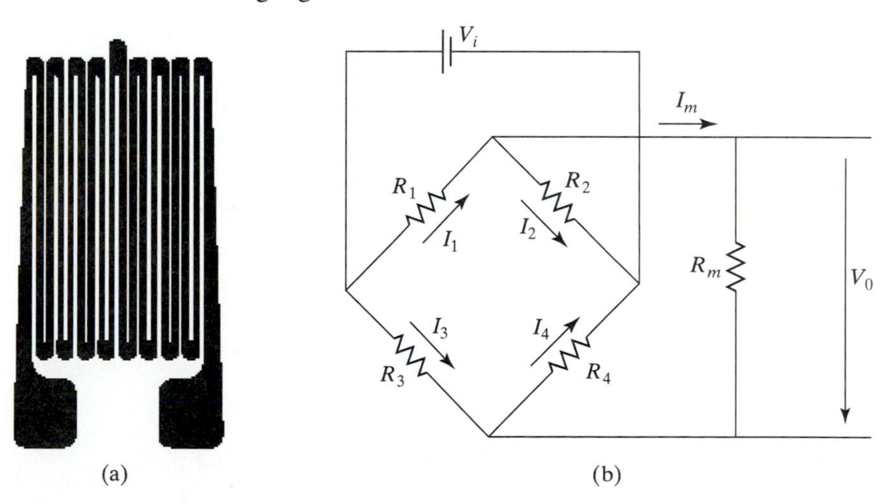

Figure 2.7

OMEGA strain gauge (*Source*: OMEGA [16]) (a) and a Wheatstone bridge (b)

This is a resistance-only bridge, which converts a relative change of resistance $\delta = \Delta R/R$ into a proportional voltage output V_o.

Resistive strain gauges usually have resistances of 120 to 700 Ω, and the bridge resistances are chosen in the same range [21]. The measuring device—a voltmeter or an ADC and a PC—has an internal resistance R_m that influences I_m and V_o. It is, however, undesirable to have the measurement voltage V_o dependent on the measuring device resistance R_m. In practice, the input resistance of the measuring device R_m is chosen larger than 1 MΩ (i.e., many orders of magnitude higher than the output impedance of the strain gauge and bridge). Such a large input impedance can be found at digital multimeters and ADC for PC-based data acquisition systems, but not at analog multimeters (which typically have 30 kΩ). As a result of this choice (for an output impedance of multimeter $R_m \geq 1$ MΩ and for a bridge of resistances of hundreds of ohms), the preceding bridge circuit can be approximated by an open circuit with $I_m = 0$ such that

$$I_1 = I_2$$

and

$$I_3 = I_4.$$

Balanced Bridge Condition If the resistances of the bridge are chosen such that $V_o = 0$, we obtain

$$I_1R_1 - I_3R_3 = 0$$

and

$$I_2R_2 - I_4R_4 = 0,$$

or

$$\frac{I_1R_1}{I_2R_2} = \frac{I_3R_3}{I_4R_4}.$$

Using this result for infinite input impedance R_m of the measuring device, when $I_m = 0$, and for

$$I_1 = I_2$$

and

$$I_3 = I_4,$$

the balanced bridge condition becomes

$$\frac{R_1}{R_2} = \frac{R_3}{R_4}.$$

When the bridge is used for a strain gauge, as the strain is applied, the strain gauge resistance changes and the bridge becomes unbalanced. Two methods can be used to evaluate this change in strain gauge resistance: the null method and the deflection method.

Null Method Assume that R_1 has the value of the nominal resistance of the strain gauge (i.e., unstressed) denoted R (Fig. 2.8). The unstressed strain gauge will give

$$R_1 = R,$$

and in this case the bridge is balanced. As a result of a strain ε, the resistance of the strain gauge becomes

$$R + \Delta R = (1 + G\varepsilon)R.$$

The null method uses the bridge balance condition

$$\frac{R_1}{R_2} = \frac{R_3}{R_4}$$

and requires us to rebalance the bridge by achieving

$$R_2 = R + \Delta R.$$

For this purpose, a galvanometer G can be used to verify that $I_m \cong 0$ as a result of using a potentiometer to modify R_2 until it becomes equal to $R + \Delta R$. This is a manual method presented only for its simplicity. A method suitable for time-varying strain measurement using a computer-based data acquisition system is the deflection method.

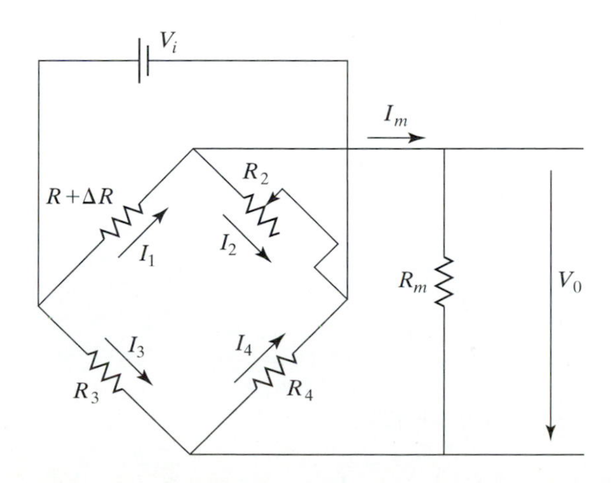

Figure 2.8

Null method for Wheatstone bridge

Deflection Method In the unbalanced case

$$R_1 = R + \Delta R,$$

and for

$$R_2 = R,$$

V_o is not 0 and can be calculated with

$$V_o = I_1 R_1 - I_3 R_3.$$

For a very large value of the input resistance R_m of the measuring device $I_m \cong 0$, such that $I_1 = I_2$ and $I_3 = I_4$, the following equations result:

$$I_1 = V_i/(R_1 + R_2)$$

and for

$$I_3 = V_i/(R_3 + R_4).$$

For an unbalanced bridge, eliminating the currents I_1 and I_3 from the preceding three equations yields

$$V_o = V_i\left(\frac{R_1}{R_1 + R_2} - \frac{R_3}{R_3 + R_4}\right).$$

Using $R_1 = R + \Delta R$ and $R_2 = R$, this equation becomes

$$V_o = V_i\left(\frac{R + \Delta R}{R + \Delta R + R} - \frac{R_3}{R_3 + R_4}\right).$$

For the case of all bridge resistances chosen equal to R, the nominal strain gauge resistance

$$V_o = V_i \frac{R\Delta R}{4R^2 + 2R\Delta R} = V_i \frac{\delta}{4 + 2\delta},$$

where $\delta = \Delta R/R$. For the frequent case of a very small value of δ

$$V_o = V_i \frac{\delta}{4},$$

or

$$\delta = 4V_o/V_i.$$

Given the value of the constant bridge supply voltage V_i, the deflection method requires the measurement of the output voltage V_o with a measuring device having a very high input resistance R_m and the calculation of

$$\delta = \Delta R/R = 4V_o/V_i$$

or

$$\Delta R = (4R/V_i)V_o.$$

This method is suitable for data acquisition systems for the frequent case in which $\delta = \Delta R/R$ has a very small value.

Example: Deflection Method for a Strain Gauge

A strain gauge with nominal resistance $R_1 = R = 600\ \Omega$ is installed in a branch of a Wheatstone bridge having for unstrained strain gauge $R_1 = R_2 = R_3 = R_4 = R$ and $V_i = 10$ Fig. 2.9). The strain gauge is subject to a strain as a result of bending the beam on which it is cemented. A digital voltmeter with input resistance $R_m = 10\ \text{M}\Omega$ gives a reading of $V_o = 5$ mV$= 5 \times 10^{-3}$ V. Calculate

(a) the change of the resistance ΔR and
(b) the strain ε for gauge factor $G = 2$.

***Solution*:**

(a) The change in resistance is given by

$$\Delta R = 4\ V_o R/V_i = 4 \cdot 5 \times 10^{-3} \cdot 600/10 = 1.2\ \Omega.$$

(b) The strain is given by

$$\varepsilon = (1/G)\ \delta = (1/G)\Delta R/R = (1/2)\,1.2/600 = 0.001\ [\text{m/m}].$$

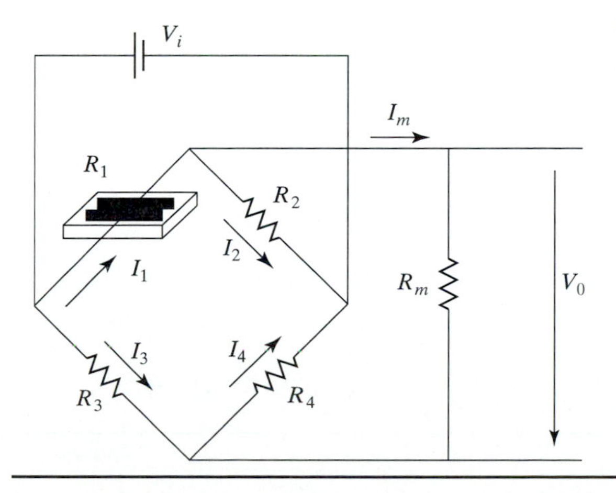

Figure 2.9
Example for the deflection method

Example: Calculation of Strain Gauge Resistance Change

A resistive strain gauge, $G = 2.1$, is cemented on a rectangular steel bar, 4 cm wide and 1 cm thick, with Young modulus $E = 200 \cdot 10^6$ kN/m^2. An axial force $F = 30$ kN is applied. Determine the change of the resistance of the strain gauge, ΔR, if the resistance of the unstressed strain gauge is $R = 350\ \Omega$.

Solution: The cross-sectional area is

$$A = 4 \cdot 1[\text{cm}^2] = 4 \cdot 10^{-4}[\text{m}^2].$$

The stress σ [kN/m^2] is

$$\sigma = \frac{F}{A} = \frac{30}{4 \cdot 10^{-4}} = 75 \cdot 10^3,$$

and the resulting strain is

$$\varepsilon = \frac{\sigma}{\mathrm{E}} = \frac{75 \cdot 10^3}{200 \cdot 10^6} = 0.375 \cdot 10^{-3}\,\text{m/m}.$$

The change in resistance is

$$\Delta R = R\varepsilon G = (350)(0.375 \cdot 10^{-3})(2.1) = 0.276\,\Omega.$$

Example: Measurement Uncertainty

The strain gauge of the previous example is connected to a resistance-measuring device having an accuracy of $\pm 0.1\ \Omega$. What is the uncertainty in determining the stress σ?

Solution: Given that $\Delta R/R = G\varepsilon$ and

$$\sigma = \varepsilon E = \frac{E}{GR}\Delta R,$$

the partial derivative of σ with regard to ΔR gives

$$\frac{\partial \sigma}{\partial(\Delta R)} = \frac{E}{GR}.$$

The uncertainty of σ, U_σ, can be determined as function of the uncertainty of ΔR, $U_{\Delta R}$, as follows [17]:

$$U_\sigma = \frac{E}{GR} U_{\Delta R} = \frac{200 \cdot 10^6}{2.1 \cdot 350} 0.1 = 27.2 \cdot 10^3 \text{ kN/m}^2.$$

2.3.4 Signal Conditioning for a Strain Gauge Installed in a Wheatstone Bridge

The voltage output of the sensor composed of a strain gauge installed in a Wheatstone bridge is compatible with the A/D converter if, for the range of the strain measured, the A/D converter receives voltages covering a full scale. If the A/D converter has an input range of 0–10 V and the sensor output is low (in the previous example, 5 mV), the sensor output voltage has to be linearly amplified for accurate digital conversion.

A solution, shown in Fig. 2.10, uses an operational amplifier [2, 6]. Various operational amplifier schemes are presented in Section 4.2. In the current section, only a specific use of an operational amplifier—an inverting amplifier for signal conditioning—is described.

Operational amplifiers have very high input impedance, such that the currents I_N and I_P are negligible:

$$I_N = I_P \cong 0.$$

Ohm's law gives

$$I_1 = (V_i - V_N)/R_1,$$

$$I_2 = (V_N - 0)/R_2,$$

$$I_3 = (V_i - V_P)/R_3,$$

and

$$I_4 = (V_P - 0)/R_4.$$

Taking into account that $I_N = I_P \cong 0$, Ohm's law also gives

$$I_C = V_P/R_o$$

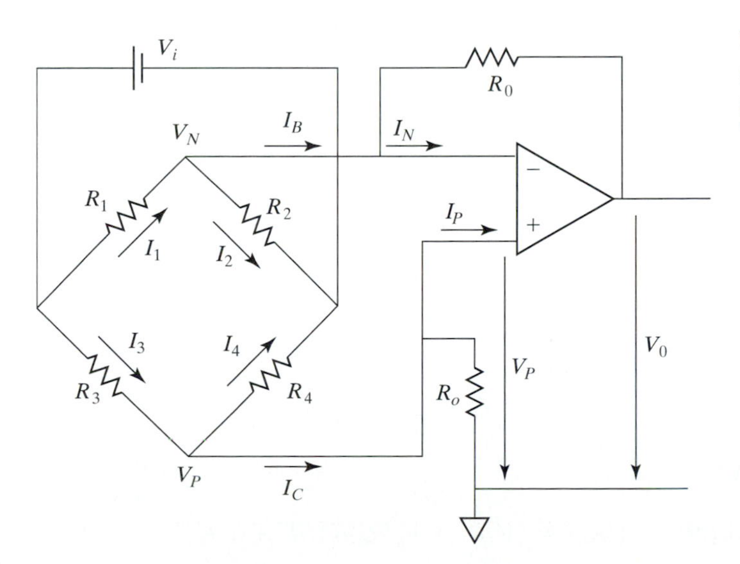

Figure 2.10

Signal conditioning for a strain gauge installed in a Wheatstone bridge

and

$$I_B = (V_N - V_o)/R_o.$$

The first Kirchhoff law gives

$$I_1 = I_2 + I_B \tag{a}$$

and

$$I_3 = I_4 + I_C. \tag{b}$$

An operational amplifier with gain K is characterized by

$$V_o = K(V_P - V_N). \tag{c}$$

These 11 equations contain 12 variables: $I_1, I_2, I_3, I_4, I_B, I_C, I_N, I_P, V_P, V_N, V_i$, and V_o. Using 10 equations to eliminate 10 variables—I_1, I_2, I_3, I_4, I_B, I_C, I_N, I_P, V_P, and V_N—a single equation with variables V_i and V_o of interest remains. First, eliminating all currents, the first two equations become

$$(V_i - V_N)/R_1 = (V_N - 0)/R_2 + (V_N - V_o)/R_o \tag{d}$$

and

$$(V_i - V_P)/R_3 = (V_P - 0)/R_4 + V_P/R_o. \tag{e}$$

Assuming that all bridge resistances are chosen equal to the nominal resistance R of the strain gauge, that is,

$$R_2 = R_3 = R_4 = R,$$

we obtain

$$V_i/R_1 + V_o/R_o = (1/R_1 + 1/R + 1/R_o)V_N \tag{f}$$

and

$$V_i/R = (2/R + 1/R_o)\,V_P. \tag{g}$$

The equation $V_o / K = V_P - V$ for a very large K gives

$$V_P \cong V_N. \tag{h}$$

The equations give

$$(V_i/R)/(2/R + 1/R_o) = (V_i/R_1 + V_o/R_o)/(1/R_1 + 1/R + 1/R_o),$$

or

$$V_i\left(\frac{1/R}{2/R + 1/R_o} - \frac{1/R_1}{1/R_1 + 1/R + 1/R_o}\right) = V_o\frac{1/R_o}{1/R_1 + 1/R + 1/R_o},$$

from which it follows that

$$V_0 = V_i R_0 \frac{(R_0 + R)(R_1 - R)}{RR_1(2R_0 + R)}.$$

For $R_0 \gg R$ and $R_1 = R + \Delta R$ (i.e., for an unbalanced bridge),

$$V_0 = V_i R_0 \frac{R_0 \Delta R}{R(R + \Delta R)2R_0}.$$

For $\Delta R = R\delta$,

$$V_0 = V_i R_0 \frac{\delta}{2R(1 + \delta)}.$$

In the frequent case that $\delta \ll 1$,

$$V_0 = \frac{V_i R_0 \delta}{2R}.$$

A strain gauge subject to a strain ε gives

$$\delta = G\varepsilon,$$

such that

$$V_0 = \frac{V_i R_0 G}{2R}\varepsilon.$$

Consequently, for a bridge voltage output V_o, the corresponding strain gauge ε can be calculated as

$$\varepsilon = \frac{2R}{V_i R_0 G} V_0.$$

For a given nominal resistance R of the strain gauge and bridge and constant supply voltage V_i, the output voltage V_o can be modified as desired by the proper choice of the operational amplifier resistance R_o.

The operational amplifier is also useful to buffer the A/D converter (which can be connected at the right-hand side of the operational amplifier in the Fig. 2.10) from high voltages and currents on the sensor side.

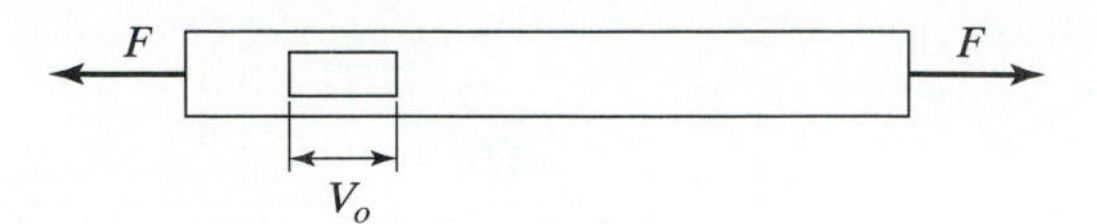

Figure 2.11

Axial force measurement with a strain gauge

2.3.5 Axial Force Transducer

A strain gauge can be installed on a bar to measure the applied axial force, as shown in Fig. 2.11. In case of a slow-varying $F(t)$, inertial and damping effects are negligible. Young's modulus E is used to define the dependence between the strain σ and the stress ε as

$$\sigma = E\varepsilon$$

where, for the area A,

$$\sigma = F/A.$$

Given that, for a strain gauge,

$$\varepsilon = \frac{2R}{V_i R_0 G} V_0,$$

it follows that

$$\varepsilon = \frac{2R}{V_i R_0 G} V_0 = \frac{\sigma}{E} = \frac{F}{EA}$$

or

$$F = \frac{2RAE}{V_i R_0 G} V_0.$$

Example: Axial Force Measurement

A steel bar with Young elastic modulus $E = 200 \cdot 10^6\,\text{kN/m}^2$ and a cross-sectional area $A = 4\ \text{cm}^2$ is subject to an axial force F. A strain gauge with voltage output V_o is used for measuring the force F, as shown in Fig. 2.11. The strain gauge, with a nominal resistance $R = 350\ \Omega$, is connected in a branch of bridge and with the inverting amplifier as shown in Fig. 2.10. The output resistance of the operational amplifier is $R_o = 1\ \text{M}\Omega$. All branches of the bridge except for the branch containing the strain gauge, have fixed resistances of $R = 350\ \Omega$. The strain gauge factor is $G = 2.1$ and the voltage $V_i = 10\ V$

Calculate the force F given a measured voltage $V_o = 5\ V$.

Solution: Using the last equation, the force is

$$F = \frac{(2)(350)(4 \cdot 10^{-4})(200 \cdot 10^9)}{(10)(10^6)(2.1)}(5) = 13{,}333\ \text{N}.$$

2.3.6 Strain Gauge-Based Accelerometer

A strain gauge mounted on a cantilever beam can be used for acceleration measurement. (See Fig. 2.12.) The beam is assumed under initial steady-state deflection due to the weight Mg, which produces a strain on the strain gauge assumed here compensated to zero (or the beam can be assumed used in space in zero gravity environment). When the accelerometer is subject to an acceleration a, an inertia force Ma results on the beam and produces an internal moment m at the point where the gauge is cemented. The moment m due to the inertia force Ma is

$$Mal = m.$$

Given this moment m, the corresponding stress σ at gauge location is $m(t/2)/I$. For the beam moment of inertia of the cross section $I = (1/12)wt^3$, the axial stress is [1, 6]

$$\sigma = \frac{Mal(t/2)}{(1/12)wt^3} = \frac{6Mal}{wt^2},$$

where w is the width of the beam.

As the result of the transversal static deflection of the beam due to the constant acceleration a applied on the mass M, the strain gauge will be subject to strain

$$\varepsilon = \sigma/E = (6\,Ml/Ewt^2)\,a.$$

When an inverting operational amplifier is connected to the bridge, its output voltage is

$$V_0 = \frac{V_i R_0 G}{2R}\varepsilon$$

or

$$V_0 = \frac{6GMlV_i R_0}{2REwt^2}a.$$

Consequently, the acceleration can be calculated as

$$a = \frac{REwt^2}{3GMlV_i R_o}V_0.$$

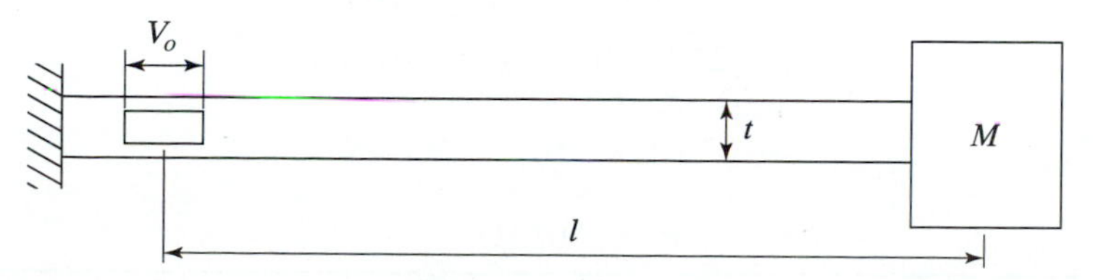

Figure 2.12
Strain gauge-based accelerometer

Example: Accelerometer Based on a Strain Gauge

An accelerometer based on a strain gauge consists in a cantilever beam of length $l = 20$ mm, width $w = 2$ mm, and thickness $t = 1$ mm, fitted with a (seismic) mass $M = 0.01$ kg. (See Fig. 2.12.) The modulus of elasticity of the beam is $E = 200 \cdot 10^9\,\mathrm{N/m^2}$. The strain gauge, cemented is at $l = 20$ mm from the free end of the beam, has $G = 2.1$ and is connected to a bridge. The bridge is interfaced with an ADC through an inverting amplifier with $R_o = 1\,\mathrm{M\Omega}$. Assume a nominal resistance of the strain gauge of $R = 350\,\Omega$ and the supply voltage of the bridge of $V_i = 10$ V.

Calculate the acceleration a that produces a voltage output $V_o = 0.1$ V.

Solution: For the given data, the acceleration is given by

$$a = \frac{(350)(200 \cdot 10^9)(0.002)(0.001)^2}{3(2.1)(0.01)(0.02)(10)(10^6)} 0.1 = 1.11 \cdot [\mathrm{m/s^2}].$$

As a fraction of gravitational acceleration $g = 9.81\,\mathrm{m/s^2}$, this represents

$$a/g = 1.11/9.81 = 0.1131$$

(i.e., 11.31% of g). Assuming that measurements up to twenty times higher than this value are expected, the voltage V_o will increase proportionally twenty times to $(0.1\,\mathrm{V})(20) = 2$ V and the range for input voltage to ADC has to be 0 to 2 V, if available as software selectable (as, for example, in the case of the DAQPad–MIO-16XE-50 of National Instruments).

Example: Design Issues in Signal Conditioning

In the previous example, the resistance R_o of the inverting amplifier had the value of $1\,\Omega$, and the voltage output from this amplifier to ADC was $V_o = 0.1$ V. In case the ADC has only the range 0 to 10 V available, calculate a value for R_o that will give the midscale 5 V for V_o, for the previous acceleration value of $1.11\,\mathrm{m/s^2}$, when all other factors being kept the same.

Solution: Given

$$a = \frac{REwt^2}{3GMlV_iR_o} V_0,$$

for the new value $V_o = 5\,\mathrm{V}$, the resistance R_o will be given by

$$1.11 = \frac{(350)(200 \cdot 10^9)(0.002)(0.001)^2}{3(2.1)(0.01)(0.02)(10)(R_0)} 5,$$

or

$$R_o = 50\,\mathrm{M\Omega}.$$

2.4 PIEZOELECTRIC STRAIN SENSORS AND ACCELEROMETERS

2.4.1 Piezoelectric Materials

The piezoelectric effect consists of generating an electric charge when a material is subject to a mechanical deformation or in producing a mechanical deformation when it is subject to an electric charge. Some natural and artificial crystals as well as polarized ceramics (piezoceramics) can have the piezoelectric effect. The piezoelectric effect, in piezoceramic materials, can be induced by applying a high electric field while the material is heated above a specific temperature called Curie temperature.

These strain transducers can also be used to measure indirectly force, torque, pressure, velocity, or acceleration.

The main parameters of a piezoceramic sensor are as follows:

Curie temperature [°C];
dielectric constant c [F/m];
Young or elastic modulus E [N/m^2] or compliance, s [m^2/N];
piezoelectric coefficients: piezoelectric charge coefficient, d_{ij}, [C/N], and piezoelectric voltage coefficient, g_{ij}, [Vm/N].

The first subscript (i) indicates the direction perpendicular to the electrodes, and the second subscript (j) indicates the direction of the applied stress or deformation.

For a sensor, the two piezoelectric coefficients are defined as

$$d_{ij} = \frac{\text{charge density produced in direction } i \text{ [C/m}^2\text{]}}{\text{mechanical stress applied in direction } j \text{ [N/m}^2\text{]}}$$

and

$$g_{ij} = \frac{\text{electric field produced in direction } i \text{ [V/m]}}{\text{mechanical stress applied in direction } j \text{ [N/m}^2\text{]}}.$$

These two coefficients have a linear dependence given by [19]

$$g_{ij} = \frac{d_{ij}}{c}.$$

These parameters are listed in manufacturer catalogues, for example [22].

In Fig. 2.13 are shown the top view (a) of a piezoelectric sensor and the diagram of the longitudinal cross section (b) of the piezoelectric sensor illustrating the piezoelectric effect. The terminals of the electrodes of the piezosensor shown in Fig. 2.13 (b) are accessible on the same side such that the other side can be properly glued to the flexible component for strain measurement.

Assuming that the piezoceramic sensor shown in Fig. 2.13 has the of length L, width W, and thickness T [m], its capacitance [F] can be calculated as

$$C = cWL/T.$$

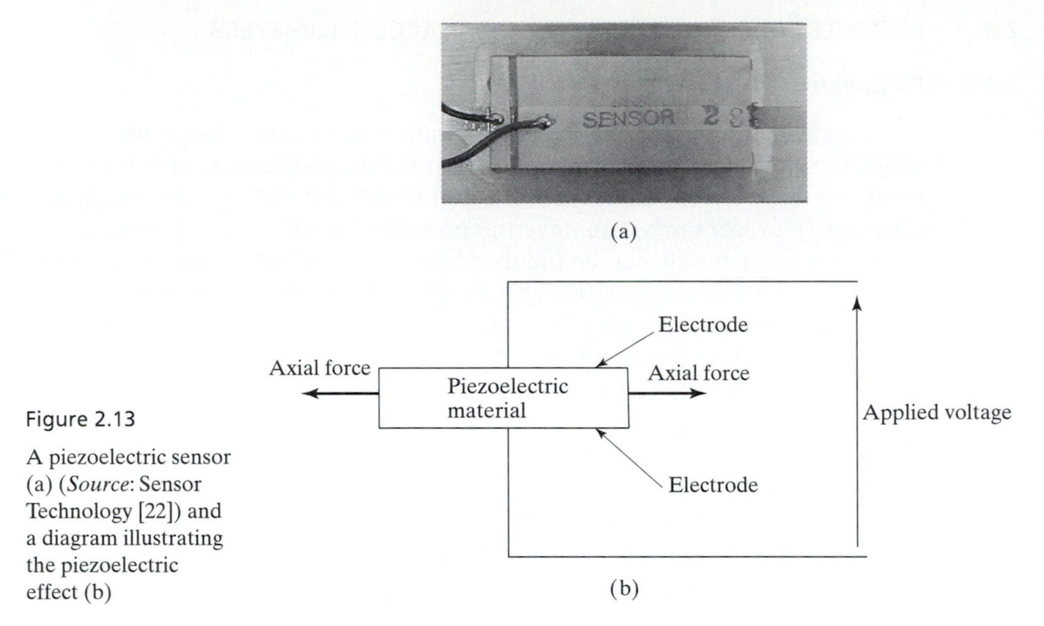

Figure 2.13

A piezoelectric sensor (a) (*Source*: Sensor Technology [22]) and a diagram illustrating the piezoelectric effect (b)

The charge $Q[C]$ of an electrically unloaded sensor of capacitance C produces the voltage V [V] across the thickness T given by

$$V = Q/C.$$

Besides the capacitance, the sensor has a leakage resistance that is of a large value and slowly discharges Q and reduces V. The measurement of the voltage V requires a measuring device with input impedance higher than the leakage resistance of the sensor.

The signal from the piezoceramic sensor can be measured with a high input impedance device such that a digital voltmeter or with a PC-based virtual instrument equipped with an I/O board for sensor connection. If a signal conditioning operational amplifier is required (e.g., a charge amplifier to convert charge Q input into a voltage V output), it also has to have high input impedance.

2.4.2 Piezoelectric Strain Sensors

The piezoceramic sensor shown in Fig. 2.14 is subject to an axial force F (along an axis denoted 1), which produces an axial deflection from L to $L + \Delta L$, or the axial strain $\Delta L/L$.

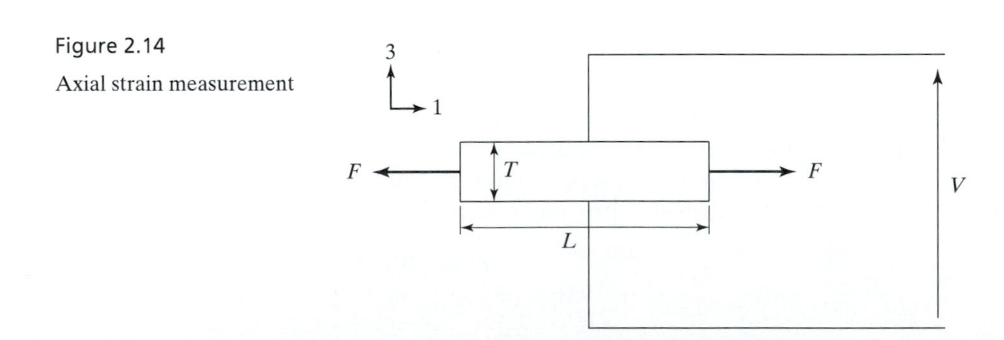

Figure 2.14

Axial strain measurement

Across the thickness T (i.e., along the axis 3) the charge Q produces a voltage $V = Q/C$ for the sensor of capacitance C. Assuming a uniform electric field between the electrodes, its value is given by V/T. The piezoelectric coefficient in this case is defined as

$$g_{31} = \frac{V/T}{\sigma_1} = \frac{\text{electric field produced in direction 3 [V/m]}}{\text{mechanical stress applied in direction 1 [N/m}^2\text{]}}.$$

Taking into account the stress (σ_1)–strain ($\Delta L/L$) relationship

$$\sigma_1 = E\Delta L/L,$$

the following voltage (V)–strain ($\Delta L/L$) relationship results:

$$V = g_{31}ET\Delta L/L.$$

Example: Strain Measurement

Assume the piezoceramic sensor BM400 Type I dimensions 0.5″ · 1.5″ · 0.020″ (12.7 mm · 38.1 mm · 0.5 mm) [22]. The catalog parameters are

$$\begin{aligned} g_{31} &= -10.5 \cdot 10^{-3}\,[\text{Vm/N}]; \\ \text{compliance} &= 12.5 \cdot 10^{-12}\,[\text{m}^2\text{/N}]; \\ E = 1/(\text{compliance}) &= 8 \cdot 10^{10}\,[\text{N/m}^2]; \\ T &= 0.5 \cdot 10^{-3}\,[\text{m}]; \\ L &= 38.1 \cdot 10^{-3}\,[\text{m}]; \end{aligned}$$

and

$$\begin{aligned} c = {} & (\text{relative dielectric constant})(\text{absolute dielectric constant of vacuum}) = \\ & (1350)(8.854 \cdot 10^{-12})[\text{ F/m}] = 11.95 \cdot 10^{-9}\,[\text{F/m}] \end{aligned}$$

Calculate

(a) the capacitance C
(b) the piezoelectric coefficient d_{31}
(c) the voltage V when the sensor is subject to a 10^{-6} [m] axial deflection.

Solution: **(a)** The capacitance of this sensor is

$$C = cWL/T = (11.95 \cdot 10^{-9})(12.7 \cdot 10^{-3})(38.1 \cdot 10^{-3})/(0.5 \cdot 10^{-3}) = 11.57 \cdot 10^{-9}\,[\text{F}].$$

(b) The piezoelectric coefficient is

$$d_{31} = cg_{31} = (11.95 \cdot 10^{-9})(-10.5 \cdot 10^{-3}) = -125.5 \cdot 10^{-12}\,[\text{C/N}].$$

(c) The voltage V is

$$V = g_{31}ET\Delta L/L = (-10.5 \cdot 10^{-3})(8 \cdot 10^{10})(0.5 \cdot 10^{-3})(10^{-6})/(38.1 \cdot 10^{-3}) = -11.02\,[\text{V}].$$

2.4.3 Piezoelectric Accelerometer

Piezoelectric sensors convert strain input into a charge output and can be used as primary sensing elements in various transducers that can have the strain as an intermediate quantity.

A piezoceramic-based accelerometer is shown in Fig. 2.15, where X is the absolute position of the case and of the right-hand-side electrode, and x is the absolute position of the left-hand-side electrode. The acceleration input applied to the case, namely,

$$a = \frac{d^2}{dt^2}X,$$

produces an inertial force F, which, in turn, produces a stress σ_3 and a strain $\Delta T/T$ applied to the piezoceramic sensor, which, finally, generates a charge output Q [1, 3]. In the process, the unstressed thickness T changes into $T + \Delta T$, where $\Delta T = x - X$ is the deflection.

A piezoceramic accelerometer can be represented in an electrical model as the capacitance C with charge Q input and voltage $V = Q/C$ and in a mechanical model as a mass–damper–spring (m–b–k) (with acceleration a input and $\Delta T/T$ strain output). These two models are linked by the piezoelectric effect of the transducer (with stress $E\Delta T/T$ input and charge electric field $V/T = Q/(CT)$ output). In this chapter, a block diagram model of the piezoceramic accelerometer, linking these three submodels, will be developed.

Piezoceramic Transducer Model The strain is $\Delta T/T$, along the same axis 3, and produces the electric field V/T as defined in the numerator of

$$g_{33} = \frac{V/T}{\sigma_3} = \frac{\text{electric field produced in direction 3 [V/m]}}{\text{mechanical stress applied in direction 3 [N/m}^2\text{]}},$$

where $\sigma_3 = E_3 = E\Delta T/T$.

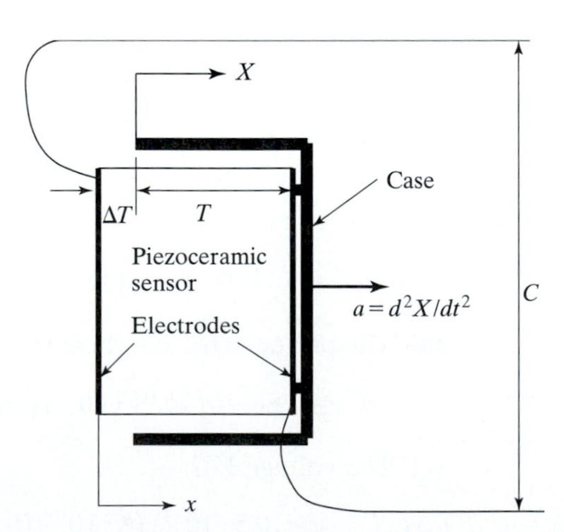

Figure 2.15

Schematic diagram of a piezoceramic accelerometer

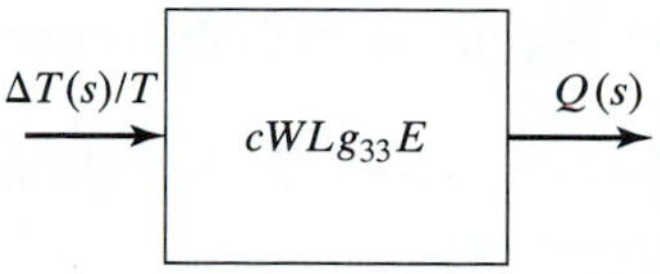

Figure 2.16

Block diagram model of piezoceramic transducing effect

The voltage output V is dependent on the input ΔT as follows:

$$V = g_{33}\sigma_3 T = g_{33} E\Delta T$$

Given the relationship

$$Q = CV,$$

or

$$Q = Cg_{33} E\Delta T,$$

and the capacitance

$$C = cWL/T,$$

the charge due to the strain $\Delta T/T$ is

$$Q = cWLg_{33} E\Delta T/T.$$

This gives the transfer function (see the block diagram in Fig. 2.16)

$$\frac{Q(s)}{\Delta T(s)/T} = cWLg_{33}E.$$

Example: Electric Charge of a Piezoceramic Sensor Produced by a Strain along Axis 3

Assume a piezoceramic sensor BM400 Type *I,* with dimensions 0.5″ · 1.5″ · 0.020″ (12.7 mm · 38.1 mm · 0.5 mm) and with the same parameters as in the previous example [22]:

$$E = 8 \cdot 10^{10} \text{ [N/m}^2\text{]};$$
$$T = 0.5 \cdot 10^{-3} \text{ [m]};$$
$$L = 38.1 \cdot 10^{-3} \text{ [m]};$$
$$c = 11.95 \cdot 10^{-9} \text{ [F/m]}.$$

Calculate the charge for a sensor strain $\Delta T/T = 10^{-6}$ [m/m].

Solution**:** In this case, the charge and the strain are both along axis 3, and the piezoelectric coefficient is given in [22] as

$$g_{33} = 25 \cdot 10^{-3} \text{ [Vm/N]},$$

such that the transfer function, from Fig. 2.16, is equivalent to the conversion gain

$$cWLg_{33}E = (11.95 \cdot 10^{-9})(12.7 \cdot 10^{-3})(38.1 \cdot 10^{-3})(25 \cdot 10^{-3})(8 \cdot 10^{10}) = 1.16 \cdot 10^{-2}\ [C/(m/m)].$$

If the sensor is subject to a 10^{-6} [m/m] strain, the charge Q is

$$Q = cWLg_{33}E\Delta T/T = 1.16 \cdot 10^{-2}(10^{-6}) = 1.16 \cdot 10^{-8}[\mathrm{C}].$$

For a given static deflection ΔT, a charge Q is generated. Given the capacitance C of the piezoelectric transducer, the corresponding voltage is $V = Q/C$. The determination of Q and ΔT from the measurement of V poses some problems. Due to the leakage resistance R_{leak} of the piezoceramic transducer and the input resistance of the voltmeter R_M, the voltage V across the capacitance C decreases in time as the initial charge $Q(0)$ is discharged by these resistances (Fig. 2.17).

R is the equivalent of the two resistances given by

$$R = 1/(1/R_{\text{leak}} + 1/R_M).$$

The current $i(t) = -dQ(t)/\mathrm{dt}$ is the same as the current in R. The voltage $V(t) = Q(t)/C$ gives

$$i(t) = V(t)/R,$$

and

$$dQ(t)/dt + Q(t)/RC = 0.$$

For initial condition $Q(0)$, the solution of this differential equation is

$$Q(t) = Q(0)\ e^{(-t/RC)}$$

and

$$V(t) = (Q(0)/C)e^{(-t/RC)}.$$

The charge $Q(t)$ decreases in time with a time constant RC. It is important to measure $Q(0)$ accurately, given its dependence on the static deflection ΔT of interest. The indirect measurement by $V(t)$ of $Q(0)$ requires a very large time constant RC to maintain $Q(t)$ close to $Q(0)$ for the duration of the measurement. Alternatively, a charge amplifier (described later in this section) has to be connected between the piezoceramic transducer and the voltmeter to convert the charge input into a proportional voltage output.

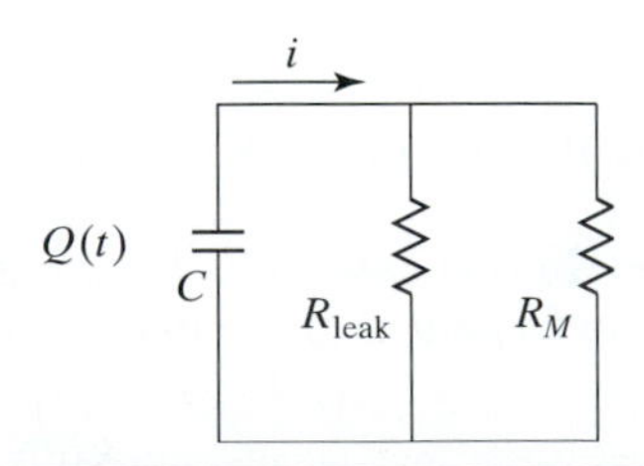

Figure 2.17
Equivalent electrical circuit

Mechanical Subsystem The accelerometer is shown in Fig. 2.18(a). Mechanically, the accelerometer model contains an equivalent mass m and a spring k in parallel to damper b (where k is the equivalent spring coefficient and b the equivalent damping coefficient of the piezoceramic part of the sensor). In Fig. 2.18(b), the mechanical model is represented as the rigid body mass m on one side and, attached to the free electrode, a spring–damper k–**b** in parallel, linked to the other electrode, solidly fixed to the case of the accelerometer.

Example: Equivalent Spring Coefficient *k* along Axis 3

Assume a piezoceramic sensor composed of a 0.5″ · 1.5″ · 0.020″ (12.7 mm · 38.1 mm · 0.5 mm) BM400 Type I with the same parameters as in the previous example:

$$E = 1/12.5 \cdot 10^{-12} = 8 \cdot 10^{10} \text{ [N/m}^2\text{]};$$
$$T = 0.5 \cdot 10^{-3} \text{ [m]};$$
$$L = 38.1 \cdot 10^{-3} \text{ [m]};$$
$$W = 12.7 \cdot 10^{-3} \text{ [m]}.$$

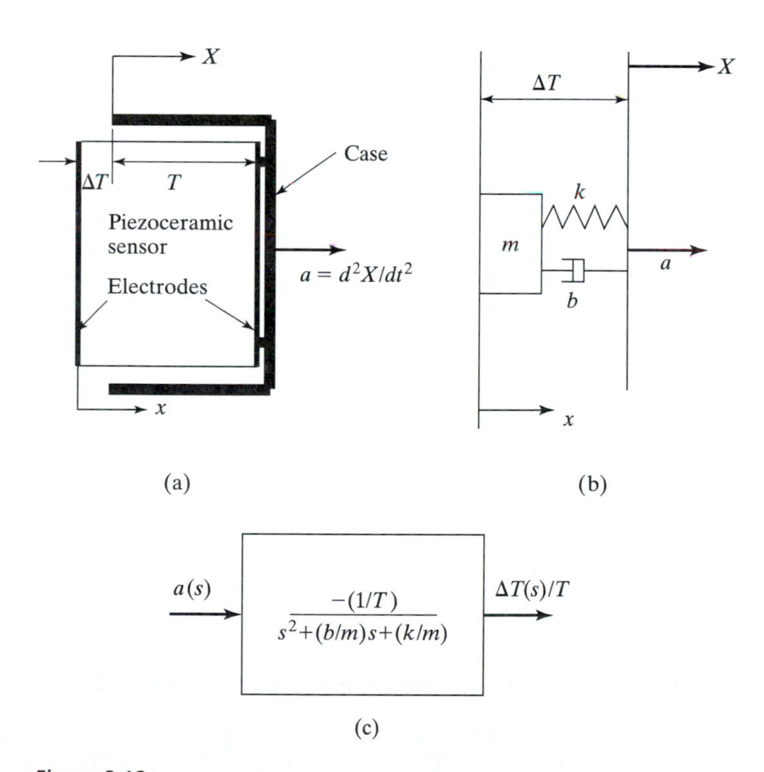

Figure 2.18

Accelerometer (a), mechanical subsystem schematic (b), and block diagram representation (c)

Assume a uniformly distributed stress

$$\sigma_3 = E\Delta T/T$$

produced by an external force applied along axis 3 ($F = F_3$)
Determine the equivalent spring coefficient k.

***Solution*:** The (force)–(strain) relationship gives

$$F_3 = \sigma_3 WL = (EWL/T)\Delta T = k\Delta T.$$

The equivalent spring coefficient is

$$k = F_3/\Delta T = EWL/T = (8\cdot 10^{10})(12.7\cdot 10^{-3})(38.1\cdot 10^{-3})/(0.5\cdot 10^{-3}) = 7.74\cdot 10^{10}\ [\text{N/m}].$$

This corresponds to a very large force 77.4 kN required for a very small deflection of 1 μm.
In Fig. 2.18 (a), the absolute coordinates are denoted as

x for the mass m;
X for the case.

For horizontal motion (no gravity effect), the mechanical model of Fig. 2.18 (b) for the piezoceramic accelerometer is based on the m–b–k model subject to the external acceleration a, which produces the inertial force ma. The relative motion of m with respect to the case is given by $x - X = \Delta T$.
Newton's second law gives

$$m\frac{d^2}{dt^2}x = -k(x - X) - b\frac{d}{dt}(x - X).$$

Given that $x - X = \Delta T$ and $dX^2/dt^2 = a$, it follows that

$$a = \frac{d^2}{dt^2}x - \frac{d^2}{dt^2}\Delta T,$$

This equation for Newton's second law results in

$$a = -\left(\frac{d^2}{dt^2}\Delta T + (b/m)\frac{d}{dt}\Delta T + (k/m)\Delta T\right).$$

The Laplace transform for zero initial conditions is

$$a(s) = -(s^2 + (b/m)s + (k/m))\Delta T(s).$$

The transfer function for acceleration $a(s)$ input and strain $\Delta T(s)/T$ output is [6]

$$\frac{\Delta T(s)/T}{a(s)} = -\frac{1/T}{s^2 + (b/m)s + (k/m)}.$$

The block diagram representation of the mechanical model is shown in Fig. 2.18(c).

Charge Amplifier The output of the sensor, the charge Q, has to be converted into a proportional voltage, easier to measure. To convert the charge Q of the electric subsystem into a voltage v output, a signal conditioning circuit is used. The conditioning of the signal is achieved with a charge amplifier based on an operational amplifier, presented in detail in Section 4.2.

The charge amplifier circuit is shown in Fig. 2.19, where the charge amplifier input current $i(t)$ comes from the piezoceramic sensor, represented by its equivalent capacitance C. A resistance R_c in parallel to a capacitance C_c form the operational amplifier feedback.

The input current in the amplifier is

$$i(t) = -\frac{d}{dt}Q(t),$$

where $Q(t)$ is the charge of the piezoceramic sensor.

The current balance, using the first Kirchhoff law, is given by

$$i = i_1 + i_2 + i_n.$$

Given that, by the design of the operational amplifier, $i_n \approx 0$, the previous equation becomes

$$i = i_1 + i_2 = -dQ(t)/dt.$$

From Section 4.2.3, $v_n \approx 0$. Hence, the currents i_1 and i_2 flow in R_c and C_c, respectively, such that

$$i = i_1 + i_2 = -\frac{v}{R_c} - C_c\frac{d}{dt}v.$$

From these equations,

$$-\frac{d}{dt}Q(t) = -\frac{v}{R_c} - C_c\frac{d}{dt}v.$$

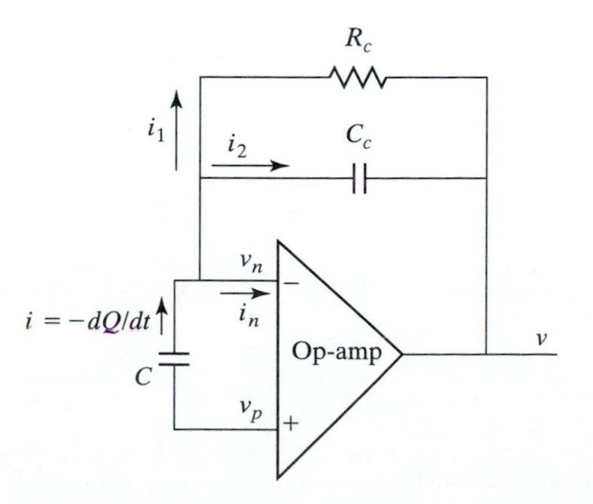

Figure 2.19

Charge amplifier for the piezoceramic sensor

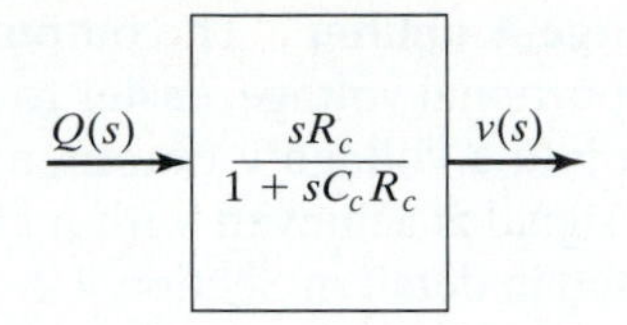

Figure 2.20

Block diagram for the charge amplifier

For zero initial conditions, the Laplace transform gives

$$sQ(s) = \left(\frac{1}{R_c} + sC_c\right)v(s).$$

Figure 2.20 shows the transfer function

$$\frac{v(s)}{Q(s)} = \frac{sR_c}{1 + sC_cR_c}.$$

Example: Charge Amplifier

Assume, for the charge amplifier shown in Fig. 2.20, that $R_c = 10^{13}\ \Omega$ and $C_c = 10^5$ pF. The transfer function of the block then becomes

$$\frac{v(s)}{Q(s)} = \frac{sR_c}{1 + sC_cR_c} = \frac{s \cdot 10^{13}}{1 + s \cdot 10^6}.$$

The time constant is in this case equal to $R_cC_c = (10^{13}\ \Omega)(10^5\ \text{pF}) = 10^6\ s$, sufficiently large for accurate measurements of $Q(0)$.

Block Diagram Representation of the Piezoceramic Accelerometer The block diagram of the accelerometer, shown in Fig. 2.21, links together the block diagrams of the mechanical model (Fig. 2.18), piezoceramic transducing effect (Fig. 2.16), and charge amplifier (Fig. 2.20).

The charge amplifier parameters, R_c and C_c can be used to obtain output voltage v in the range suitable for a data acquisition board.

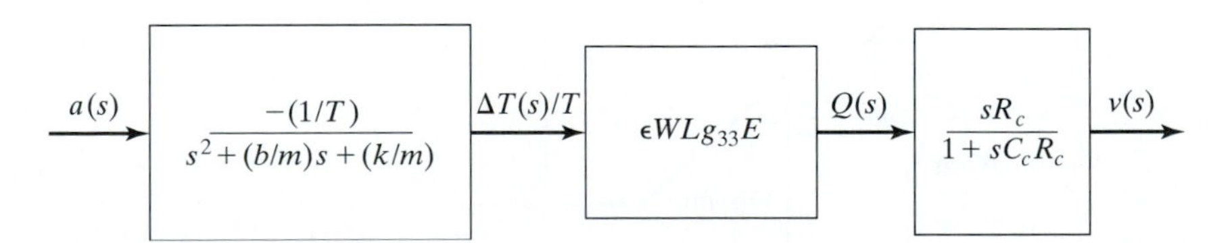

Figure 2.21

Block Diagram of the piezoceramic accelerometer

Example: Accelerometer Transducer

Assuming the same data as in the previous three examples, viz., that $k = 7.74 \cdot 10^{10}$ [N/m], $cWLg_{33}\,E = 1.16 \cdot 10^{-2}$ [(C/m)], and given that $m = 10^{-3}$ Kg, $b = 0$, $R_c = 10^{10}\ \Omega$, and $C_c = 10^3$ pF, calculate

(a) the steady-state voltage v response to an acceleration step input of magnitude $a = 2$ [m/s²];

(b) the time constant.

Solution:

(a) For the given values, the transfer functions of the three blocks, shown in Fig. 2.21, become

$$\frac{\Delta T(s)/T}{a(s)} = -\frac{1/T}{s^2 + (b/m)s + (k/m)} = \frac{2000}{s^2 + 7.74(10^{13})},$$

$$\frac{Q(s)}{\Delta T(s)/T} = cWLg_{33}E = 1.16(10^{-2}),$$

and

$$\frac{v(s)}{Q(s)} = \frac{s \cdot R_c}{1 + sC_cR_c} = \frac{s \cdot 10^{10}}{1 + s10}.$$

An acceleration step input of magnitude $a = 2$ [m/s²] has the Laplace transform 2/s. The steady-state output voltage can be obtained with the finite value theorem as the limit for $s \rightarrow 0$ of

$$v = \lim_{s\rightarrow 0} \frac{s \cdot 2000}{s^2 + 7.74(10^{13})} 1.16(10^{-2}) \frac{s \cdot 10^{13}}{1 + s \cdot 10^6} \frac{2}{s} = 0 \text{ V}.$$

This means that the reading of the measurement has to be fast; how fast the reading should be depends on the time constant. Longer time constants permit lower frequencies and static measurements because the decay of the voltage output v towards zero steady-state value is slow.

(b) The time constant is given by the product of feedback resistance and capacitance $R_cC_c = (10^{10}\ \Omega) \cdot (10^3 \text{ pF}) = 10$ s. This corresponds to an exponential decay $e^{-\{t/10\}}$. Given the value for $t = 0$, $e^{-\{0/10\}} = 1$, and the decay of 3% to 0.97 of the value at $t = 0$ gives

$$e^{-\{t/10\}} = 0.97,$$

or

$$-t/10 = \ln 0.97 = -0.03,$$

or

$$t = 0.3 \text{ [sec]}.$$

This result shows that in 0.3 [sec], a decay to 97% of the initial value takes place. This makes accurate visual reading difficult. A very high input impedance transistor (e.g., a field effect transistor, presented in Section 4.2) is an alternative to a charge amplifier for signal conditioning of piezoceramic transducers [19].

2.5 ANALOG POSITION MEASUREMENT: POTENTIOMETERS

A potentiometer can be used as a basic analog position transducer for measuring displacements. Figure 2.22 shows a potentiometer for rectilinear displacement measurement, consisting of a wire coil, a moving contact or a wiper and direct current (DC) voltage source of constant value v_i.

The potentiometer has a resistance R and its wiper can move from the position $y = 0$ to $y = y_{max}$. The resistance R_y corresponds to the position y of the wiper such that

$$R_y/y = R/y_{max}$$

or

$$R_y = (R/y_{max})y.$$

This linear relationship between R_y and y suggests a displacement sensor; however, the practical transducer requires an output voltage dependent of y. The measurement of the output voltage v_o with a voltmeter is affected by the voltmeter input resistance R_v, and the dependence between v_o and y can be derived from the following equations:

$$i = i_y + i_v;$$

$$v_i = i R_{eq};$$

$$v_o = i_v R_v = i_y R_y.$$

Figure 2.22

Linear potentiometer for rectilinear displacement measurement

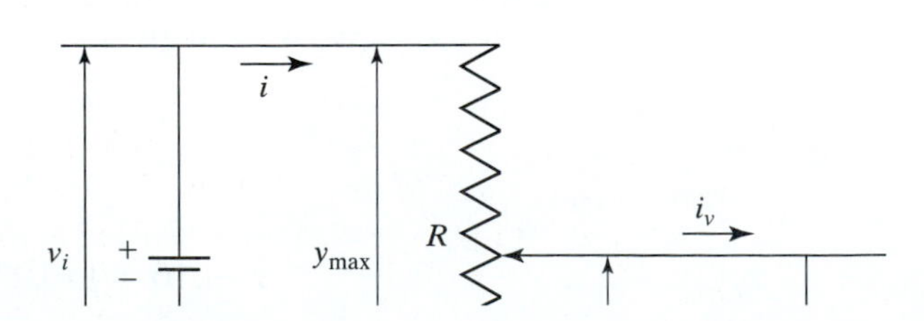

Here,

$$R_y = (R/y_{max})y$$

and

$$R_{eq} = R - R_y + 1/(1/R_y + 1R_v).$$

These equations contain the variables of interest v_o, y, and v_i. After eliminating i, i_y, and i_v, we obtain

$$v_i/R_{eq} = v_o/R_v + v_o/R_y,$$

or

$$v_i = v_o R_{eq}(1/R_v + 1/R_y).$$

Replacing R_{eq} yields

$$v_i = v_o(R - R_y + 1/(1/R_y + 1/R_v))(1/R_v + 1/R_y) =$$
$$v_o((R - R_y)(1/R_v + 1/R_y) + 1).$$

Eliminating R_y gives

$$v_o = v_i\left[1 + \left(\frac{R}{R_v} + \frac{y_{max}}{y}\right)\left(1 - \frac{y}{y_{max}}\right)\right]^{-1}.$$

This equation is nonlinear in y, but for $R_v >> R$, the following approximate linear relationship can be obtained [6]:

$$v_o = \left(\frac{v_i}{y_{max}}\right)y.$$

This linear dependence between v_o and y for $R_v \gg R$ is the approximate model of the linear potentiometer as a displacement transducer. Preferably, current i has to be small in order to produce only small heat loss in the potentiometer, and consequently, R is chosen to have a large value. Analog voltmeters have typically $R_v = 30\ \mathrm{k\Omega}$ and the condition $R_v > R$ might not be satisfied. Digital voltmeters are preferable, given that they have typically $R_v = 10\ \mathrm{M\Omega}$.

The resolution of this potentiometer transducer for displacement depends on the number of turns N of the wire coil. The smallest displacement that can be measured, or the resolution is y_{max}/N. Angular displacements can be sensed by rotary potentiometers (Fig. 2.23).

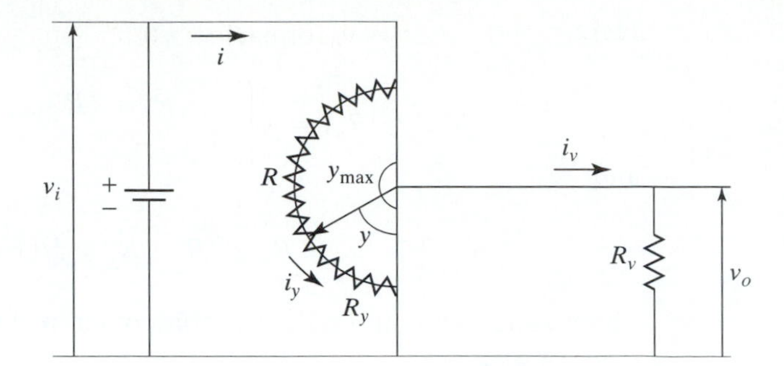

Figure 2.23

Rotary potentiometer for angular displacement measurement

The preceding two equations are valid for rotary potentiometer also, but in this case y is an angular displacement. The maximum angular displacement that can be measured with the rotary potentiometer shown in Fig. 2.23 is $y_{max} = 180$, but there are rotary potentiometers designed to measure angular displacements of multiple turns.

Moving wiper, wiper–coil contact, and the resolution (depending on the number of turns of the coil) are limiting factors for this displacement transducer performance. Also, this resistive potentiometer converts displacement into voltage, but time variation $y(t)$ is sensed only for slow varying displacements, a fact that translates into poor dynamic performance. In particular, the contact wiper–coil wire is a limiting factor for dynamic measurement as it can result in unstable contact and can produce parasitic electric noise in the output voltage.

More advanced designs of potentiometers address these limitations [1].

Example: Rotary Potentiometer Resolution

Assume a rotary potentiometer of the type shown in Fig. 2.23 with $y = 180°$ and $N = 1000$ turns. Calculate the resolution r.

***Solution*:**

$$r = y_{max}/N = 180°/1000 = 0.18°.$$

Example: Measurement Error Due to Voltmeter Input Resistance

For the rotary potentiometer from the preceding example, one of the following voltmeters has to be selected for measuring v_o:

(a) an analog voltmeter with $R_v = 30\ k\Omega$;

(b) a digital voltmeter with $R_v = 10\ M\Omega$. The resistance of the potentiometer is $R = 1\ k\Omega$ and $v_i = 10$ V.

Select the voltmeter such that the error due to R_v is smaller than the resolution when measuring an angle of $y = 50$.

***Solution*:** The calculation is based on the following steps:

For $y = 50°$, calculate v_o for each voltmeter, taking into account its R_v in kΩ:

$$v_o(R_v) = v_i\left[1 + \left(\frac{R}{R_v} + \frac{y_{max}}{y}\right)\left(1 - \frac{y}{y_{max}}\right)\right]^{-1} =$$

$$10\left[1 + \left(\frac{1}{R_v} + \frac{180}{50}\right)\left(1 - \frac{50}{180}\right)\right]^{-1}.$$

For this value of $v_o(R_v)$, calculate the linear approximation of the angle

$$y(R_v) = \left(\frac{y_{max}}{v_i}\right)v_o(R_v) = \left(\frac{180}{10}\right)v_o(R_v).$$

The angular measurement error $e(R_v)$ due to R_v is

$$e(R_v) = 50 - y(R_v),$$

and this error can be compared to the resolution of $r = 0.18°$.

(a) For $R_v = 30$ kΩ,

$$v_o(30) = 10\left[1 + \left(\frac{1}{30} + \frac{180}{50}\right)\left(1 - \frac{50}{180}\right)\right]^{-1} = 2.759 \text{ V}$$

$$y(30) = \left(\frac{180}{10}\right)2.759 = 49.67,$$

and

$$e(30) = 50 - 49.67 = 0.33° > 0.18°.$$

Consequently, when the effect of its input resistance is ignored and the angle is calculated with the linear approximation $y = (y_{max}/v_i)v_o$, the analog voltmeter introduces an error of 0.33°. This error is larger than the resolution of the potentiometer $r = 0.18°$ and the analog voltmeter is rejected.

(b) For $R_v = 10\ M\Omega = 10{,}000\ k\Omega$,

$$v_o(10{,}000) = 10\left[1 + \left(\frac{1}{10000} + \frac{180}{50}\right)\left(1 - \frac{50}{180}\right)\right]^{-1} = 2.777772 \text{ V}$$

$$y(10{,}000) = \left(\frac{180}{10}\right)2.777772 = 49.99999°,$$

and

$$e(10{,}000) = 50 - 49.99999 = 0.00001° < 0.18°.$$

The digital voltmeter introduces an error of 0.00001° (when the effect of its input resistance is ignored), which is much smaller than the resolution of the potentiometer $r = 0.18°$, and the digital voltmeter is chosen.

Example: Analog-to-Digital Converter (ADC) Selection

Analog-to-digital converters (ADCs) will be presented in detail in Section 4.2. In this example, for interfacing the potentiometer from the previous example with a PC, the converter word length of the ADC will have to be selected from the following list:

(a) 8-bit converter;
(b) 12-bit converter.

Select the ADC word length that has a better resolution than the potentiometer resolution of $r = 0.18°$.

Solution:

(a) An 8-bit converter permits $2^8 = 256$ distinct digital readings.
For the full-scale $y_{max} = 180°$, the digital resolution is $180°/256 = 0.7° > 0.18°$
The 8-bit converter is rejected because the digital resolution is poorer than the potentiometer resolution $r = 0.18°$.

(b) A 12-bit converter permits $2^{12} = 4096$ distinct digital readings.
For the full-scale $y_{max} = 180°$, the digital resolution is $180°/4096 = 0.044° < 0.18°$.
The 12-bit converter is chosen because the digital resolution is better than the potentiometer resolution r = 0.18°.

2.6 DIGITAL POSITION MEASUREMENT: OPTICAL ENCODERS

2.6.1 Digital Transducers

Analog transducers (e.g., the potentiometers) have a voltage output proportional to the measured physical variable. Digital transducers (e.g., optical encoders, digital resolvers, digital tachometers, limit switches, etc.) produce an output of digital signals, normally a train of constant magnitude voltage pulses [1]. Digital transducer output does not maintain the dependence between the magnitude of the voltage output and the magnitude of the input signal; this dependence is contained in the pulse duration or in a coded form of a binary word.

The output of pulses that is not coded for direct computer reading is processed by a counter and can eventually be coded in a computer readable form.

This chapter will focus on optical encoders. Optical encoders are either incremental optical encoders or absolute optical encoders. Incremental encoders give a measurement relative to an initial position of the rotating component. Absolute encoders give the absolute displacement of the rotating component relative to a non-moving reference.

The incremental optical encoder generates pulse signals with the measurement information encoded in the pulse duration and frequency. This information is converted into coded form. The pulse voltage is compatible for direct digital reading as a serial input to the computer by complying with the transistor-to-transistor logic (TTL). In this case, for voltages below 0.8, the assigned binary value is 0 and for over 2 V, the binary value is 1. The pulses are supposed to have no values between 0.8 V and 2 V, and their frequency and duration are supposed to be compatible with the characteristics of the computer input reading device.

Absolute optical encoders produce n-bit parallel signals that encode the displacement in a particular coded form. The coded form can be straight binary code (SBC), binary coded decimal (BCD), or gray code. These codes will be presented in Section 2.6.3.

2.6.2 Incremental Optical Encoders

Basic Principle An incremental encoder for measuring angular displacement in only one direction is shown in Fig. 2.24. This encoder has four windows.

The photosensor output is a voltage U that has a high value during the passage of light from the light source through the window (represented by a black circle in Fig. 2.24) to the photosensor and a low value when facing the opaque area. A window followed by an opaque area form a sector. Figure 2.25 shows the ideal pulse voltage output (i.e., with no transient variation and no noise) for a constant velocity rotation.

For an incremental encoder with M windows per circumference, the physical resolution is [1]

$$r_M = \frac{360°}{M}.$$

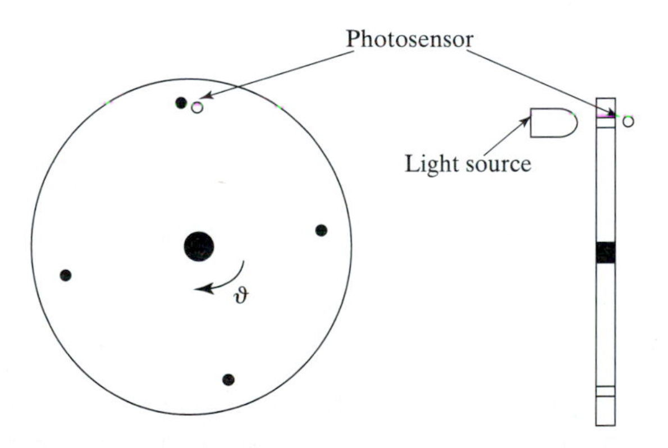

Figure 2.24
An incremental encoder

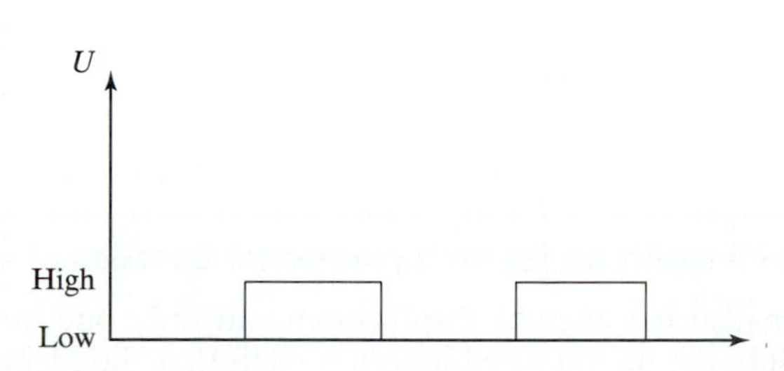

Figure 2.25
Voltage output of the photosensor for constant angular velocity

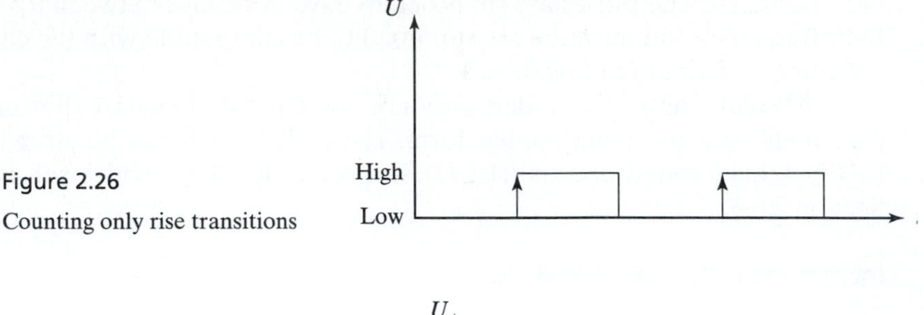

Figure 2.26
Counting only rise transitions

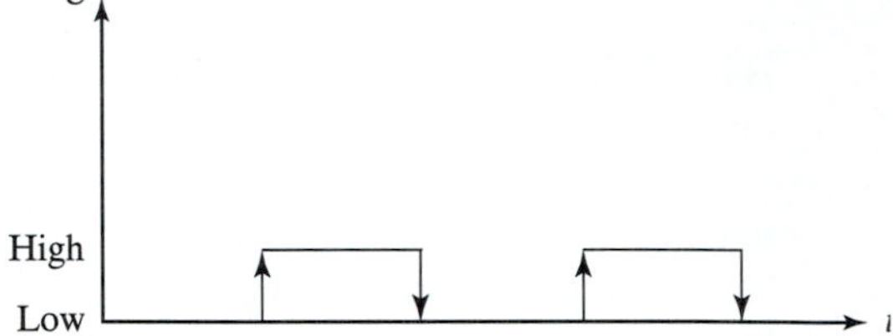

Figure 2.27
Counting both rise and fall transitions

Figure 2.26 shows, by upward arrows, the case in which only the rise from low to high voltage is counted (i.e., one digital reading per pulse cycle or per sector). This resolution r_M corresponds to the pitch.

Figure 2.27 shows, by upward and downward arrows, the case in which both rise and fall of the voltage are counted (i.e., two digital readings, from one rise transition to the next, for one pulse cycle of the encoder signal output).

In this case, the digital resolution is

$$r_M = \frac{360^\circ}{2M}.$$

This resolution corresponds to a half pitch.

If the measurement of the maximum angular displacement of a full revolution (360º) is represented in a digital counting by a n-bit word, the digital resolution is

$$r_n = \frac{360^\circ}{2^n}.$$

In this case, the angles 0° and 360° are counted twice. To avoid this, the reading for 0° is excluded and only $2^n - 1$ values are retained such that in this case the digital resolution is [1]

$$r_n = \frac{360^\circ}{2^n - 1}.$$

Example: Physical and Digital Resolution for an Incremental Encoder

An incremental encoder, for measuring angular displacement in only one direction, has $M = 1000$ windows. Calculate the digital word length n such that the ideal design (the same physical and digital resolution) is achieved. Assume that both rise and fall transitions are counted and that reading for 0° is excluded.

***Solution*:** Physical resolution is given by

$$r_M = \frac{360°}{2M} = \frac{360°}{2(1000)} = 0.18°.$$

Ideal design requires that

$$r_M = r_n,$$

or

$$\frac{360°}{2M} = \frac{360°}{2^n - 1}.$$

Thus,

$$2M = 2^n - 1,$$

or

$$M = 2^{n-1} - 0.5.$$

In order to obtain an integer number solution for n, the last equation can be solved by trial and error to determine n such that at least $M = 1000$ distinct values can be represented digitally. For $n - 1 = 10$ the value of

$$2^{n-1} - 0.5 = 2^{10} - 0.5 = 1023.5$$

(i.e., the digital word length has to be $n = 11$).

Direction of Motion Detection Most applications require not only unidirectional, but also bidirectional angular displacement measurement over time. Incremental optical encoders can achieve direction of motion detection either with the help of a second concentric circular track and identical set of windows placed in quadrature (shifted by a quarter pitch from the other set of windows) or by installing a second photosensor shifted again by a quarter pitch from the other photosensor. The latter solution is shown in the front and side views from Fig. 2.28.

The two photosensors—Photosensor 1 and Photosensor 2—are placed apart and generate voltage outputs U_1 and U_2, respectively, which differ by a quarter of pulse cycle.

Figure 2.29 displays the voltage outputs U_1 and U_2 of the photosensors when the incremental encoder moves clockwise as shown in Fig. 2.28. In this case, at the end of the opaque zone, the window first opens to Photosensor 1 for one-half pitch and, after one quarter of a pitch, the Photosensor 2 reaches the window. A high-frequency clock

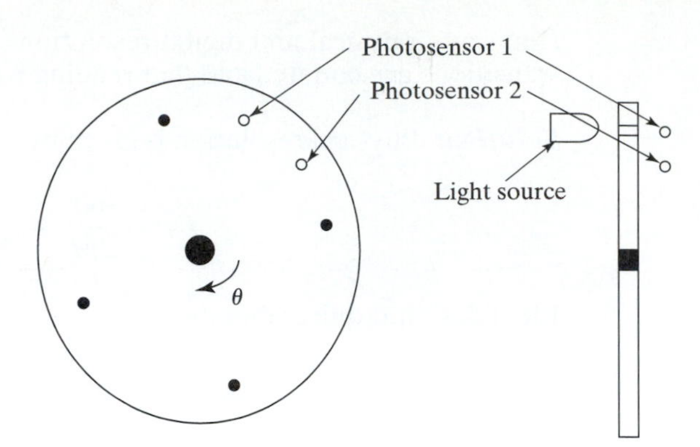

Figure 2.28

Incremental encoder with detection of the direction of motion

is used and N clock cycles are counted for the encoder cycle. For example, for a clock frequency of 1 MHz (i.e., a clock cycle of 1 μs) and, in the case of a constant encoder cycle of 1 ms (i.e., encoder window frequency of 1 kHz), $N = 1000$ clock cycles count/encoder cycle will result.

Assuming the clock cycles count N_1 from the rise of U_1 until the rise of U_2 (i.e., along a $\frac{1}{4}$ encoder cycle or pulse-to-pulse period, as in the case shown in Fig. 2.29), the result in this case verifies that $N_1 < N - N_1$.

Figure 2.30 displays the voltage outputs U_1 and U_2 of the photosensors when the incremental encoder moves counterclockwise. In this case, at the end of the opaque zone, the window first opens to Photosensor 2 for one-half pitch and, after one quarter of the encoder cycle, the Photosensor 1 reaches the window. Counting the clock cycles N_1 from the rise of U_1 until the rise of U_2 (i.e., along $\frac{3}{4}$ of the encoder cycle), the result in this case verifies that $N_1 > N - N_1$.

In order to account for the direction of motion, one bit can be assigned the value 1 for clockwise motion and the value 0 for counterclockwise motion.

Figure 2.29

Voltage outputs U_1 and U_2 of the photosensors for clockwise motion

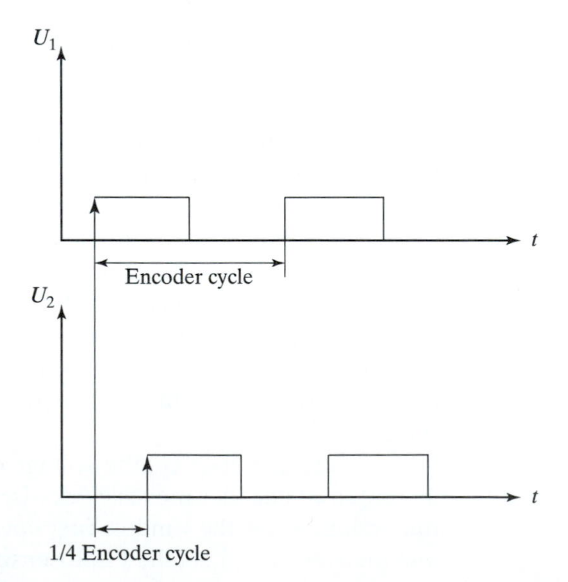

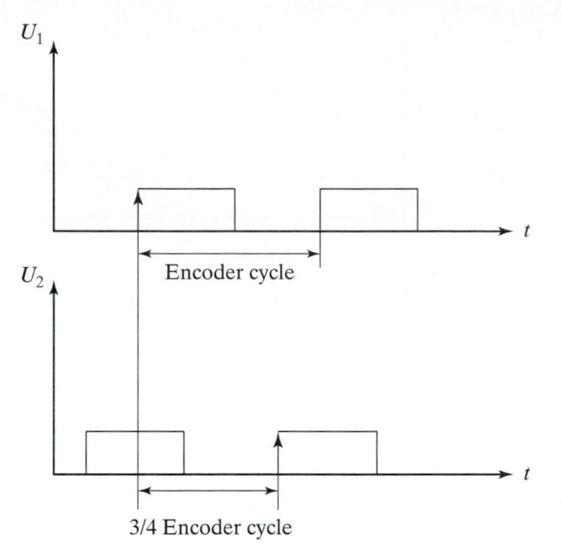

Figure 2.30
Voltage outputs U_1 and U_2 of the photosensors for counterclockwise motion

In case of n-bit word digital counting, retaining one bit for the direction of motion, $n - 1$ bits will be assigned to the value of the angular displacement. The digital resolution, when the reading for 0° is excluded, is

$$r_n = \frac{360°}{2^{n-1} - 1}.$$

Example: Physical and Digital Resolution for an Incremental Encoder with Direction of Motion Detection

An incremental encoder, for measuring angular displacement, has $M = 1000$ windows and is equipped with direction of motion detection. Assume that both rise and fall transitions are counted and that reading for 0° is excluded. Calculate the digital word length n such that the ideal design (the same physical and digital resolution) is achieved.

***Solution*:** The physical resolution is given by

$$r_M = \frac{360°}{2M} = \frac{360°}{2(1000)} = 0.18°.$$

The ideal design requires that

$$r_M = r_n,$$

or

$$\frac{360°}{2M} = \frac{360°}{2^{n-1} - 1}.$$

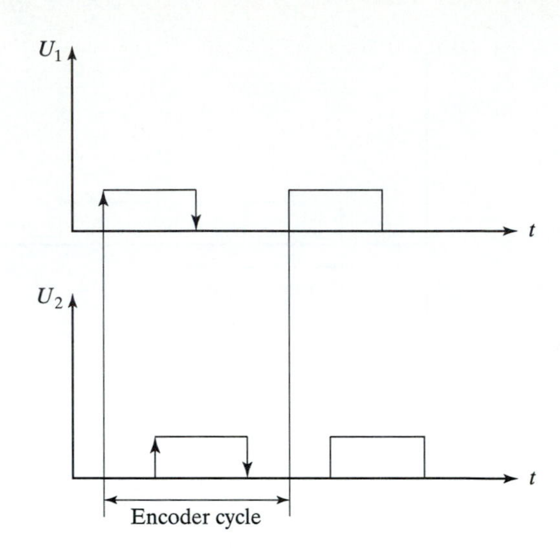

Figure 2.31
Four transitions per encoder cycle reading

Hence,

$$2M = 2^{n-1} - 1,$$

or

$$M = 2^{n-2} - 0.5.$$

The last equation can be used by trial and error to determine an integer value for n such that at least $M = 1000$ distinct values can be represented digitally. For $n - 2 = 10$, the value of

$$2^{n-2} - 0.5 = 2^{10} - 0.5 = 1023.5$$

(i.e., the digital word length has to be $n = 12$ bits).

Physical resolution in this case can be further improved by counting all four transitions over the incremental encoder cycle, as shown in Fig. 2.31. The physical resolution of this incremental encoder is

$$r_M = \frac{360^\circ}{4M}.$$

Example: Physical and Digital Resolution for an Incremental Encoder with Direction of Motion Detection and Four Readings per Encoder Cycle

The incremental encoder of the preceding example is now modified to read four transitions per encoder cycle. Recalculate the digital word length n for an ideal design.

***Solution*:** The ideal design requires that

$$\frac{360^\circ}{4M} = \frac{360^\circ}{2^{n-1} - 1}$$

or

$$M = 2^{n-3} - 0.25.$$

An integer value for n can be obtained by trial and error such that at least $M = 1000$ distinct values can be represented digitally. For $n - 3 = 10$, the value of

$$2^{n-3} - 0.25 = 2^{10} - 0.25 = 1023.75$$

(i.e., the digital word length has to be $n = 13$ bits).

Incremental encoders can be built to measure multirevolution angular motion by adding an extra track, with only one window, and the corresponding light source and sensor. The digital word has to be increased with sufficient bits to record the number of revolutions. For example, counting the maximum 16 revolutions requires four bits.

2.6.3 Absolute Optical Encoders

An absolute encoder has several concentric circular tracks that are divided into sectors. Figure 2.32 shows a simplified example for $\mu = 3$ concentric tracks divided in sectors:

$$v = 2^{\mu} = 2^3 = 8.$$

In Fig. 2.32, each sector has opaque areas (represented as shaded areas) and windows that correspond to the straight binary code (SBC). The opaque areas are coded as 0

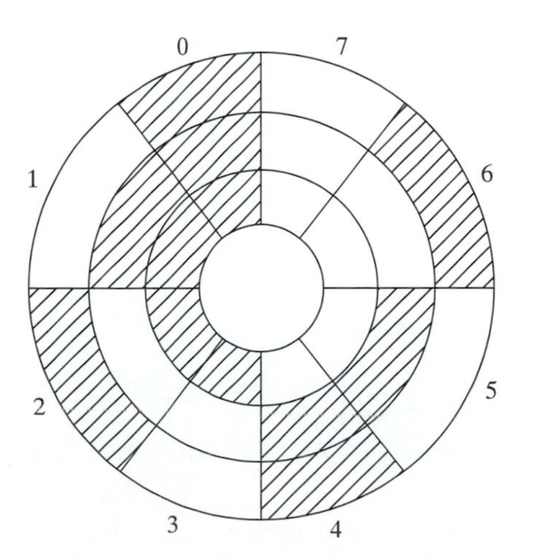

Figure 2.32

Absolute encoder using straight binary code (SBC)

and the windows are coded as 1. The following mapping of SBC and decimal sector number results:

Decimal number	0	1	2	3	4	5	6	7
SBC	000	001	010	011	100	101	110	111

Figure 2.33 shows the absolute encoder with the binary number associated with each track such that the outer track contains the least significant bit (LSB), while the inner track contains the most significant bit (MSB). The reading of binary numbers starts with the MSB (i.e., from the inner track).

The advantage of absolute encoders is that they indicate the correct angular position with regard to the fixed reference frame of the light source and sensor at any time, even after a power supply interruption.

In SBC, the transitions from sectors 3 to 4 and 5 to 6 contain two digit changes. In the case of an error in reading SBC numbers—for example if only one bit is read correctly and the other not—the error in identifying the sector can be important. For example, if at the transition from sector 3 to 4 only the second bit change is read correctly, the result is not 100, but 001 that identifies sector 1. For absolute encoders with a large number of tracks, this might become a serious problem. It is preferable that in binary form, adjacent sectors differ by only one bit. This is achieved in gray code:

Decimal number	0	1	2	3	4	5	6	7
Gray code	000	001	011	010	110	111	101	100

Using the gray code requires eventual conversion to SBC.

Physical resolution and digital resolution are the same in the case of absolute encoders:

$$r = \frac{360^\circ}{2^\mu}.$$

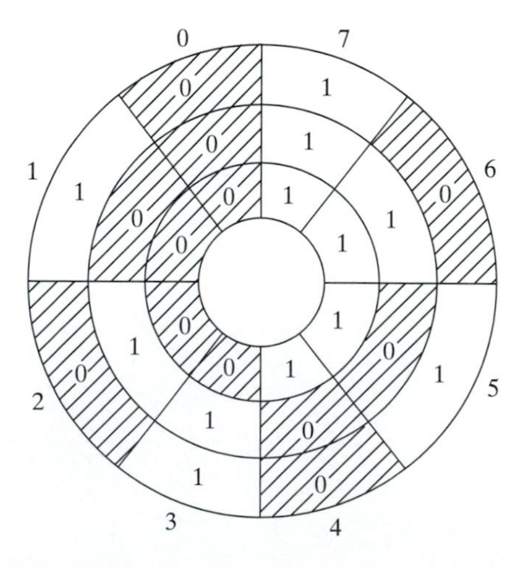

Figure 2.33

Absolute encoder with binary numbers inscribed

Example: The Resolution of an Absolute Encoder

An absolute encoder has $\mu = 3$ tracks. Calculate the number ν of sectors and the resolution.

***Solution*:**

For $\mu = 3$,

$$v = 2^3 = 8 \text{ [sectors]}$$

and

$$r = \frac{360°}{8} = 45°.$$

An absolute encoder with three tracks has a too low resolution for practical applications. Optical encoders are built with resolutions up to $360°/2^{22} = 0.000086°$. They achieve better resolutions than rotary potentiometers that are limited in practice to $360°/2^{10} = 0.35°$ [19].

2.7 VELOCITY MEASUREMENT: TACHOMETERS

Analog Measurement of Velocity: Tachometers A permanent magnet DC generator , shown in Fig. 2.34, can be used for analog measurement of angular velocity. In this figure, ω is the angular velocity to be measured, T is the torque required to drive the PM–DC generator, L and R are the inductance and the capacitance of the rotor, I is the current in the rotor windings, and U is the voltage output at the rotor windings terminals, while U_r is the voltage induced by the stator in the rotor.

The design requirements for a velocity transducer are:

minimum mechanical load in order to have a negligible T;

minimum influence of the voltage-measuring device, achieved in the case of a very high input impedance.

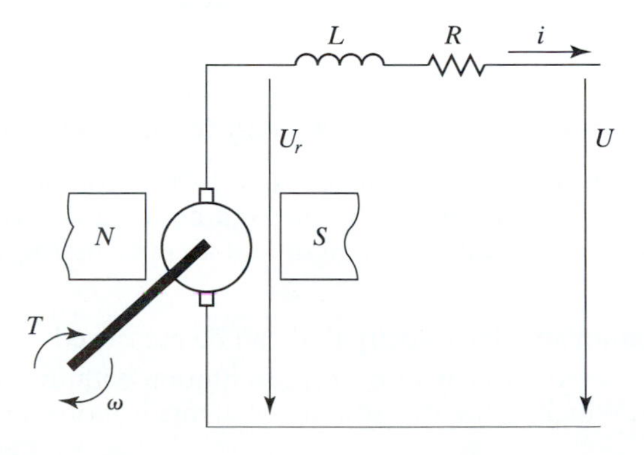

Figure 2.34

Permanent magnet DC generator

The low torque T is obtained in case of a low moment of inertia of the rotor and a low magnetic torque accounting for the load on the electric side. This means a small-size rotor and a high-impedance electrical load.

The only phenomenon that remains significant is the generation of the induced voltage in the rotor U_r due to the motion of the rotor winding with angular velocity ω in the magnetic field of the permanent magnet, such that

$$U_r = K\omega.$$

The electric circuit of the DC generator has the following voltage drop equation:

$$U_r = Ldi/dt + Ri + U.$$

A very high input impedance of the voltage-measuring device results in $I \approx 0$, and, in this case, the previous two equations give the model of the ideal tachometer:

$$U = K\omega.$$

This model is valid for a limited frequency domain due to the design requirements listed earlier that cannot be satisfied at very fast variations of the angular velocity ω. Indeed, at high angular accelerations, even a rotor with low moment of inertia J produces a significant inertia torque

$$Jd\omega/dt.$$

The requirement of minimum mechanical load, achieved by a small size rotor, also results in a rotor winding with a small number of conductors and this produces a geometrically significant variation of the distance in the air gap between the rotor and the permanent magnet. This generates a ripple voltage U_{ripple} in the output voltage U, in accordance with the formula

$$U = K\omega + U_{\text{ripple}}.$$

This results in an error of measurement of ω that can amount, even for high accuracy tachometers, up to 2% of the average value of U for $\omega > 5$ radians/sec [19].

Example: Compatibility of Tachometer with an Analog-to-Digital Converter

Assume a tachometer with a conversion gain of 6 V per 1000 revolutions/min or 6 [V/krpm]. This tachometer is interfaced with a PC through an analog-to-digital converter (ADC). The ADC has 12-bit digital word length and accepts analog voltage inputs of a maximum of 10 V.

(a) Calculate the maximum angular velocity that can be measured.

(b) Compare the voltage resolution at 12-bit representation with the ripple voltage assumed as 2%. Which of the two sources of errors is more significant?

Solution:

(a) The maximum voltage output acceptable for the ADC is U_{max}. The maximum angular velocity is given by

$$\omega_{max} = U_{max}/K,$$

where K has the units volts per radians/sec.

For this problem, it is more convenient to use angular velocity as n in [1000 revolutions/min] or [krpm] such that

$$n_{max} = U_{max}/K_n,$$

where

$$K_n = 6 \text{ [V/krpm]}.$$

The maximum angular velocity that can be measured is then

$$n_{max} = 10 \text{ V} / (6 \text{ V/krpm}) = 1.667 \text{ krpm} = 1667 \text{ revolutions per minute.}$$

(b) The voltage resolution at 12-bit is

$$r_d = 10 \text{ V}/2^{12} = 10/4096 = 0.00244 \text{ V}.$$

Ripple voltage of 2% of 10 V gives 0.2 V, an error about 80 times higher than the digital resolution. Reduction of ripple voltage effect is required for a more accurate angular velocity measurement [19].

Digital Measurement of Velocity Optical encoders can be used not only for angular displacement, but also for angular velocity measurement [1, 6].

Assume an incremental encoder equipped with a digital clock of frequency f (clock cycles/sec). A counter gives the number m of clock cycles per sector (window-to-adjacent window). A high-frequency digital clock has to be used to generate a large number m of clock cycles per sector (i.e., clock frequency has to be much higher than the angular velocity measured). The cycle time of the digital clock is

$$T_c = 1/f.$$

The time T_e for the motion of the encoder over a sector (i.e., the encoder cycle time) counted in m digital clock cycles of duration T_c is

$$T_e = mT_c = m/f.$$

For an incremental encoder with N sectors per revolution, the time for a complete revolution is $NT_e = NmT_c = Nm/\text{f}$ in (sec/rev). The angular velocity in revolutions per second is

$$n = \frac{1}{NmT_c} = \frac{f}{Nm},$$

or, in radians per second [1],

$$\omega = \frac{2f\pi}{Nm}.$$

Example: Incremental Optical Encoder for Velocity Measurement

An incremental encoder with $N1000$ sectors is equipped with a digital clock with $f = 1$ MHz. The counter gives a reading of $m = 2000$ digital clock cycles per sector. Calculate the angular velocity.

Solution:

$$n = \frac{f}{Nm} = \frac{10^6 \qquad [\text{cycles/sec}]}{(1000)(2000)\,[\text{sectors/rev}][\text{cycles/sector}]} = 0.5\,\text{rps},$$

or

$$2(3.14)(0.5) = 3.14 \text{ rad/sec}.$$

PROBLEMS

2.1. A K thermocouple produces a voltage that is measured by the potentiometer as 25 mV. Determine the temperature T when the reference junction isothermal block is indicated by a thermistor as 0°C. Use the Seebeck coefficient for 20°C.

2.2. A PC-based data acquisition system is considered for a K-type thermocouple. Two limit designs will have to be evaluated.

Design A—a cheap and less accurate design with

a four-bit A/D converter;

a standard electronic reference junction block with a temperature uncertainty of ±0.57°C at 0°C;

a standard thermocouple with ±2.3°C error limit.

Design B—an expensive and more accurate design with

a eight-bit A/D converter;

a special electronic reference junction block with a temperature uncertainty of ±0.15°C at 0°C;

a special thermocouple with ±1.15°C error limit.

Compare the two designs based on the following calculations:

1. Determine the scale range for the analog input voltage in A/D converter.
2. Calculate the resolution.
3. Calculate the combined uncertainty (of the A/D converter, reference junction, and thermocouple) regarding the temperature measurement.

2.3. A strain gauge with nominal resistance $R = 120\ \Omega$ is installed in a branch of a Wheatstone bridge having for unstrained strain gauge $R_1 = R_2 = R_3 = R_4 = R$ and $V_i = 10$ V.

As a result of bending the beam, on which it is cemented, the strain gauge is subject to a strain. A digital voltmeter with input resistance $R_m = 10\ M\Omega$ gives a reading of $V_o = 5\ mV = 5 \cdot 10^{-3}\ V$. Calculate

(a) the change in the resistance ΔR;

(b) the strain ε for gauge factor $G = 2$.

2.4. A resistive strain gauge, $G = 2.2$, is cemented on a rectangular steel bar with the elastic modulus $E = 205 \times 10^6\ kN/m^2$, width 3.5 cm, and thickness 0.55 cm. An axial force of 12 kN is applied. Determine the change of the resistance of the strain gauge, ΔR, if the normal resistance of the gauge is $R = 100\ \Omega$.

2.5. The strain gauge of the previous problem is connected to a resistance device having an accuracy $\pm 0.25\ \Omega$. What is the uncertainty in determining the stress σ?

2.6. A steel bar with an elastic modulus $E = 205 \cdot 10^6\ kN/m^2$ and a cross section area $A = 6.5\ cm^2$ is subject to an axial force F. For measuring this force, a strain gauge is cemented on the bar. The nominal resistance of the strain gauge is $R = 100\ \Omega$. The strain gauge is connected in a branch of Wheatstone bridge with all other branches with resistances equal to $R = 100\ \Omega$. Wheatstone bridge voltage output is conditioned, as shown in Fig. 2.10, using an inverting amplifier with a resistance $R_o = 1\ M\Omega$. The strain gauge factor is $G = 2.1$ and the voltage $V_i = 8.5\ V$. Calculate the force F given a measured voltage $V_o = 6.5\ V$.

2.7. An accelerometer based on a strain gauge, shown in Fig. 2.12, consists in a cantilever beam of length $L = 27$ mm, width $w = 2.5$ mm, and thickness $t = 1.5$ mm, fitted with a (seismic) mass $M = 0.017$ kg. The modulus of elasticity of the beam is $E = 205 \times 10^9\ N/m^2$. The strain gauge cemented at $l = 23$ mm from the free end of the beam and having $G = 2.05$ is connected to a bridge which is interfaced to an ADC, as shown in Fig. 2.10, through an operational amplifier with $R_o = 1.15\ M\Omega$. Assume nominal resistance of the strain gauge of $R = 100$ and the supply voltage of the bridge is $V_i = 8.5\ V$. Calculate the acceleration a which produces a voltage output $V_o = 0.155\ V$.

2.8. In the previous problem, the resistance R_o of the operational amplifier was given as $1.15\ M\Omega$ and the voltage output from this amplifier to ADC was $V_o = 0.155\ V$. In case the ADC has only the range 0 to 10 V available, calculate a value for R_o, which will give the midscale 5 V for V_o, for an acceleration value and all other factors being kept the same.

2.9. Assume that a positioning measuring potentiometer (with total resistance $R = 15.5\ k\Omega$, total length $L = 12$ cm, and input voltage $V_i = 10\ V$) has the wiper at $x = 12$ cm. The voltage V_o of the potentiometer is measured with an analog voltmeter with input resistance $R_o = 35\ k\Omega$ connected as the load of a potentiometer.

(a) Calculate the voltage V_o measured by the analog voltmeter.

(b) Calculate the error in analog voltmeter measurement of V_o with regard to the V_o measured by a digital voltmeter with $R_o = 12\ M\Omega$.

2.10. A tachometer has the tachometer constant $K = 6$ V/krpm (krpm = 1000 revolutions per minute) and is connected to an eight-bit ADC that has input voltage range from 0 to 10 V.

(a) Calculate the maximum acceptable velocity, which can be measured by the tachometer in this configuration.

(b) Calculate the velocity measurement resolution of the tachometer with ADC.

2.11. A two-track incremental encoder installed on a rotating shaft has 1000 slots evenly distributed along the circumference and has the output sampled at 10 MHz. If the photocell for the outer track is counted "on" for 10,000 sampling pulses, calculate the angular velocity of the rotating shaft.

2.12. An incremental encoder with 1° slots evenly distributed along the circumference of 1 cm radius is installed on a shaft rotating at 1000 rpm. Calculate the frequency of the pulse output (in pulses per second) of the encoder before sampling.

CHAPTER 3

Actuator Modeling

3.1 DIRECT CURRENT MOTORS

3.1.1 Introduction

Faraday demonstrated in 1841 that a coil conducting a current rotates when placed in the magnetic field produced by a permanent magnet. This discovery enabled the design of simple power actuators, direct current (DC) motors. DC motors require DC power supply, often available for other purposes such as analog electronic and digital devices. DC motors are often used in precise position servomotors and for driving loads at variable speed. When the applications consist of driving loads at a given speed, open-loop control can be used. Position control is achieved with servomotors that use closed-loop position control.

In this chapter, the focus is on interfacing issues of actuators with a mechanical system and the digital controller. Permanent magnet (PM) and shunt DC motors will be presented. The detailed presentation of various other DC actuators can be found in books dedicated to actuators [1, 23, 28, 137].

3.1.2 PM–DC Motor

PM–DC Motor Model The development of new magnetic materials led to a new generation of permanent magnet DC motors with improved performance, which are cheaper and easier to control. The simplified diagram of a PM–DC motor is shown in Fig. 3.1. The stator consists of a pair of magnetic poles, N–S. The rotor consists of coils of conducting wires connected through the segments of a collector to a DC power supply.

Figure 3.1 shows the cut from a mechanical load (with cut variables torque T and angular velocity ω) as well as the cut from a DC power supply (with cut variables voltage U and current i). The rotor is modeled mechanically as a rigid body with a moment of inertia J and a viscous friction coefficient b, accounting for the air drag and viscous

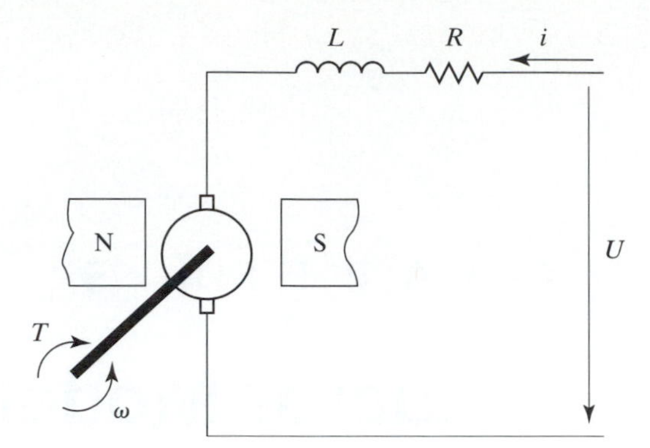

Figure 3.1

The diagram of a permanent magnet–DC motor

friction in the lubricated bearings. The electric model of the rotor is given by the lumped parameters R and L, the rotor winding circuit resistance and inductance, respectively. For a motor, the power will flow from the electrical cut (U, i) towards the mechanical cut (T, ω).

The conversion of the electrical energy from the DC power supply into the mechanical energy supplied to the load takes place in the DC motor, in particular in the air gap between the stator and the rotor. Forces applied on rotor coils are generated as a result of the current i flowing through the rotor winding surrounded by the magnetic field produced by the PM of the stator. At the same time, voltages are induced in the moving rotor winding as well due to the magnetic field, generating the so-called back electromotive force (back e.m.f.). These two effects in a PM–DC motor can be modeled by separating the mechanical subsystem and the electrical subsystem, each being modeled by two port elements, as shown in Figs. 3.2 and 3.3, respectively.

In the left-hand side of Fig. 3.2, torque components are represented around a cross section of the shaft. T_r denotes the (magnetic) torque acting on the rotor, while U_r represents the back electromagnetic force (back e.m.f.) induced by the magnetic field in the rotor winding, opposite to the supply voltage U. The torque T and angular veloc-

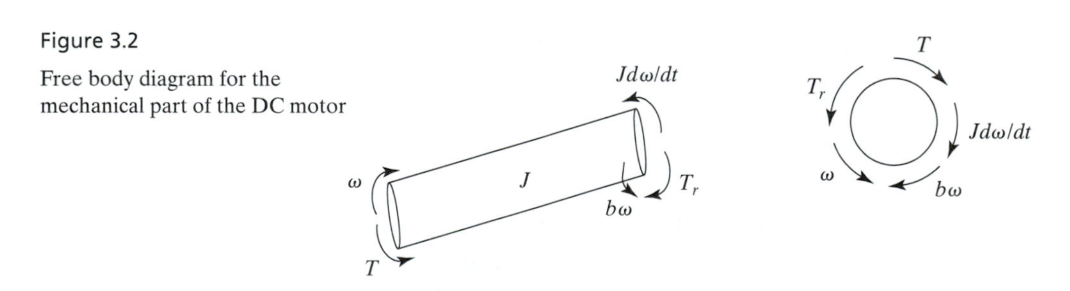

Figure 3.2

Free body diagram for the mechanical part of the DC motor

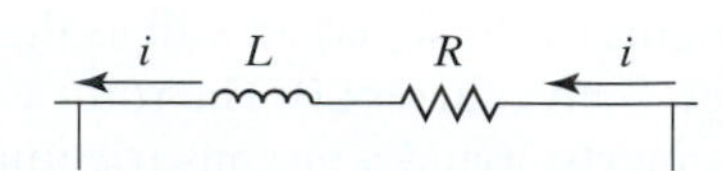

Figure 3.3

Two-port $[(U_r, i)$ and $(U, i)]$ circuit of the electrical part of the DC motor

ity ω are the cut variables toward the mechanical load, while the voltage U and the current i are the cut variables toward the DC power supply.

Two algebraic equations model the electromechanic conversion phenomena, viz.,

$$T_r = Ki$$

and

$$U_r = K_e \omega,$$

where K [Nm/A] is the torque constant and K_e [Vs/rad] is the electrical constant.

These two algebraic equations describe the electromechanical energy conversion. Assuming the conversion efficiency, $\eta \leq 1$, power transfer, from electrical power $U_r * i$ into mechanical power $T_r\ \omega$, is given by

$$T_r * \omega = \eta * U_r * i.$$

Using the foregoing two algebraic equations, we obtain the relationship

$$K = \eta K_e.$$

Power losses occur due to winding resistance, magnetic losses, friction, etc. In the case of negligible losses, ideal power conversion can be assumed ($\eta = 1$), such that for

$$K \text{ in [Nm/A] and } K_e \text{ in [Vs/rad]},$$
$$K = K_e.$$

For the mechanical part, shown in the free body diagram of Fig. 3.2, Newton's second law gives

$$J d\omega/dt = T_r - b\omega - T.$$

For the electrical part shown in Fig. 3.3, the voltage drop equation gives

$$U = L di/dt + Ri + U_r.$$

The last two differential equations and the two algebraic equations regarding the electromechanic conversion form a system of four differential-algebraic equations containing six variables T, ω, T_r, U_r, i, and U. This system of four differential-algebraic equations represents the analytical model of the PM–DC motor.

The elimination of internal variables T_r and U_r results in a model reduced to two differential equations with four variables of the two cuts (T, ω) and (U, i):

$$Ki = J d\omega/dt + b\omega + T;$$
$$U = L di/dt + Ri + K_e \omega.$$

Most DC motors have negligible L, such that the model for $L = 0$ is reduced to

$$J d\omega/dt = Ki - b\omega - T;$$
$$U = Ri + K_e\omega.$$

Solving for the cut variables (U, i) on electrical side and (T, ω) on the mechanical side, we obtain

$$\omega = (1/K_e)U - (R/K_e)i;$$
$$T = -(J/K_e)dU/dt - (b/K_e)U + (JR/K_e)di/dt + (K + bR/K_e)i.$$

This form is used later in this chapter for block diagram representation and in Section 5.4 for effort-flow modeling. This form will permit us to show how the block diagram language can incorporate features of modularity using the cuts (U, i) and (T, ω).

For Simulink modeling based on the data flow approach, the model is cast as follows:

$$i = (1/R)U - (K_e/R)\omega;$$
$$d\omega/dt = (K/J)i - (b/J)\omega - (1/J)T.$$

Simulink is a block diagram simulator based on data flow approach. In data flow approach, each block is executed immediately after all input values are available. The equations of the model were rearranged in order to satisfy the data flow requirement and to create a model that can be solved numerically using integration only and avoid numerical derivations. Numerical derivation can result in numerical computation difficulties in case of discontinuous or noisy signals. The examples of Simulink programs presented in this book can be easily understood after following the Simulink tutorial [35].

Example: Simulink Model of a PM DC Motor

The last two differential equations are used in the Simulink model with the following notation:

$$\begin{aligned} i &= (1/R)U - (K_e/R)\omega \\ &= A_u U - A_o\omega; \\ d\omega/dt &= (K/J)i - (b/J)\omega - (1/J)T \\ &= B_i\, i - B_o\omega - B_T\, T. \end{aligned}$$

The model, represented using Simulink graphical conventions, is shown in Fig. 3.4. For a small PM–DC motor, typical values for the parameters are $K = 0.015$ Nm/A, $K_e = 0.066$ Vs/rad, $R = 3.3\ \Omega$, $b = 0$, and $J = 30 \cdot 10^{-6}$ kg m^2 such that $A_u = 1/R = 0.303$, $A_o = K_e/R = 0.02$, $B_i = K/J = 500$, $B_o = b/J = 0$, and $B_T = 1/J = 33333$. In Simulink models, triangles represent multiplication of the input with the value assigned to the gain. Only three positions are visible in the display of the gain value. As a consequence, in Fig. 3.4, $A_o = 0.02$ appears truncated as 0.0, $B_i = 500$, appears correctly as 500, and $B_T = 33333$ as a generic gain K and not as a numerical value.

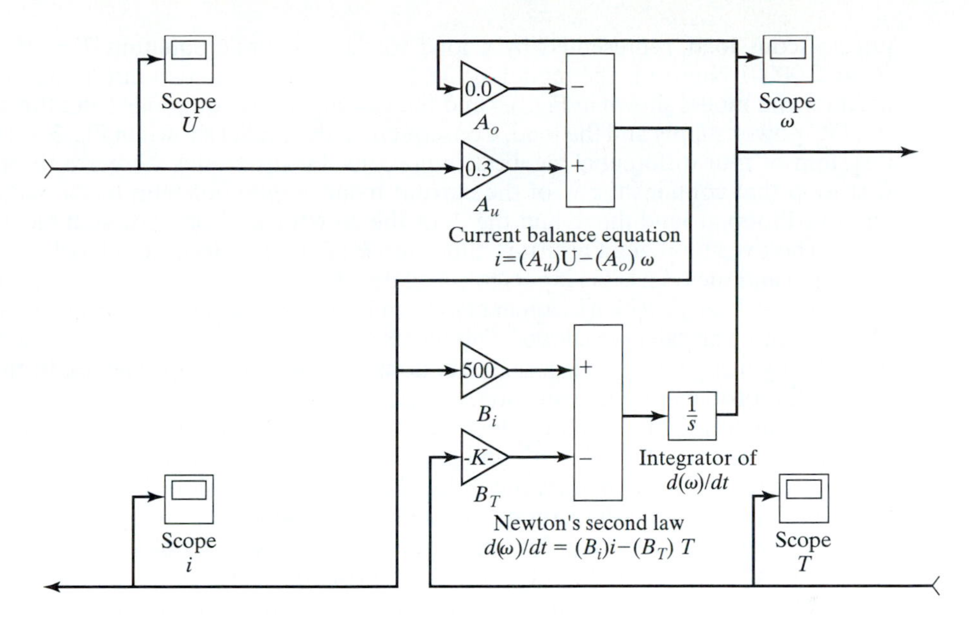

Figure 3.4

Simulink model of a PM–DC motor subsystem

The model of Fig. 3.4 requires interfacing of cut variables (U, i) and (T, ω) with the submodels of the other components.

In Fig. 3.5, the motor is assumed supplied from a voltage source, modeled by the voltage drop algebraic equation $U = U_s + R_s i$, where $U_s = 10$ V and $R_s = 0.3\ \Omega$ are the constant DC source voltage and internal resistance, respectively. The load is assumed a

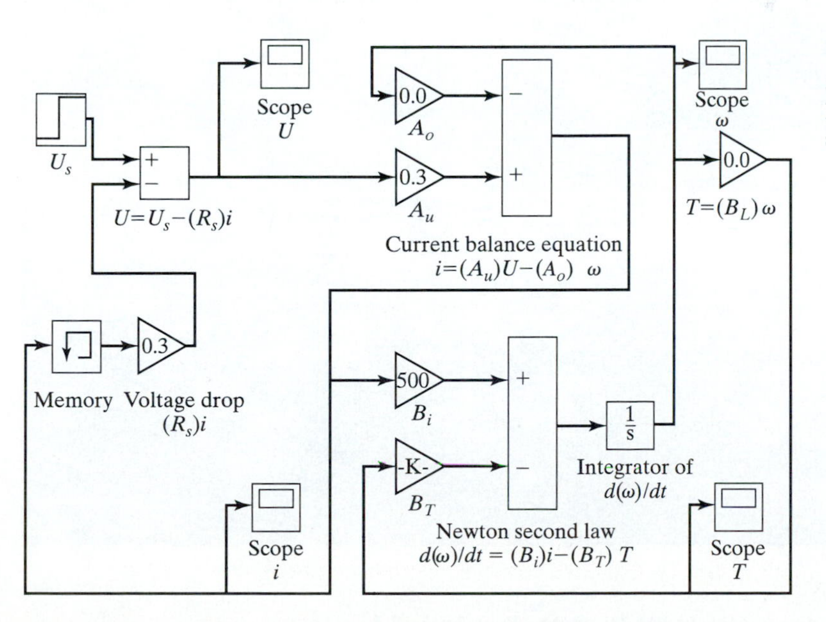

Figure 3.5

Simulink model of the PM–DC motor with DC supply and load

pure viscous load, represented by a load torque algebraic equation $T = B_L\omega$, where $B_L = 0.00001$ Nms/rad. The model in Fig. 3.5 is based on the two differential equations used for the model shown in Fig. 3.4 and the two algebraic equations from the models of the DC power supply and the load. Consequently, the model shown in Fig. 3.5 is based on a system of four differential algebraic equations. In this model there are three loops, a first loop that contains the + of the current balance equation sum block and a second and third loop around the + and the − of the Newton's second law sum block, respectively. The execution at an arbitrary time step k of the first loop sum block requires the $(A_u)U(k)$ and the $(A_o)\ \omega\ (k)$ inputs to calculate $i(k)$ output. Given that $U(k)$ is calculated with $U(k) = U_s + R_s\, i(k), i(k)$ appears as both input and output at time step k of the sum block, creating an "algebraic loop." This algebraic loop problem can be easily solved symbolically by using the two algebraic equations to eliminate two variables. In the case of solving the problem numerically, special algorithms are required. In Simulink, the solution is the inclusion of a memory block in the current i. The memory block is a block that delays the output $i(k) = i(k - 1)$, and this breaks the algebraic loop on the expense-reduced accuracy in the dynamics computation. For relatively slow dynamics compared to the max step size of the simulation, this can be acceptable [73].

The second and the third loops contain an integrator $1/s$ block, and this does not permit the appearance of the algebraic loop problem. For example, numerical computation of the integral using the Euler method is based on earlier time step values for the current time step and has the same effect as the memory block.

Figure 3.6 shows the simulation results. As a result of the voltage drop on $R_s = 0.3\ \Omega$, the voltage U differs from the step input $U_s = 10$ V, more significantly during the time

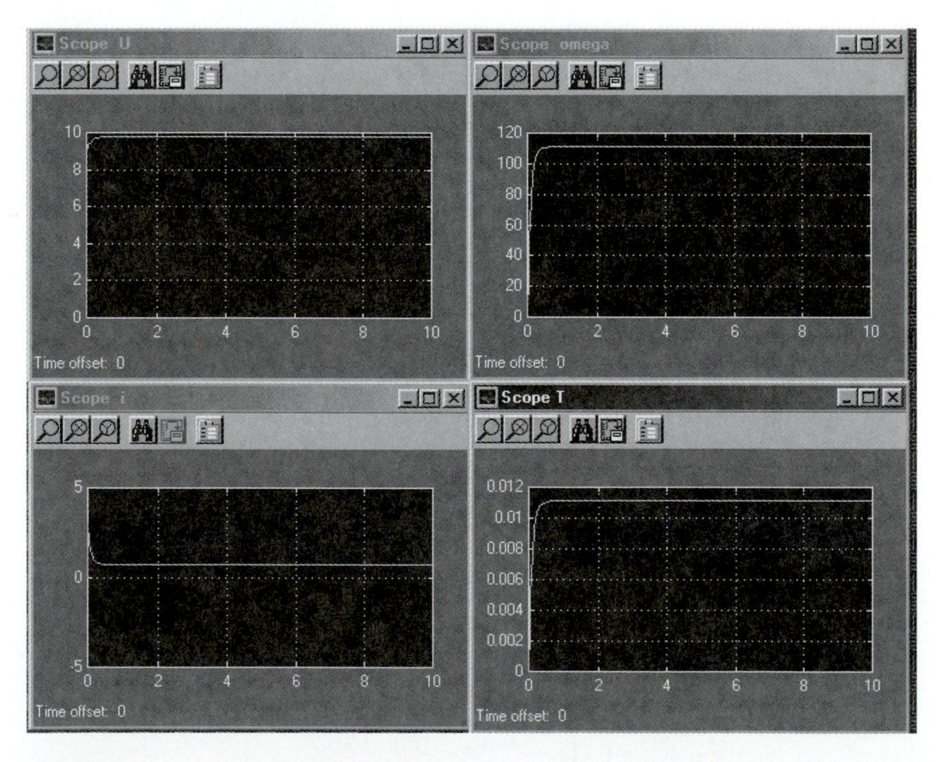

Figure 3.6

Simulation results for the PM–DC motor shown in Fig. 3.5

of high current i and then it stabilizes at a steady-state value of 9.7 V. The angular velocity reaches a steady-state value of about 116 rad/s and the torque reaches a steady-state value of about 0.011 Nm while the current stabilizes at about 1 A.

The Simulink model shown in Fig. 3.5 maintains explicitly the cut variables (U, i) and (T, ω), but due to the data flow approach of Simulink, an assignment of U and T as input variables and of i and ω as output variables was required. Other assignments are possible, but for interfacing, modules on the supply and on the load side have to be given compatible assignments (i.e., outputs have to match inputs). For example, a module on the supply side has to have U as output and i as input, as shown in Fig. 3.5.

The results of the simulation can be displayed on four scopes giving the time variation of the four cut variables for the step input voltage $U_s = 10$ V. This modular representation permits easy replacements of the subsystems of the model; for example, on the DC supply side, to consider a power supply located at some distance from the motor, the effect of the resistance r and inductance l of the wire connection needs to be accounted for:

$$U = U_s - R_s i - ri - l\,di/dt.$$

Also, a more complex load can be modeled by a moment of inertia J_L, besides B_L.

The use of this modularity property of the effort-flow models will be illustrated in more detail in Sections 5.3 and 5.4.

Block Diagram Representation Block diagram representation is often used in modeling and control. For block diagram representation of linear systems, the input–output relationship of each block is derived using the Laplace transform.

The PM–DC motor model for the two cuts (T, ω) and (U, i), when $\eta = 1$ (i.e., for $K_e = K$), is

$$Ki = J d\omega/dt + b\omega + T;$$
$$U = L di/dt + Ri + K\omega.$$

The Laplace transform for zero initial conditions gives

$$Ki(s) = (Js + b)\omega(s) + T(s);$$
$$U(s) = (Ls + R)i(s) + K\omega(s),$$

or

$$[Ki(s) - T(s)]/(Js + b) = \omega(s);$$
$$[U(s) - K\omega(s)]/(Ls + R) = i(s).$$

Block diagram representation requires the assignment of an input signal and of an output signal. This implies, for the output cut, a known relationship between the cut variables (T, ω).

In the case of a linear relationship, its Laplace transform gives

$$T(s) = f(s)\omega(s),$$

where $f(s)$ is the transfer function between the input $\omega(s)$ and the output $T(s)$. This equation, plus two equations with four variables representing the model of a DC motor, allows us to eliminate two variables such that for the desired variation of one output variable, a solution for one input cut variable can be obtained. This approach permits the formulation of open-loop control using the inverse dynamics solution linking the desired output variable to the required input variable.

Either U or i can be chosen as input variable and either T or ω as output variable. Four different combinations can be considered:

	Input signal	Output signal
(a)	U	ω
(b)	U	T
(c)	i	ω
(d)	i	T

Case (a) will be illustrated in the following example.

Example: Block Diagram of a DC Motor with Inertial-Viscous Load

A DC motor used for stirring the liquid in a tank can be assumed to have a load modeled by the lumped parameters J_L (moment of inertia) and b_L (viscous friction coefficient). Figure 3.7 shows the DC motor and the free-body diagram of the inertial-viscous load.

A free-body diagram of the load gives

$$T = J_L d\omega/dt + b_L\omega.$$

After applying the Laplace transform for zero initial conditions, the form $T(s) = f(s)\,\omega\,(s)$ is obtained:

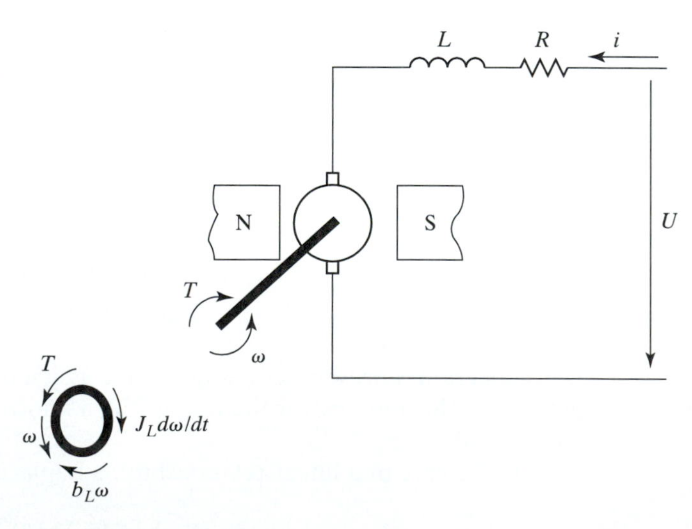

Figure 3.7

DC motor and an inertial-viscous load

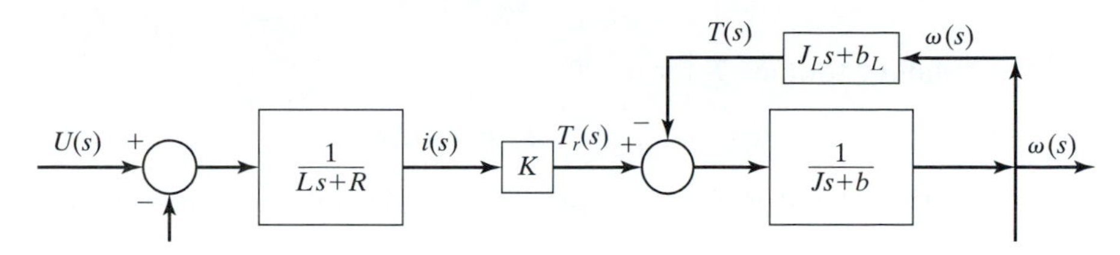

Figure 3.8

Block diagram of the DC motor with inertial-viscous load

$$T(s) = (J_L s + b_L)\omega(s).$$

This load equation plus the two equations of the PM–DC motor model

$$[U(s) - K\omega(s)]/(Ls + R) = i(s)$$

and

$$[Ki(s) - T(s)]/(Js + b) = \omega(s)$$

can be used to obtain the block diagram from Fig. 3.8, where U is the input and ω is the output.

This block diagram has the same structure as the Simulink model from Fig. 3.5. This block diagram is based on transfer functions, available only for linear systems. In general, block diagrams lack the modularity feature, a problem that will be addressed in Section 5.4.

Steady-State Characteristics of a DC Motor The steady-state characteristics of a DC motor are based on the steady-state dependence between the motor torque $T_r = Ki$ and the motor velocity ω.

For the DC motor, the voltage drop equation is

$$U = Ldi/dt + Ri + K\omega.$$

In the case that the DC motor is in a steady state, the derivative vanishes:

$$U = Ri + K\omega.$$

For $i = T_r/K$,

$$U = RT_r/K + K\omega.$$

Two limit cases can be defined:

no-load case $T_r = 0$, in which the "no load speed" is

$$\omega_0 = U/K;$$

"stalling torque" T_s for $\omega = 0$:

$$T_s = KU/R.$$

Substituting U and R/K in the preceding steady-state equation, the following equation results:

$$T_r/T_s + \omega/\omega_0 = 1.$$

Using the notations ω_0 and T_s, this equation shows that the motor torque T_r is not independent from the angular velocity ω, such that its simulation has to be based on the reproduction of the linear dependence between T and ω. The supply voltage U can, however, be modeled as independent from the current i if the electric energy source voltage U is practically insensitive to the current i variation for normal electric loads.

3.1.3 Shunt DC Motor

A simplified diagram of a shunt DC motor is shown in Fig. 3.9. The permanent magnets of the PM–DC motor are replaced in this case, by a field winding on the stator with inductance L_f, resistance R_f, and current i_f. The rotor consists of coils of conducting wires connected through the segments of a collector to a DC power supply.

The motor is modeled by separating the mechanical subsystem and the electrical subsystem, each modeled by two port elements shown in Figs. 3.10 and 3.11, respectively.

Two algebraic equations model electromechanical conversion phenomena:

$$T_r = K_a i_a K_f i_f$$

and

$$U_r = K_e \omega.$$

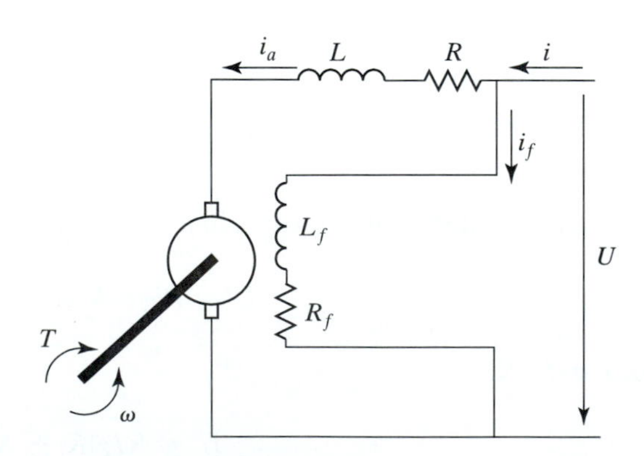

Figure 3.9
A shunt DC motor diagram

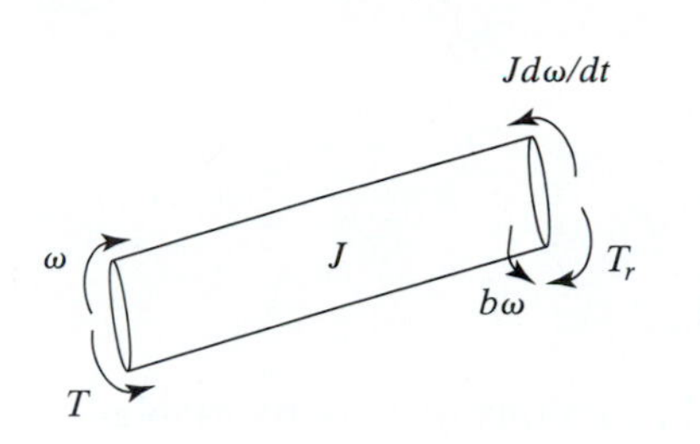

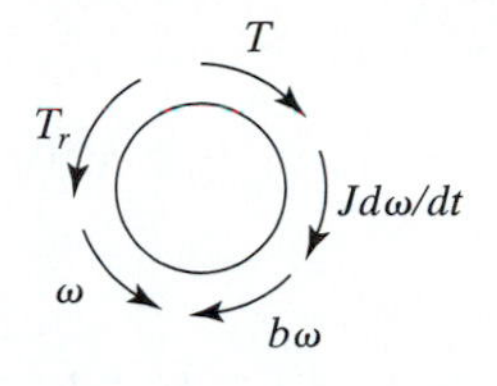

Figure 3.10

Free-body diagram for the mechanical part of the DC motor

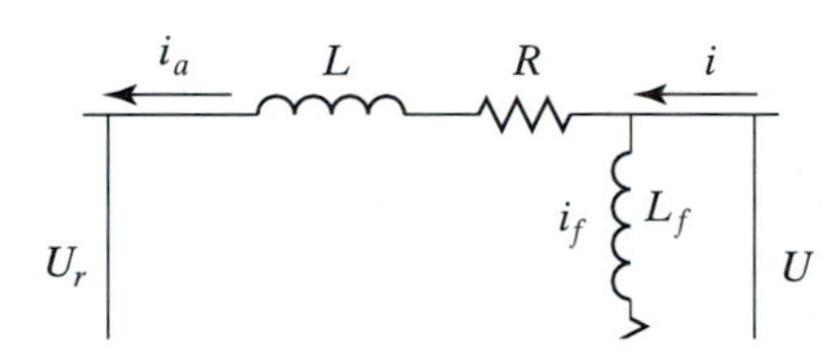

Figure 3.11

Two-port $[(U_r, i_a)$ and $(U, i)]$ circuit of the electrical part of the DC motor

For the mechanical part, shown in the free-body diagram of Fig. 3.10, Newton's second law gives

$$J d\omega/dt = T_r - b\omega - T.$$

For the electrical part shown in Fig. 3.11,

$$U = L di_a/dt + Ri_a + U_r,$$
$$U = L_f di_f/dt + R_f i_f,$$

and

$$i = i_a + i_f.$$

A system of six algebraic-differential equations containing eight variables T, ω, T_r, U_r, i, i_a, i_f, and U was obtained. Most DC motors have negligible L, and L_f such that $L = 0$ and $L_f = 0$ will be considered further on.

The elimination of internal variables T_r and U_r results in a model reduced to four equations:

$$K_a i_a K_f i_f = J d\omega/dt + b\omega + T,$$
$$U = Ri_a + K_e\omega,$$
$$U = R_f i_f,$$

and

$$i = i_a + i_f.$$

Eliminating

$$i_f = U/R_f$$

and

$$i_a = i - i_f = i - U/R_f,$$

the model is reduced to the following two differential equations with four variables of the two cuts $(T, \omega)$and (U, i):

$$K_a(i - U/R_f)K_f(U/R_f) = Jd\omega/dt + b\omega + T$$

and

$$U = R(i - U/R_f) + K_e\omega.$$

Solving the last equation for ω gives

$$\omega = (1 + R/R_f)(1/K_e)U - (R/K_e)i.$$

Replacing ω in the torque equation gives

$$K_a(i - U/R_f)K_f(U/R_f) = (J/K_e)(1 + R/R_f)dU/dt + (b/K_e)(1 + R/R_f)U - (JR/K_e)di/dt + b(1 + R/R_f)(1/K_e)U - (bR/K_e)i + T.$$

Solving for T yields

$$T = -(J/K_e)(1 + R/R_f)dU/dt + (JR/K_e)di/dt - (b/K_e)(1 + R/R_f)U + (bR/K_e)i + (K_aK_f/R_f)Ui - (K_aK_f/R_f{}^2)U^2.$$

This model is nonlinear due to the terms Ui and U^2. The Laplace transform cannot be applied to eliminate variables U and i that appear also in differential form. This representation, based on electrical side cut (U, i) and mechanical side cut (T, ω) can be used for a Simulink model.

Example: Simulink Model of a Shunt DC Motor

The equations, represented in the Simulink model shown in Fig. 3.12, were derived following the same procedure as in the case of the PM–DC motor (for U and T as inputs and ω and i as outputs, and the cuts (U, i) and (T, ω), resulting in

$$i = (1/R)(1 + R/R_f)U - (K_e/R)\omega = A_uU - A_o\omega$$

and

$$dw/dt = ((K_aK_f)/(JR_f))Ui - ((K_aK_f)/(JR_f^2))U^2 - (1/J)T = B_{ui}Ui - B_{uu}U^2 - B_TT,$$

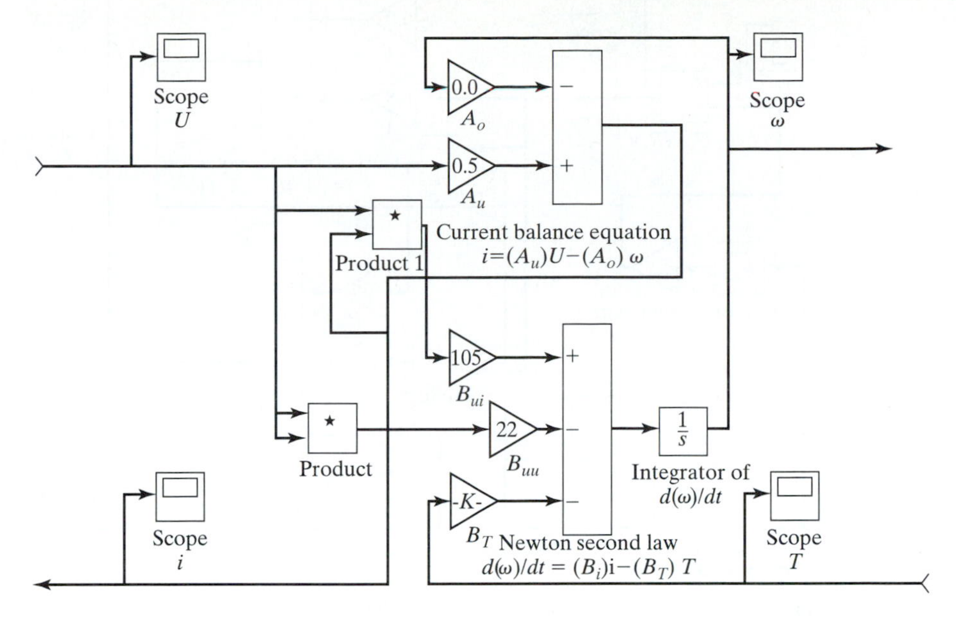

Figure 3.12

Simulink model of a shunt DC motor

where

$$
\begin{aligned}
A_u &= (1/R)(1 + R/R_f),\\
A_o &= K_e/R,\\
B_{ui} &= (K_a K_f)/(JR_f),\\
B_{uu} &= (K_a K_f)/(JR_f^{\,2}),
\end{aligned}
$$

and

$$B_T = 1/J.$$

For $K = K_a K_f = 0.015$ Nm/A, $K_e = 0.066$ Vs/rad, $R = 3.3\ \Omega$, $b = 0$, $R_f = 4.75\ \Omega$, and $J = 30 \cdot 10^{-6}$ kg m^2 such that $A_u = (1/R)\ (1 + R/R_f) = 0.5135$, $A_o = K_e/R = 0.02$, $B_{ui} = (K_a K_f)/(JR_f) = 105$, $B_{uu} = (K_a K_f)/(JR_f^{\,2}) = 22.16$, $B_o = b/J = 0$, and $B_T = 1/J = 33{,}333$.

The Simulink model for the shunt motor subsystem is shown in Fig. 3.12.

This Simulink model for the nonlinear model of the shunt DC motor differs from the PM–DC model of Fig. 3.4 only due to the nonlinear terms $(B_{ui}U_i)$ and $(-B_{uu}U^2)$. Similarly to the PM–DC motor model, the model of Fig. 3.12 requires interfacing of cut variables (U, i) and (T, ω) with the models of the other components in order to be executable. In Fig. 3.13, the shunt DC motor model is shown with the same DC power supply and load as in Fig. 3.5 Figure 3.14 shows the simulation results for the shunt DC motor. The results are similar in shape with the results for PM–DC motor shown in Fig. 3.6, but with different steady-state values for current i (2.5 A), voltage U (9.25 V), angular velocity ω (140), and load torque T (0.014 Nm).

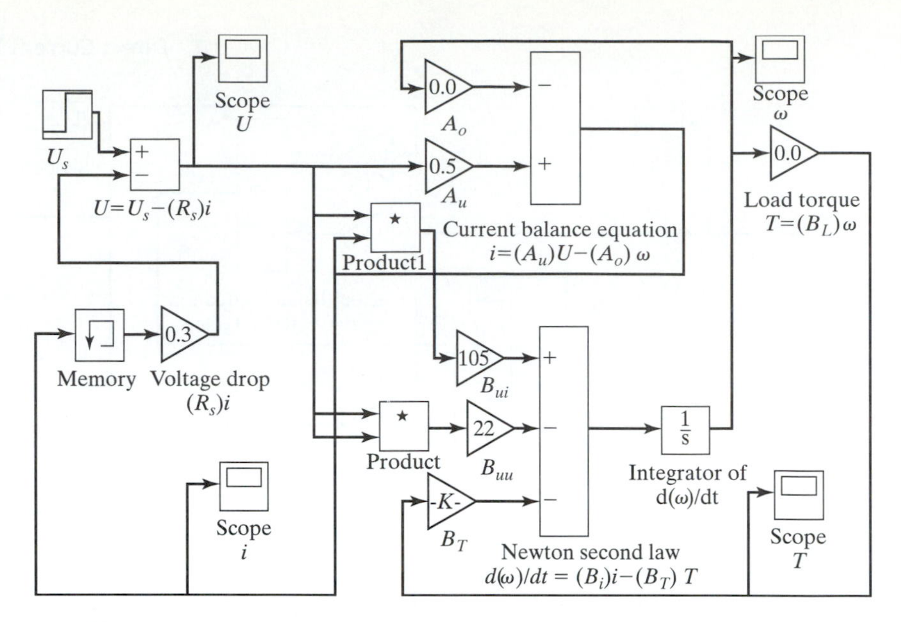

Figure 3.13

Shunt DC motor with DC power supply and viscous load

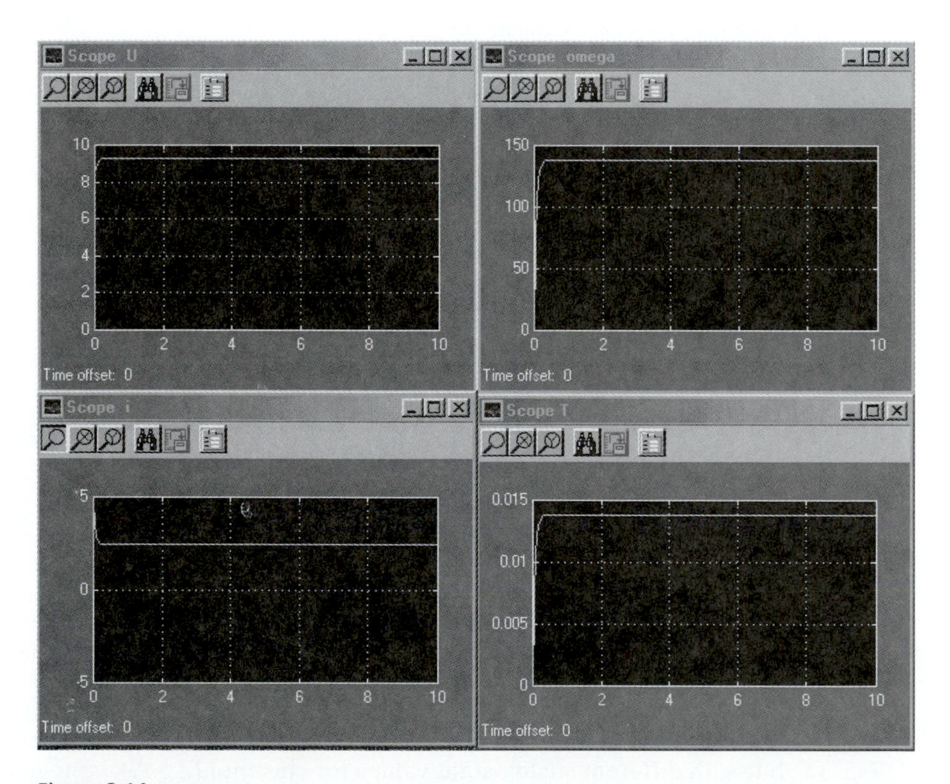

Figure 3.14

Simulation results for the shunt DC motor shown in Fig. 3.13

3.2 STEPPER MOTORS

Stepper motors are frequently used due to their high ratio of torque versus size, easy computer control, good performance in open-loop control, etc. The basic types are permanent magnet and variable reluctance stepper motors. A hybrid type is most frequently used in industry and will be presented here.

Hybrid stepper motors have a permanent magnet rotor magnetized along the rotor shaft axis. As shown in Fig. 3.15, the same number of teeth fitted as end plates of steel along the rotor are allocated to the north (N) pole and to the south (S) pole of the permanent magnet. As shown in Fig. 3.15, the N and S adjacent teeth are offset by 60°, and the end plates constitute a variable reluctance path for the stator windings.

In Fig. 3.16, a hybrid stepper motor with a six-toothed-rotor and a two-phase stator is shown. The stator has four windings, 1, 1′, 2, and 2′. Windings 1 and 1′ are con-

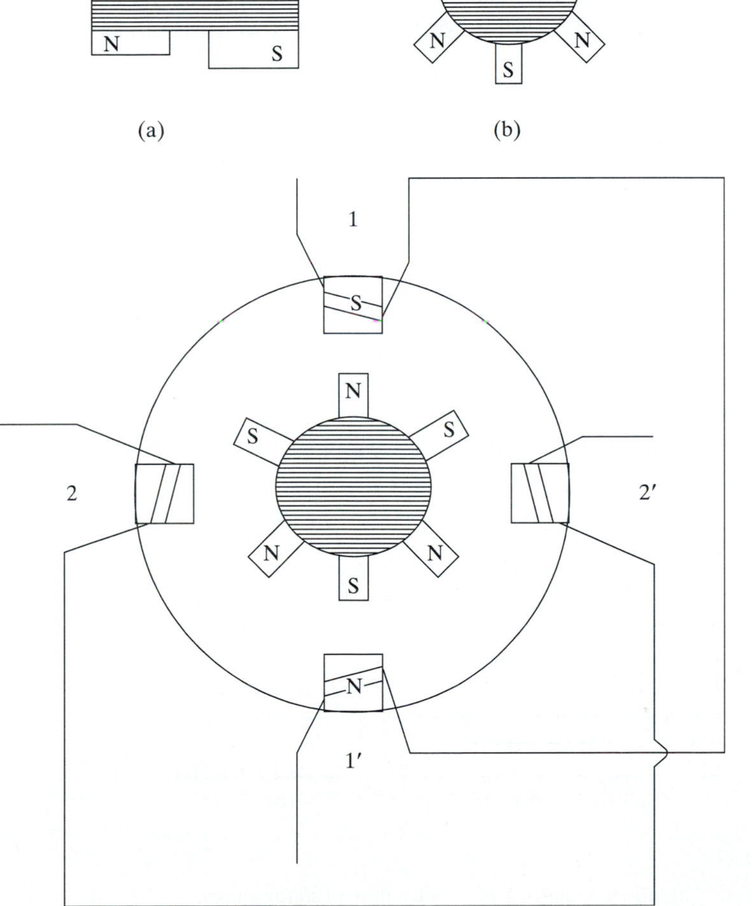

Figure 3.15

Rotor teeth arrangement vertical-longitudinal (a) and vertical-lateral (b) cross sections

Figure 3.16

Hybrid stepper motor with six rotor teeth and two-phase stator

nected in series to form phase 11′, and windings 2 and 2′ are also connected in series to form phase 22′. The teeth of the rotor will tend to find a minimum reluctance (or magnetic resistance) path along stator and will align along one or another of the stator winding cores, called a détente position [139]. The rotor is held in the equilibrium position by a small détente torque due to the attractive forces generated by the permanent magnets. With six rotor teeth and a two-phase stator, there are 12 equilibrium positions. In Fig. 3.16 an equilibrium position with N, rotor teeth aligned with the core of the phase 1 is shown.

The stepper motor rotor turns as a result of applying excitation current to the stator phases in a particular sequence, which changes the attractive forces produced by the interaction of the permanent magnet of the rotor with stator windings. In Fig. 3.17, a simple sequential excitation of the stator phases called full stepping is shown. In this case, at any time only one stator phase is powered [139]. Figure 3.18 shows a sequence of four steps of 30° each, as a result of exciting stator phases as shown in Fig. 3.17.

In case (a), only phase 11′ is energized. For the given direction of the current, winding 1 produces the pole S and winding 1 produces pole N. The N–S poles, of the rotor shown in Fig. 3.16, are attracted by the equilibrium position 11′. The position shown in Fig. 3.16 is maintained in this case by a holding torque due to the interaction of phase 11′ and the rotor teeth. The holding torque is much greater than the détente torque.

In case (b), only phase 22′ is energized and pole S results from winding 2 and N from 2′. The rotor turns clockwise by 30° to reach a new equilibrium position along 22′.

In case (c), only phase 11′ is energized such that pole N results from winding 1 and S from 1′. The rotor turns further clockwise by 30° to reach another equilibrium position.

In case (d), only phase 22′ is energized such that pole N results from winding 2 and from 2′. The rotor turns further by another 30° to reach the next position.

The sequence of four phase excitations shown in Fig. 3.17 resulted in four angular steps of 30° in clockwise rotation of the rotor (i.e., a quarter of a revolution). A repetition of the 30° phase excitations keeps the motor turning clockwise. A full revolution requires 360°/30° = 12 steps.

Counterclockwise motion can be obtained by applying the reverse sequence of excitations shown in Fig. 3.17 (i.e., (d), (c), (b), and (a)).

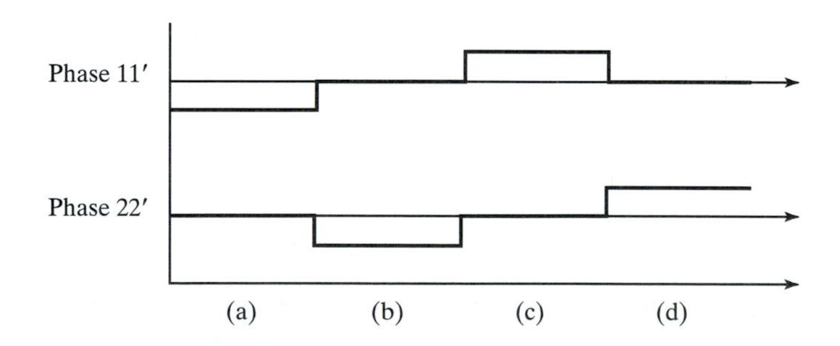

Figure 3.17

Time diagram of full-stepping, one-phase excitation of stator phases

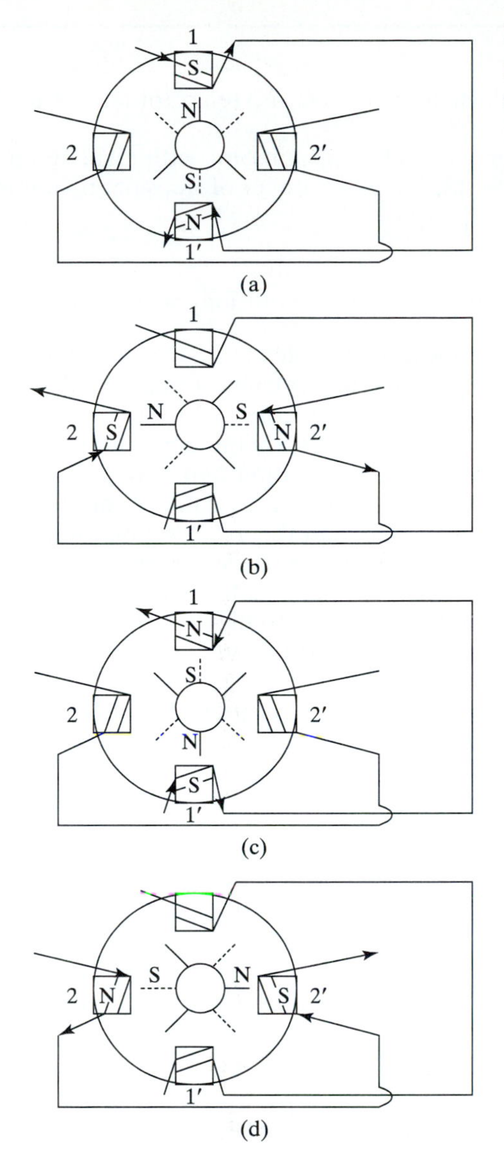

Figure 3.18

Full-stepping, one-phase excitation of stator phases

For a hybrid stepper motor with n phases on the stator and m teeth on the rotor, the total number of steps per revolution N is [32]:

$$N = nm.$$

In the case of the stepper motor shown in Fig. 3.18, $n = 2$ phases and $m = 6$ rotor teeth, such that $N = 12$ [steps/revolution].

The step angle, or the angular resolution r is [140] r [°/step] = 360 [°/revolution] / N [steps/revolution] and $r = 360°/12 = 30$[°/step], as shown in Fig. 3.18.

Example

A hybrid stepper motor has eight stator windings and ten rotor teeth. Calculate the angular resolution.

The number of phases for a hybrid stepper motor with eight stator windings is $n = 4$ [phases]. Given $m = 10$ [teeth], the total number of steps per revolution is $N = 40$ [steps/revolution].

The resolution is

$$r = 360[°/\text{revolution}]/40[\text{steps/revolution}] = 9[°/\text{step}].$$

Often, stepper motors are designed with step angles of 1.8 or 0.9 [°/step].

The interfacing of a computer generating a train of pulses and a stepper motor requires a translator and a drive, as shown in Fig. 3.19 [32, 139].

Digital output of a pulse train and the direction bit (e.g., 0 for clockwise and 1 for counterclockwise motion) from a PC have to be converted into the required sequence of phase excitations to achieve stepwise rotation of the rotor of the stepper motor [32, 140]. The translator converts the pulse train and the direction bit information in commands to the driver. Typically, a drive uses power transistors with base signals coming from the translator. These power transistors have as loads the windings of the stator [140]. In Fig. 3.19, the translator sends four signals to the drive for sequential excitation of the four windings of a typical two-phase hybrid stepper motor. Drives for actuators are presented in more detail in Section 4.6.

In open-loop velocity control, the PC has to generate a constant frequency pulse train for the desired velocity v_d [revolutions/sec] over a given duration of time t_d [sec].

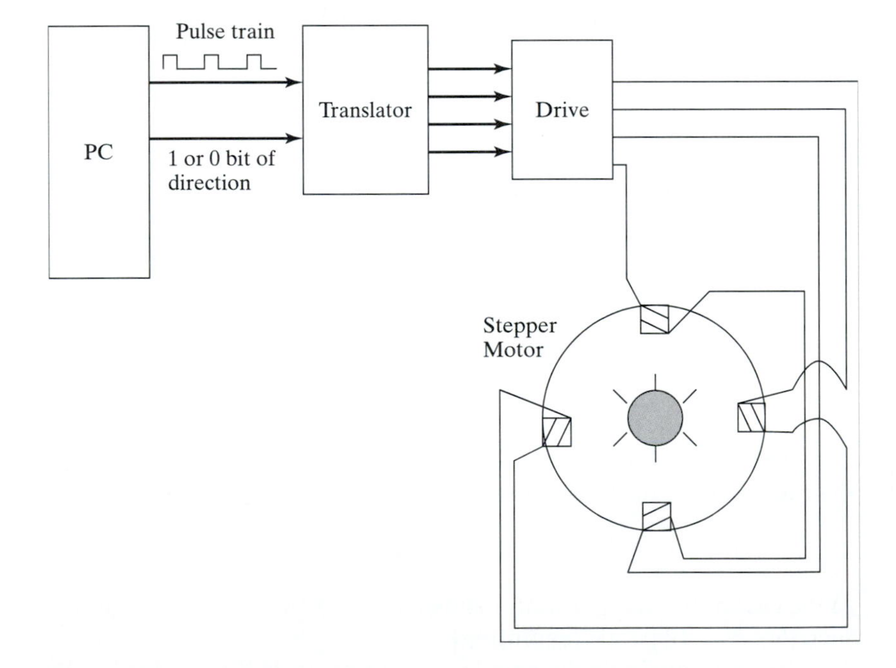

Figure 3.19

Translator and drive for a stepper motor PC-based controller

Assuming a full-step excitation of an N [steps/revolution] stepper motor, this corresponds to a desired velocity V_d [steps/sec] given by

$$V_d = v_d N.$$

For a rotation of T_d [steps],

$$T_d = V_d t_d.$$

The generation of a total number of steps T_d [steps] with a step frequency of V_d [steps/sec] can be obtained by a PC using a variety of methods, such as the digital differential analysis (DDA) method [32]. In Fig. 3.20, the block diagram of the DDA method is shown. Assume a PC clock frequency F [cycles/sec] and 1 [addition/cycle] (i.e., addition frequency f [addition/sec] equal to PC clock frequency). Also, it is assumed that the change from one value of the constant velocity to another is small, the acceleration or deceleration durations are negligible, and the word length M [bit] of the PC is large enough to have negligible effect on numerical computation errors.

For the desired velocity V_d [steps/sec] and the addition frequency f [additions/sec], the DDA algorithm calculates VV [steps/addition]:

$$VV = V_d/f.$$

The binary computation uses M-bit binary representation $\theta\theta$ of the current angular position θ.

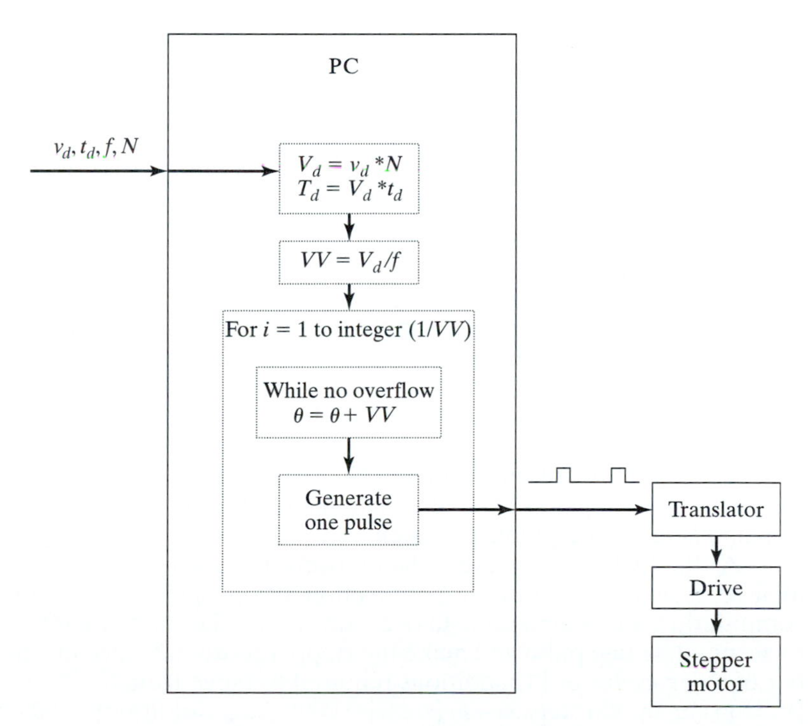

Figure 3.20

Block diagram of DDA method of generating a pulse train

At each PC clock cycle, DDA algorithm calculates $\theta\theta + VV$ from the value $\theta\theta$ at the beginning of the clock cycle and the value VV. This computation is repeated until an overflow occurs (i.e., until the repeated addition of VV value reaches at least one step value). At overflow, a new pulse is generated to make the stepper motor advance by one step. This method permits us to synchronize in real time the pulse train generation using the PC clock. An accurate pulse frequency generation requires that the PC interrupts do not affect real-time pulse generation. The accuracy of pulse frequency generation can be guaranteed by the real-time control solutions presented in Chapter 7.

Example

Assume that a PC has to generate a constant frequency pulse train for the desired velocity $v_d = 0.25$ [revolutions/sec] to change the angular position of the rotor from $\theta =$ 0.5 [revolution] to $\theta = 2.5$ [revolutions]. The stepper motor has $N = 200$ [steps/revolution] and uses full step excitation. PC has a clock frequency of $F = 100{,}000$ [cycles/sec] and M64 bit.

Calculate the desired velocity V_d [steps/sec], T_d [steps], and the number of clock cycles required to using DDA method.

***Solution*:** The desired velocity is

$$V_d = 0.25 \cdot 200 = 50 \text{ steps/sec.}$$

To move from $\theta = 0.5$ [revolution] to $\theta = 2.5$ [revolutions] a stepper motor with $N =$ 200 [steps/revolution] requires a pulse train that moves the rotor from $\theta = 0.5 \cdot 200 =$ 100 [step] to $2.5 \cdot 200 = 500$ [step] (i.e., by 400 steps).

At a constant velocity of 0.25 [revolutions/sec], the motion from $\theta = 0.5$ [revolution] to $\theta = 2.5$ [revolutions] requires $t_d =$ (2 [revolutions]) / 0.25 [revolutions/sec] = 8 [sec] or, as computed before, $T_d = 50 \cdot 8 = 400$ [steps].

Assuming that 1 [addition/cycle], the PC addition frequency is $f = 100{,}000$ [addition/sec]. For the binary computation using 64-bit binary representation, 1 bit is for sign and 63 bits for the binary representation of numbers. The error due to 63-bit binary representation is $1 - 1/2^{63} = 1 - 1.08 \cdot 10^{-19}$ and, in this case, it is acceptable to ignore the error due to binary representation VV of the V_d and the binary representation $\theta\theta$ of θ.

The DDA algorithm calculates $VV = 50/100{,}000 = 0{,}0005$ [steps/addition]. The value of VV is supposed to be less than one in order to have several PC additions per step and achieve acceptable precision.

At the first PC clock cycle, the first addition adds 0.0005 [steps/addition], in decimal representation, to the initial position 100 [steps] and no overflow occurs. This computation is repeated until an overflow occurs (i.e., after 1/0.0005 = 2000 additions to accumulate one pulse and make the stepper motor advance by one step). The number of clock cycles or PC additions required to move from $\theta = 100$ [steps] to $\theta = 500$ [steps] (i.e., by 400 steps is 400 [steps}/0.0005 [steps/addition] = 800,000 additions). It can be verified that the PC clock will count for 800,000 [additions] /100,000[additions/sec] = 8 [sec].

3.3 HYDRAULIC MOTORS

Hydraulic motors are heavy-duty actuators that can produce high actuating forces. They are often used for construction machines, airplane direction and attitude control, machine tools, vehicle steering , etc. Hydraulic motors use high-pressure oil for power transmission and can produce high forces in a rapidly changing pattern. The use of high pressure oil requires pumps and hydraulic pipes, which increase the size and operational difficulties of the system.

Figure 3.21 shows a cross section of the Medium Duty Hydraulic Cylinder, Parker Hannifin [139]. The two ports of the hydraulic cylinder are used for interfacing with the valve controlling the flow of oil under pressure, and the piston rod is used to transmit mechanical power to the load to be moved.

Figure 3.22 shows the diagram of the complete hydraulic actuator with the valve and the mass-spring-damper, $m_L - k_L - b_L$ load [1, 30]. The input variable is the posi-

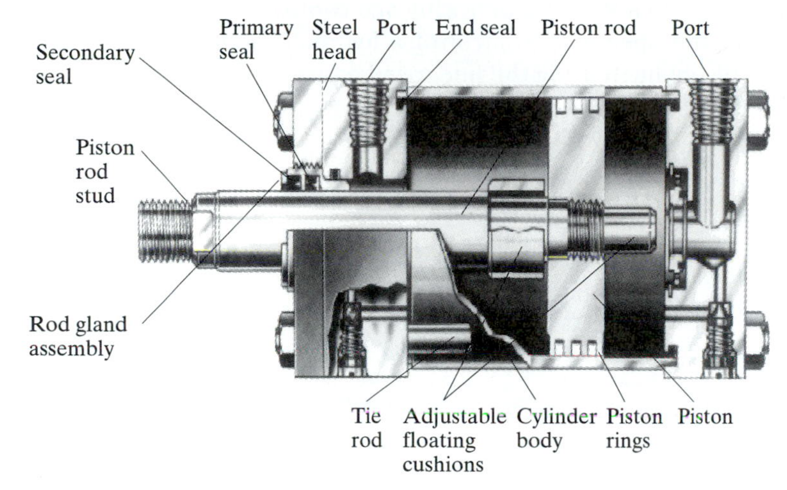

Figure 3.21

Cross section of Medium Duty Hydraulic Cylinder, (*Source*: Parker Hannifin [139])

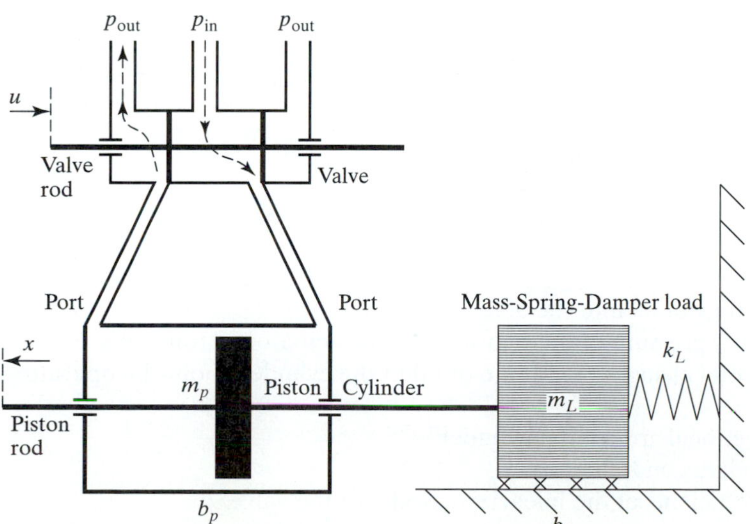

Figure 3.22

Open-loop control of a hydraulic actuator with mass-spring-damper load

tion u of the valve rod, which controls the motion of the piston rod. The output variable is the position x of the piston rod and of the mass m_L of the load.

The piston is modeled by the mass m_p and the viscous friction coefficient b_p. The oil under pressure comes at a constant high pressure p_{in} and leave the valve at low pressure p_{out}. The valve pistons are shown in Fig. 3.22 displaced slightly to the right such that the right-hand side port of the valve is open and permits the flow of oil to the right-hand side port of the cylinder. As a result, the piston is pushed leftwards and the oil in the left-hand side of the piston is pushed out at p_{out}. If the valve rod moves leftwards and opens the left-hand side port, the motion of the piston is rightwards. As a result, the displacement u of the valve rod is transformed in opposite direction displacements of the x of the cylinder piston rod. The displacement u of the valve rod can be controlled by a solenoid actuator and requires reduced power, while the displacement u of the valve rod use oil under high pressure and can produce high-power motion. The solenoid actuator is controlled by the voltage supply of the solenoid coil and can be easily interfaced with a PC for monitoring and control.

The linear model of the open-loop control of a hydraulic actuator with mass-spring-damper load can be obtained using the linearized equations of the valve, cylinder, and load [1, 30, 141]:

valve model

$$q = k_u u - k_p p$$

hydraulic piston-cylinder actuator model

$$q = A\,dx/dt + (V/2\beta)\,dp/dt$$

load

$$md^2x/dt^2 + bdx/dt + kx = Ap,$$

where

$$m = m_L + m_p,$$
$$b = b_L + b_p,$$
$$k = k_L,$$

q is the average flow of oil into the actuator,
p is the differential pressure of the two sides of the actuator piston,
V is the incremental changes of volume of oil in the cylinder about the operating point,
A is the cross-sectional area of the cylinder,
β is the bulk modulus, and
k_u and k_p are coefficients of the linearized model of the valve.

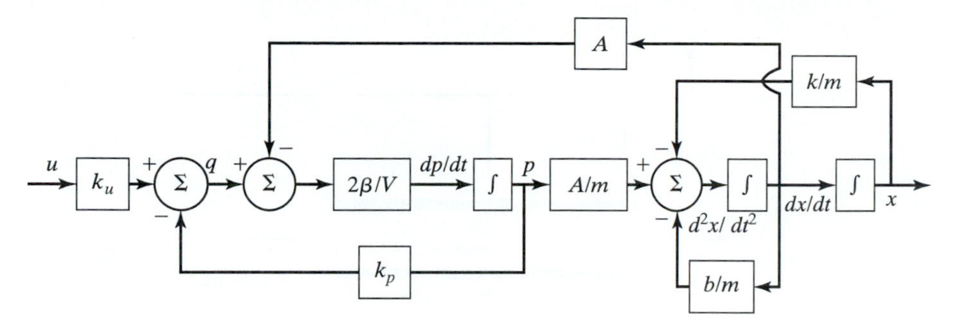

Figure 3.23

Block diagram of a hydraulic actuator with mass-spring-damper load

For obtaining the block diagram representation of the system, the last two equations are rewritten such that the highest derivative with a unitary coefficient appears on the left-hand side:

$$dp/dt = (2\beta/V)(q - A\,dx/dt);$$
$$d^2x/dt^2 = (A/m)p - (b/m)\,dx/dt - (k/m)x.$$

The valve model equation and the rewritten equations of the hydraulic piston-cylinder actuator and of the load are used for the three summators Σ shown in the block diagram of the hydraulic actuator with mass-spring-damper load, shown in Fig. 3.23. For given values of the hydraulic actuator and the mass-spring-damper load, the block diagram can be used for obtaining the Simulink model of the system [141].

3.4 PIEZOELECTRIC ACTUATORS

Section 2.4 presented the piezoelectric effect, which consists of generating an electric charge when a material is subject to a mechanical deformation or producing a mechanical deformation when subject to electric charge. The former is used for piezoelectric sensors, while the latter is used for piezoelectric actuators.

For explaining the use of piezoelectric effect in piezoelectric actuators, the coefficient g_{31} [m²/C] will be used. The first subscript (3) indicates the direction perpendicular to the electrodes, and the second subscript (1) indicates the direction parallel to the electrodes, as shown in Fig. 3.24 for an piezoelement of length L, width W, and thickness T [m].

For an actuator, the piezoelectric coefficient is defined as follows [19, 22]:

$$g_{31} = \frac{\text{mechanical strain produced in direction 1 [m/m]}}{\text{electric charge density applied in direction 3 [C/m}^2\text{]}}.$$

The coefficient g_{31} is defined here differently from the definition of the same coefficient for a sensor in Section 2.4.2. The units of these coefficients, [m/m]/[C/m²] and

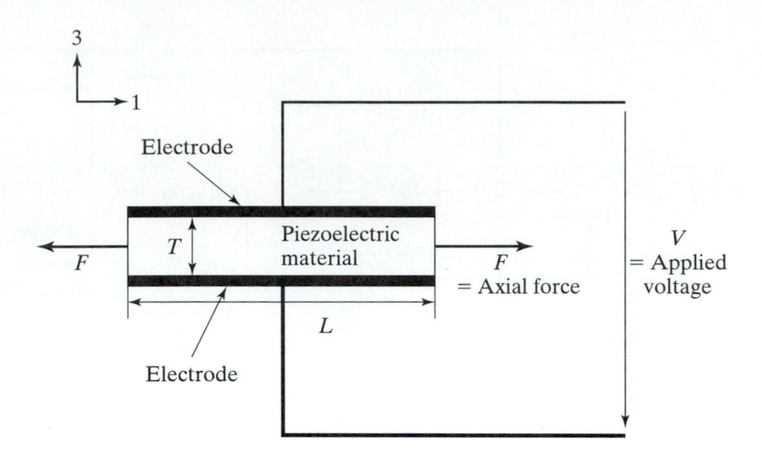

Figure 3.24

Piezoelectric effect used for actuators

[V/m]/[N/m²], are the same, given that both [CV] and [Nm] are energy units. The strain produced is $\Delta L/L$ and depends on the axial stress (σ_1) according to the formula

$$\Delta L/L = \sigma_1/E,$$

where E is the Young modulus [N/m²]. For a uniform stress distribution on the piezoelement transversal cross-sectional area WT, the axial force F, produced in direction 1, depends on σ_1 as follows:

$$\sigma_1 = F/(WT),$$

such that the mechanical strain $\Delta L/L$, produced in direction 1, depends on F as follows:

$$\Delta L/L = F/(EWT).$$

In the case of a piezoelement of capacitance C subject to an applied voltage V, the electric charge Q [C] is

$$Q = CV.$$

For an axial cross-sectional area WL, the electric charge density [C/m²], applied in direction 3, is $CV/(WL)$. The definition of the piezoelectric coefficient g_{31} gives

$$g_{31} = \frac{\frac{F}{EWT}}{\frac{CV}{WL}} = \frac{(L)F}{(CET)V}.$$

The capacitance C [F] of the piezoelement shown in Fig. 3.24, in case of an uniform charge distribution, can be calculated as

$$C = cWL/T,$$

where c [F/m] is the dielectric constant.

The last two equations give a simplified model for a piezoactuator:

$$F = (g_{31} cEW)V.$$

The units of the right-hand side are [m^2/C] [F/m] [N/m^2][m][V] =[FNV/C]. Given the equation $Q = CV$, with units [C] = [F][V], the unit [FNV/C] is equivalent to the units of F [N].

Using $\Delta L/L = F/(EWT)$, the axial strain is given by

$$\Delta L/L = V(g_{31} cEW)/(EWT) = (g_{31} c/T)V.$$

Example: Strain Produced by an Applied Voltage

Assume piezoceramic sensors (BM400 Type I [22]) of dimensions 0.5″ · 1.5″ · 0.020″ (12.7 mm · 38.1 mm · 0.5 mm). The catalog parameters are

$$g_{31} = -10.5 \cdot 10^{-3} \text{ [Vm/N]},$$

$$E = 1/12.5 \cdot 10^{-12} = 8 \cdot 10^{10} \text{ [N/m}^2\text{]},$$

$$T = 0.5 \cdot 10^{-3} \text{ [m]},$$

$$L = 38.1 \cdot 10^{-3} \text{ [m]},$$

and

$$c = 11.95 \cdot 10^{-9} \text{ [F/m]}.$$

The capacitance C, calculated in Section 2.4, is $C = cWL/T = (11.95 \cdot 10^{-9})(12.7 \cdot 10^{-3})(38.1 \cdot 10^{-3})/(0.5 \cdot 10^{-3}) = 11.57 \cdot 10^{-9}$[F].

Calculate axial strain $\Delta L/L$ for the value $V = 10$ [V] of the applied voltage.

***Solution*:**

$$\Delta L/L = (g^{31}\, c/T)\, V = ((-10.5 \cdot 10^{-3})(11.95 \cdot 10^{-9})/(0.5 \cdot 10^{-3})(10) = 2.51 \cdot 10^{-6} \text{ [m/m]}.$$

Piezoelements are used in a variety of custom and generic piezoelectric actuators:

audible output devices [118, 105]

linear micrometer displacement actuators [22]

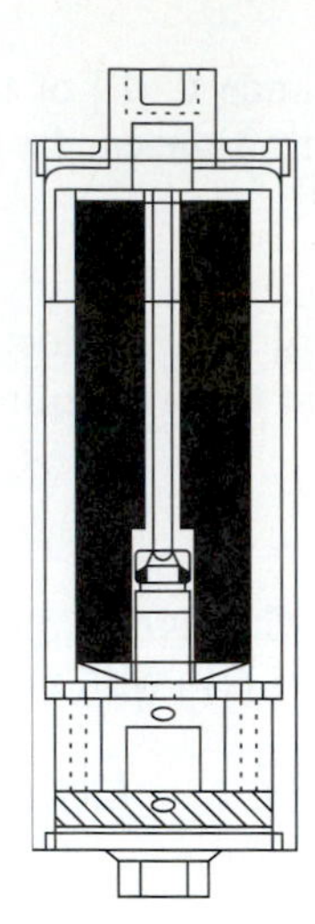

Figure 3.25

Active strut member cross section (*Source*: Sensor Technology [22])

active structural modules [22]

linear motors [142, 145]

rotary motors, etc. [144].

An example of a linear micrometer displacement actuator is the active strut member, shown in Fig. 3.25; it is designed to replace a conventional strut member for providing active control of the truss structure. It is composed of a cylindrical housing containing a piezoeletric stack, which generates forces and motion for active control of the structure. It also contains a linear displacement sensor for position feedback control and signal conditioning electronics. This actuator provides a ±40-μm displacement and maximum 1-kN pull force and 10-kN push force. The power supply is rated at ±300 [V] maximum [22].

An example of a linear motor is HR-1, from Nanomotion Ltd [142]. This motor uses piezoceramic elements coupled to a precision stage to produce both longitudinal extension and transversal bending resulting in a small elliptical trajectory of the ceramic edge. The generated driving force causes stage motion. A periodic driving force at frequencies higher than mechanical system resonance frequency allows for continuous motion over an unlimited travel. Maximum velocity at negligible load is 0.25 [m/sec]; stall force is 4 [N]. Power supply is a DC voltage of 48 V at a maximum current of 0.5 [A] [142].

The trend in piezoelectric actuators is to produce integrated systems containing the actuator, power electronics, and a real-time digital controller in a single miniaturized module [146].

PROBLEMS

3.1. Consider a more precise model of the PM–DC motor in which the effect of rotor inductance $L = 0.002$ [H] is not negligible. Obtain the corresponding Simulink model with a DC supply and a pure viscous load. For the same data used for the simulation results shown in Fig. 3.6, compare the results obtained for nonnegligible rotor inductance.

3.2. Consider the PM–DC motor model in which the load is not a pure viscous load, but an inertial-viscous load with a load moment of inertia of $90 \cdot 10^{-6}$ 10 kg m^2. Obtain the corresponding Simulink model and, for the same data used for the simulation results shown in Fig. 3.6, compare the results obtained for an inertial-viscous load.

3.3. Obtain the block diagram representation of a PM–DC motor that has nonnegligible viscous friction in the rotor bearings b_R. Compare with the block diagram shown in Fig. 3.8.

3.4. Modify the Simulink model of a PM–DC motor from Fig. 3.5 to account for nonnegligible viscous friction in the rotor bearings $b_R = 0.00003$ Nms/rad. For the same data used for the simulation results shown in Fig. 3.6, compare the results obtained for nonnegligible viscous friction in the rotor bearings.

3.5. Modify the Simulink model of a shunt DC motor from Fig. 3.13 to account for nonnegligible viscous friction in the rotor bearings $b_R = 0.00005$ Nms/rad. For the same data used for the simulation results shown in Fig. 3.14, compare the results obtained for non-negligible viscous friction in the rotor bearings.

3.6. Choose DC Gearmotor Series GM 8000 LO-COG from the *Pittman Catalogue* on the Internet [138]. Use the catalogue data for the model developed in Problem 3.2 and compare the simulation results with those displayed in Fig. 3.6.

3.7. A two-phase hybrid stepper motor is considered for driving a load. The application requires an angular step of 0.9 [°/step]. Determine the number of rotor teeth needed in this case.

3.8. Choose a hybrid stepper motor from the PDS or PDX Series in the *Parker Hannifin Catalogue* on the Internet [139]. For a typical step angle, determine the number of rotor teeth required for that motor.

3.9. The PDS or PDX Series, in the *Parker Hannifin Catalogue* on the Internet [139], offers various optional ministepping resolutions in steps/rev. A standard two-phase hybrid stepper motor, listed in the catalogue, has 50 teeth for the N section and 50 teeth for the S section of the rotor. Calculate the resolution with full stepping. How many times does the ministepping driver improve the resolution of the motor using full stepping?

3.10. Assume that a hybrid stepper motor using full stepping, has $N = 400$ [steps/revolution]. The DDA method is employed for a PC with $M = 64$ [bit] to generate a pulse train to achieve a constant velocity of 0.5 [revolutions/sec] for 2 [sec] and then, 0.55 [revolutions/sec] for 1 [sec]. Ignoring the duration of acceleration and deceleration, calculate number and the frequency of the steps to be generated.

3.11. For the case given in the previous problem, calculate the error due to binary representation in case the digital representation is reduced to $M = 8$ [bit].

3.12. For the block diagram of a hydraulic actuator with mass-spring-damper load shown in Fig. 3.23, the following values are given: $m_L = 2$ [Kg], $m_p = 1$ [Kg], $b_L = 0.1$ [Ns /m], $b_p = 0$, $k_L =$ 10 [N/m], $V/2\beta = 0.01$ [m^{4}s/Kg], $A = 0.005$ [m^2], $k_u = 0.2$ [kg/ms] , and $k_p = 0.01$ [m^4/Kg]. Obtain the Simulink model and simulate system dynamics for a step input with magnitude of 0.01 [m]. Is the response $x(t)$ overdamped or underdamped? Change the value of the damping coefficient b_L from 0 to 3 and observe the change in the response $x(t)$.

3.13. Modify the diagram in Fig. 3.22, considering that between the piston and the load, the shaft is flexible and has a spring coefficient k_s. Modify the load model to include the flexible shaft and obtain the new block diagram.

3.14. For a piezoceramic element BM500 Type I [22] of thickness 0.04″, width 1.0″, and length 3.0″, the catalog parameters are

$$g_{31} = -11.5 \cdot 10^{-3} \text{ [Vm/N]},$$

$$\text{compliance} = 15.5 \cdot 10^{-12} \text{ [m}^2\text{/N]},$$

and

$$\text{relative dielectric constant} = 1750.$$

Calculate the capacitance and the axial strain $\Delta L/L$ for the value $V = 5$ [V] of the applied voltage.

CHAPTER 4

Interfacing

4.1 COMPUTER INTERFACE REQUIREMENTS

In mechatronic systems, the PC is interfaced with signal and power transmission components. Signal transmission components have analog outputs (Fig. 4.1) or digital outputs (Fig. 4.2). A conditioned analog signal has typically 0–10 V and 0–20 mA. The conditioned digital signal has to be transistor-to-transistor (TTL) compatible (i.e., 0–0.6 V for binary value 0 and 2.4–5.5 V for binary value 1).

Normally, signal transmission components interfaced to the PC are transducers monitored by a data acquisition system, while power transmission components interfaced to a PC are actuators receiving command signals from the PC.

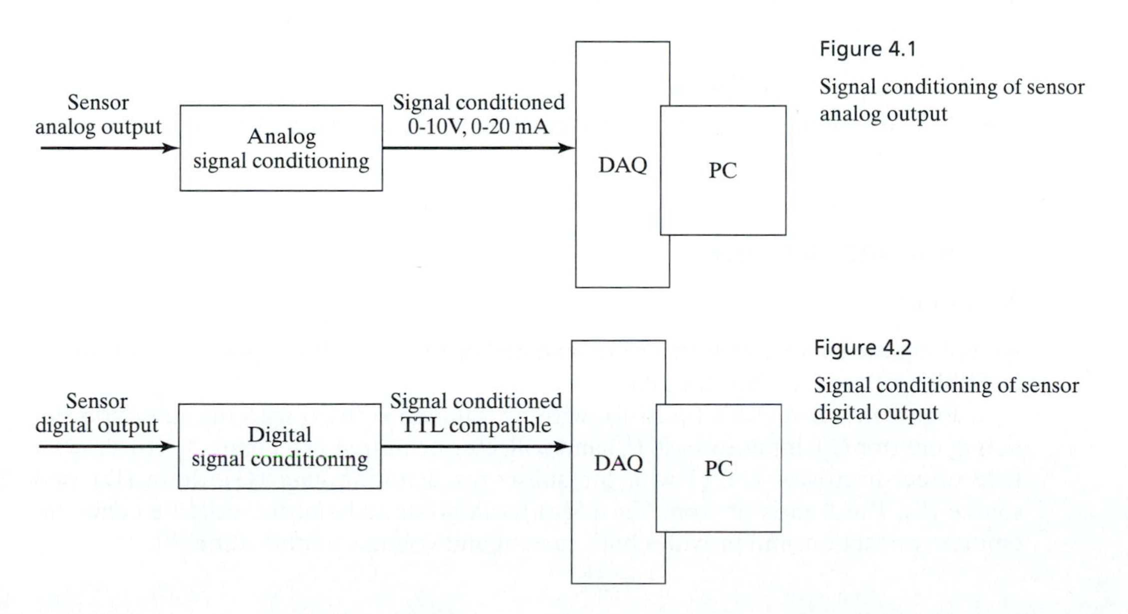

Figure 4.1

Signal conditioning of sensor analog output

Figure 4.2

Signal conditioning of sensor digital output

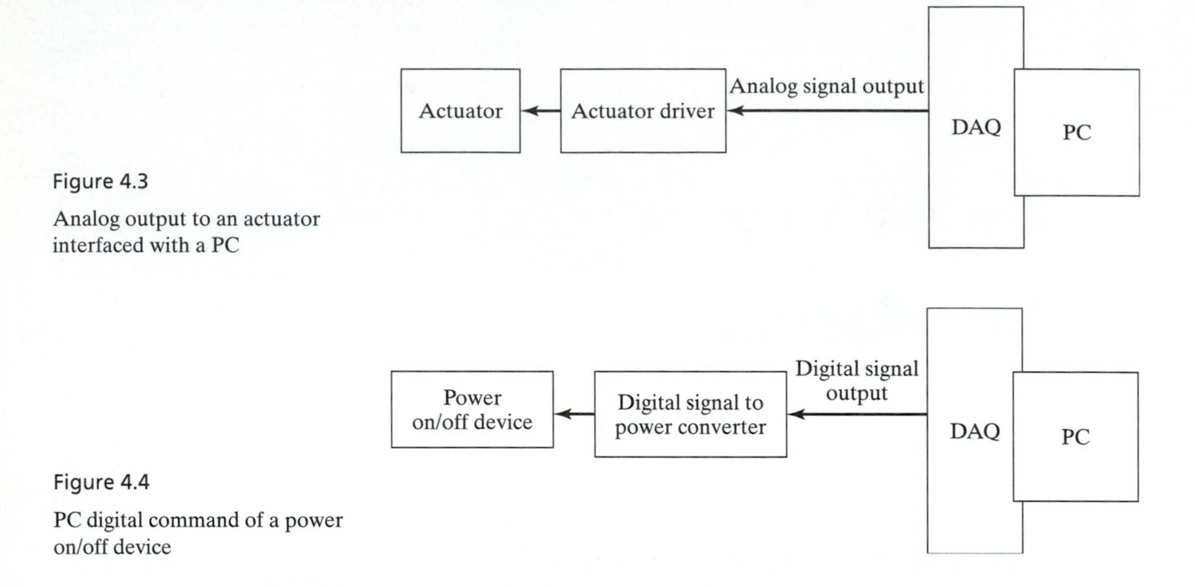

Figure 4.3

Analog output to an actuator interfaced with a PC

Figure 4.4

PC digital command of a power on/off device

Power transmission components can receive analog commands (Fig. 4.3) or digital commands (Fig. 4.4) from a PC. Output signals from the PC have to be made compatible with the actuator's or relay's input requirements. For actuators and power on/off devices this usually requires a signal conversion to a higher voltage and any current demanded by the actuator (or a higher current and any voltage demanded by the actuator).

Signal conditioning has already been used in Chapter 2 as specific solutions for thermocouples (junction compensation), strain gauges (Wheatstone bridge, amplification), piezoelectric sensors (charge amplifier), etc. Often, signal conditioning requires operational amplifiers. In what follows operational amplifiers (Section 4.2) and common types of signal conditioning (Section 4.3) are presented. Digital-to-analog conversion is the topic of Section 4.4, while analog signal digitizing using analog-to-digital converters is presented in Section 4.5. Actuators interfacing with PCs will be presented in Section 4.6.

4.2 OPERATIONAL AMPLIFIERS

4.2.1 Amplifiers

Amplifiers use transistors which were invented in 1947 at Bell Telephone Laboratories by Shokley, Bardeen, and Brattain.

Figure 4.5(a) shows a transistor with a grounded emitter, with the subscripts denoting emitter (e), input or base (b) and collector or output (s). Figure 4.5(b) shows a field effect transistor (FET) with the subscripts denoting gate (G), drain (D), and source (S). The transistor from Fig. 4.5(a) has a grounded emitter, called a common-emitter connection, and provides both current and voltage amplification [9].

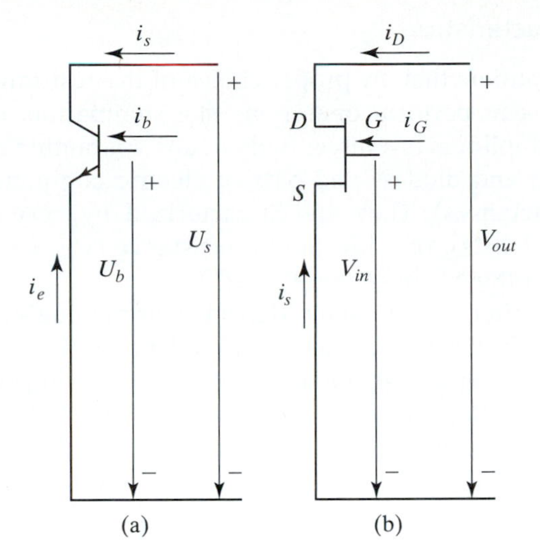

Figure 4.5
Amplifiers with a transistor having a common emitter configuration (a) and a FET (b)

The transistor, being a nonlinear device, has its operation defined using changes of current Δi instead of the current i. The current transfer ratio is defined as

$$\alpha = -\Delta i_s / \Delta i_e.$$

Kirchhoff 's current law gives

$$\Delta i_s + \Delta i_e + \Delta i_b = 0.$$

After eliminating Δi_e , the relationship between Δi_b and Δi_s is

$$\Delta i_s = \alpha \Delta i_b / (1 - \alpha) = \beta \Delta i_b,$$

where the current amplification ratio is defined as

$$\beta = \alpha / (1 - \alpha).$$

Normally, $0.9 < \alpha < 0.999$, such that $10 < \beta < 1000$. The variation of β with the current i_s is significant; the value of β at high i_s can be the double of the value of low i_s. This problem is even more significant in the case of power transistors, which operate at a much larger range of variation of i_s. Operational amplifiers with feedback solve this transistor gain variability problem. The input impedance of the transistor is in the range 0.3 to 3 MΩ and this poses a problem in the case of a comparable output impedance of the circuit connected to the base. A higher input impedance, in the range 0.1 to 1000 GΩ is available from a FET [2]. In Fig. 4.5(b), a FET with voltage gain defined by $\Delta V_{out}/\Delta V_{in}$ is shown.

4.2.2 Operational Amplifiers Characteristics

Operational amplifiers are amplifiers that, by proper choice of the resistances and capacitances for their feedback, can perform operations like summation, integration, subtraction, etc. Operational amplifiers are integrated circuits containing active electronic components (transistors and diodes) and passive electric components (resistances, capacitances, and inductances). They are characterized by very high input resistance (R_{iA} of the order of 1 MΩ), very low output resistance (R_{oA} less than hundreds of ohms), and very high voltage gain ($A = 10^4$ to 10^6).

A typical operational amplifier is manufactured as an eight-pin dual-in-line package. In Fig. 4.6(a), only the pins 2, 3, 4, 6, and 7 are identified, being the pins to which further reference will be made. In Fig. 4.6(b), the schematic diagram is represented.

The dual DC supply shown often has $V^+ = +15$ V and $V^- = -15$ V with regard to the ground or the differential voltage $V^+ - V^- = 30$ V.

The model of the open loop operational amplifier is shown in Fig. 4.7. Cut variables are

(a) (Vi, ii), where the differential voltage Vi is

$$V_i = V_p - V_n;$$

(b) (V_s, i_s), where the current i_s flows from V^+ to V^- and the differential voltage is

$$V_s = V^+ - V^-; \text{ and}$$

(c) (V_o, i_o), where the output voltage V_o is referenced to the ground.

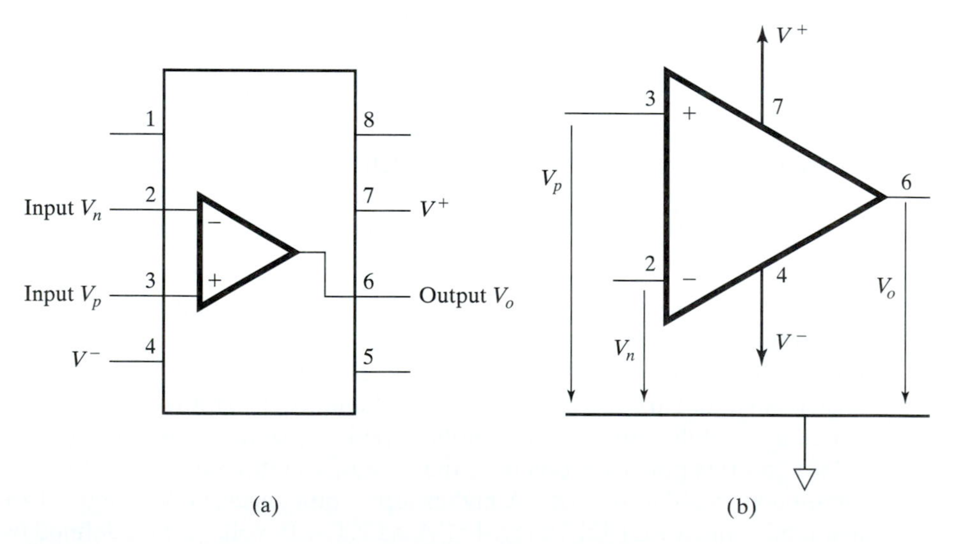

Figure 4.6

Pin assignment (a) and schematic diagram (b) of an operational amplifier

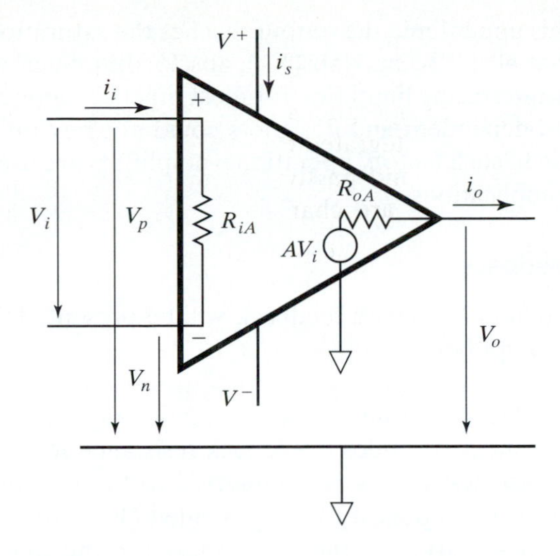

Figure 4.7

Three-port model of an open loop operational amplifier

The model in Fig. 4.7 is a three-port model. The supply voltages V^+ and V^- (as well as the differential voltage V_s) are assumed constant and the current i_s is assumed available as requested by any normal load of the operational amplifier. The voltage V_o has an amplitude lower than max{abs(V^+), abs(V^-)} (i.e., normally V_o has an amplitude lower than 15 V).

Given that the supply voltage is constant, the three-port model is often reduced to a two-port model with cut variables (V_i, i_i) and (V_o, i_o) shown in Fig. 4.8. The representation of power transfer requires, however, the three-port model.

The output voltage for no load ($i_o = 0$) is $V_o = AV_i = (10^4 \text{ to } 10^6)V_i$. For a V_o less than 15 V, this gives

$$V_i \leq 15/(10^4 \text{ to } 10^6)$$

or

$$V_i \leq 15 \text{ to } 1500\,\mu\text{V}$$

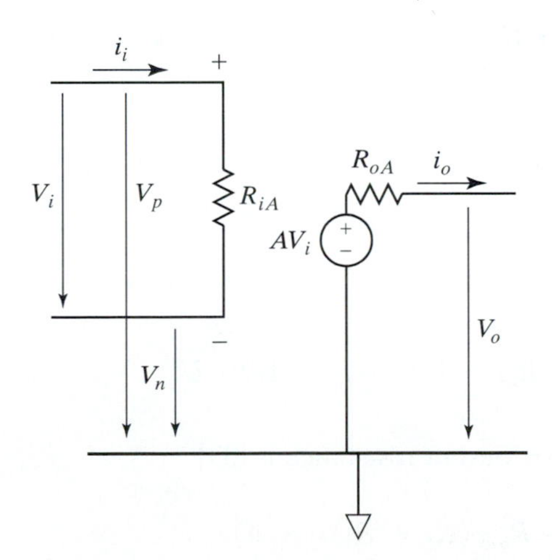

Figure 4.8

Two-port model of an open loop operational amplifier

As the input voltage V_i reaches this upper limit, the output reaches the saturation voltage 15 V due to the condition that abs(V_o)< max{abs(V^+), abs(V^-)}[9]. Such a small input voltage V_i results in a major interfacing limitation for most practical applications. Moreover, the high gain A is load-dependent and designers could not rely on a constant value. As a matter of fact, due to such factors, operational amplifiers are used with a feedback circuit for interfacing applications.

4.2.3 Operational Amplifiers with Feedback

Two common types of operational amplifiers with feedback will be presented: inverting and non-inverting operational amplifiers.

Inverting Operational Amplifiers Figure 4.9 shows the schematic diagram of an inverting operational amplifier. This scheme includes a feedback resistance R_F and a resistance R_I in the input circuit. These resistances are connected to the negative (−) polarity V_n of the input, while positive (+) polarity V_p is grounded ($V_p = 0$). As a result, the output voltage $A(V_p - V_n) = -AV_n$ has the sign opposite to the sign of the input V_n, such that the amplifier is called an inverting operational amplifier. Figure 4.10 shows the two-port model of the inverting operational amplifier.

The following six equations with eight variables ($i^{(c)}$, i_i, i_F, V_n, $U^{(c)}$, i, i_o, and U) represent the analytical model of the inverting operational amplifier:

$$
\begin{aligned}
i^{(c)} &= i_i + i_F; \\
i_i &= V_n / R_{iA}; \\
V_n &= U^{(c)} - R_1\, i^{(c)}; \\
i &= i_F + i_o; \\
i_F &= (V_n - U)/R_F; \\
i_o &= (-A\, V_n - U)/R_{oA}.
\end{aligned}
$$

After eliminating three variables (i_i, i_F, and i_o), three equations with five variables (V_n, $U^{(c)}$, $i^{(c)}$, i, and U) remain:

$$
\begin{aligned}
V_n &= (i + (1/R_{oA} + 1/R_F)U)/(1/R_F - A/R_{oA}); \\
U_{(c)} &= (1 + R_1/R_{iA} + R_1/R_F)V_n - (R_1/R_F)U;
\end{aligned}
$$

and

$$
i_{(c)} = (1/R_{iA} + 1/R_F)V_n - (1/R_F)U.
$$

The inverting operational amplifier output resistance is [33]

$$
R_o = R_{oA}\ (R_F + R_1)/(R_1 A).
$$

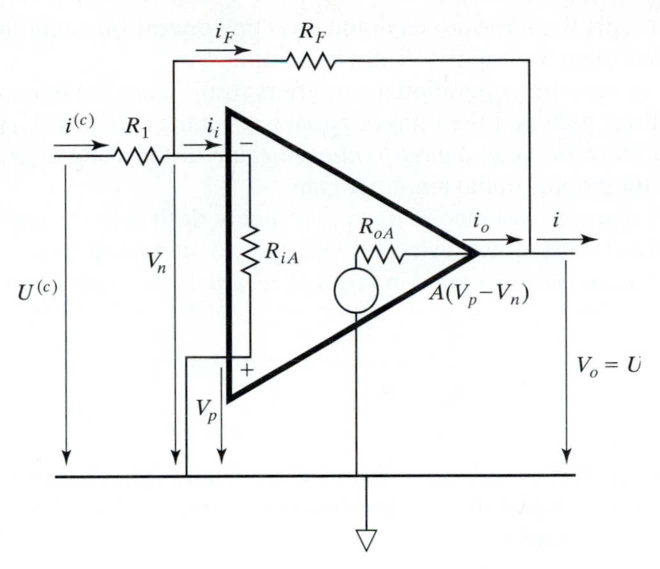

Figure 4.9

Inverting operational amplifier

For the very high value of the gain $A = 10^4$ to 10^6,

$$V_n \cong 0,$$
$$U_{(c)} = -(R_1/R_F)U,$$
$$i_{(c)} = -(1/R_F)U,$$

and

$$R_o \cong 0$$

or

$$i_{(c)} = -(1/R_1)U_{(c)}$$

and

$$U = -(R_F/R_1)U_{(c)}.$$

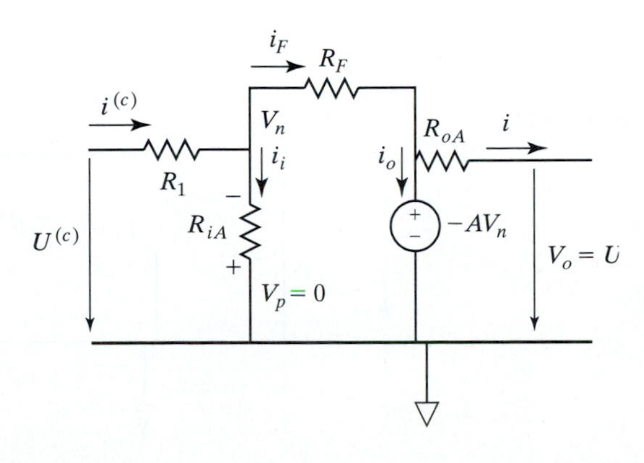

Figure 4.10

Two-port model of the inverting amplifier

The last equation represents the basic model of the inverting operational amplifier and gives the relationship between the input $U^{(c)}$ and the output U.

The applications of inverting operational amplifiers result from the linear dependence between the voltage gain and the ratio of passive resistances R_F / R_1. Calibrated resitances give a much more stable and easy-to-choose value of the voltage gain when compared to the open loop operational amplifier gain.

The fact that the output resistance is $R_o \cong 0$, is highly desirable in signal transmission from the cut (U, i) to another device, for example, to an analog-to-digital converter. The equivalent circuit is represented in Fig. 4.11 in which the input resistance R_I is given by

$$R_i = R_1.$$

The value of R_1 is lower than R_{iA}, and this might be undesirable when a sensor is connected to the cut $(U^{(c)}, i^{(c)})$ because it increases the current input $i^{(c)}$. In this case, further signal conditioning is required.

Noninverting Operational Amplifiers Figure 4.12 shows the schematic diagram of a noninverting operational amplifier. The input $U^{(c)}$ is applied at V_p (i.e., $V_p = U^{(c)}$). Given that the input resistance R_{iA} has a very high value, $i_i = 0$ and consequently, $V_n = V_p = U^{(c)}$. The model is given by five equations with seven unknowns (i_1, i_F, i_o, V_n, U, i, and $U^{(c)}$):

$$\begin{aligned}
i_1 &= i_F;\\
i_1 &= -V_n/R_1 = U^{(c)}/R_1;\\
i_F &= (V_n - U)/R_F = (U^{(c)} - U)/R_F;\\
i &= i_F + i_o;\\
i_o &= (A(V_p - V_n) - U)/R_{oA}.
\end{aligned}$$

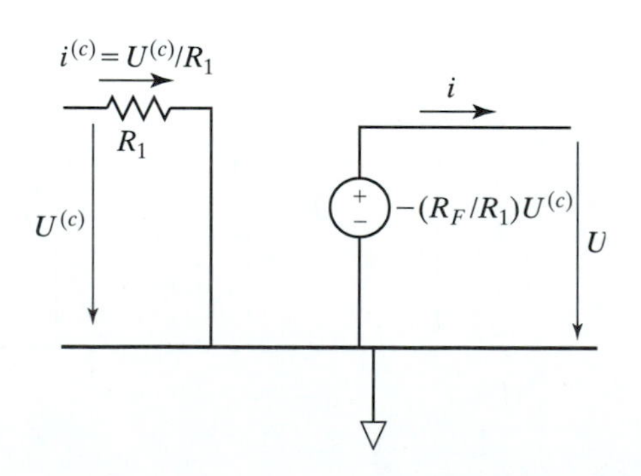

Figure 4.11

Equivalent circuit of the inverting amplifier

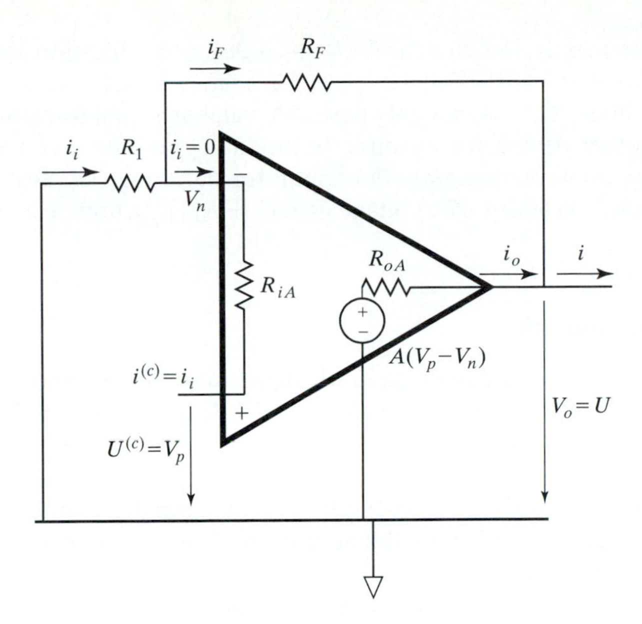

Figure 4.12

Schematic diagram of an noninverting operational amplifier

Eliminating three variables (i_1, i_F, and i_o) leaves two equations with four variables (V_n, U, i, and $U^{(c)}$):

$$V_n = (i + (1/R_{oA} + 1/R_F)U - (A/R_{oA})U^{(c)})/(1/R_F - A/R_{oA}) =$$
$$(i + (1/R_{oA} + 1/R_F)U)/(1/R_F - A/R_{oA}) - (A/R_{oA})U^{(c)}/(1/R_F - A/R_{oA}) =$$
$$(i + (1/R_{oA} + 1/R_F)U)/(1/R_F - A/R_{oA}) - U^{(c)}/(R_{oA}/AR_F - 1);$$
$$U = (1 + R_F/R_1)\, U^{(c)}.$$

For the very high value of the gain A,

$$V_n \cong U^{(c)}.$$

The equivalent circuit is represented in Fig. 4.13. In this case, $i^{(c)} = i_i = 0$. The input resistance R_i, for the cut ($U^{(c)}$, $i^{(c)}$), is equal to the input resistance R_{iA}, which has a value

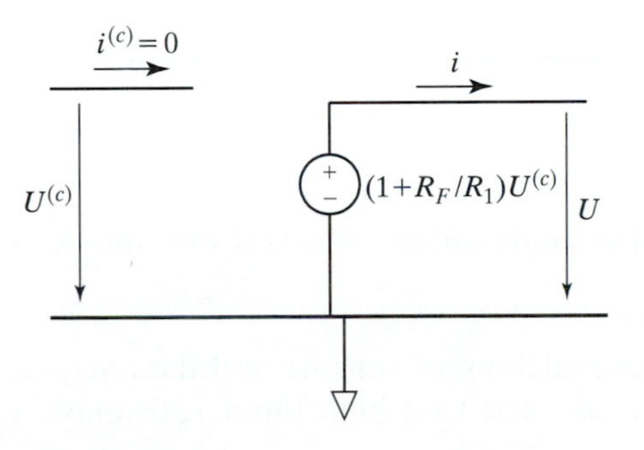

Figure 4.13

Equivalent circuit for the noninverting operational amplifier of more than $10^4\ \Omega$. Consequently, the current $i^{(c)} \cong 0$—a desirable situation when a sensor is connected to this cut $(U^{(c)}, i^{(c)})$ [33]. Output resistance is, as in the case of the inverting operational amplifier, $R_o \cong 0$—a highly desirable value in signal transmission from the cut (U, i) to another device, for example to an analog-to-digital converter. These properties make the noninverting amplifier useful for buffering (i.e., isolating) sensors from the data acquisition board. The voltage gain $(1 + R_F/R_I)$, however, is limited to values over 1.

4.2.4 Operational Amplifier Bandwidth

The bandwidth is defined as the frequency range in which the ratio of the magnitude of the output/input drops 3 decibels (dB). The value of -3 decibels corresponds to -3 dB = 20 $\log_{10}$ (output/input) and gives output/input = $10^{-(3/20)} = 0.71$ (i.e., output = 71% (input)).

The bandwidth defines the useful frequency range of the amplifier. For an input signal with frequencies outside the bandwidth, the outputs for those frequencies are reduced to less than 71% of the inputs.

Operational amplifiers are characterized by a (gain)(bandwidth) product GBP. Typical values of GBP for open loop operational amplifiers are from 10^6 to 10^7Hz. Open loop amplifiers have maximum gain $A = 10^4$ to 10^6 (which, using 20 $\log_{10}$A, gives 80 to 120 dB). The maximum gain often drops -3 dB at about 10 Hz (i.e., a limited bandwidth) [33]. Operational amplifiers with feedback have a lower maximum gain and this extends their bandwidth.

Example: Bandwidth of an Inverting Operational Amplifier

Assume, for the inverting operational amplifier shown in Fig. 4. 11, the following values: $R_F = 5\ \text{k}\Omega$, $R_I = 1\ \text{k}\Omega$, and GBP = 10^6 Hz. Determine its bandwidth.

Solution: The gain $G = R_F / R_I = 5$.

The bandwidth is

$$B = (GBP)/G = 10^6/5 = 0.2 \cdot 10^6 \text{ Hz}.$$

It can be observed that the inverting amplifier has a much larger bandwidth than the open loop operational amplifier.

4.3 SIGNAL CONDITIONING

Various signal conditioning operations can be achieved with the inverting and noninverting amplifiers, as follows:

(a) Isolation and impedance matching of sensors and data acquisition boards require that sensors are connected to a high input resistance to reduce sensor

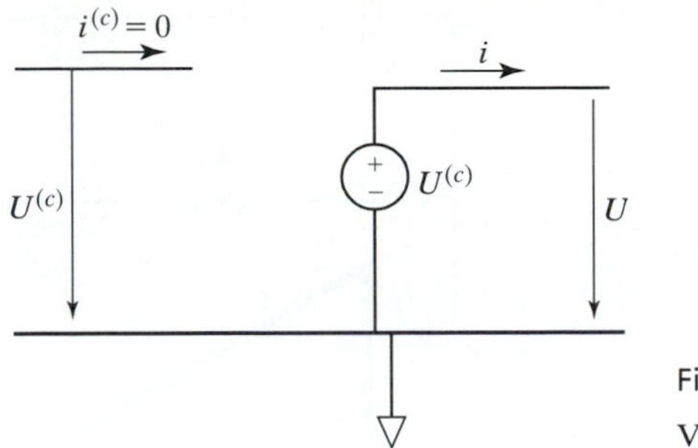

Figure 4.14

Voltage follower

loading and their power consumption. Also, data acquisition boards require very low output resistance for interfacing with other devices to reduce voltage drop and minimize the modification of the output voltage carrying the measurement information.

(b) Amplification of a sensor signal is often required to match signal data acquisition board characteristics, which often include a maximum input voltage of 10 V. An inverting amplifier can be used to satisfy this requirement by a proper choice of the voltage gain $R_F/R_{I.}$

(c) Filtering the noise from the signal received by a data acquisition system reduces the noise and increases the signal/noise ratio.

Isolation can be achieved by inserting a noninverting amplifier between the sensor and the data acquisition board with a unity gain. The schematic diagram shown in Fig. 4.12 can achieve this for $R_F = 0$, which gives the unity gain $(1 + R_F/R_I) = 1$. This is called a voltage follower (Fig. 4.14).

Amplification of a voltage signal can be achieved with an inverting or a noninverting amplifier.

Example: Amplification of Sensor Signal

Assume a transducer that produces a positive voltage output of maximum 10 mV. Choose the inverting amplifier gain to make the signal compatible with the 0 to 10 V range of the analog input to a data acquisition board.

***Solution*:** For 10 mV = 0.01 V, the required gain is:

$$10 \text{ V}/0.01 \text{ V} = 1000.$$

Filtering with operational amplifiers is frequently used to remove the noise from signals. This is possible only for the noise with frequencies outside the range of useful frequencies of the signal. Often, this means the removal from a noisy signal of the high frequency noise, while retaining a lower frequency measurement signal. For this purpose, a low-pass analog filter is used.

A low-pass active filter built with an inverting amplifier is shown in Fig. 4.15.

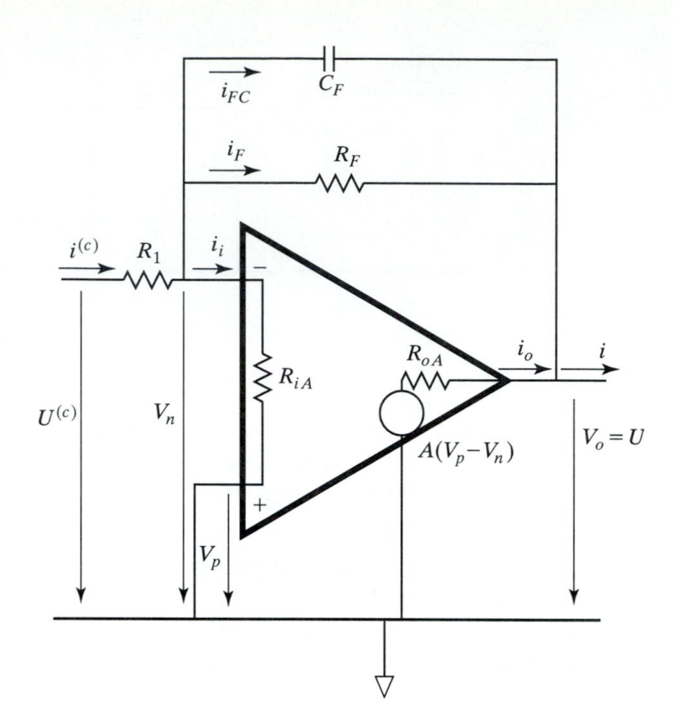

Figure 4.15

Low-pass filter

It is desired that the sensor has a much larger signal amplitude than the amplitude of the noise accumulated in the transmission from the sensor to the data acquisition system (i.e., a high signal/noise ratio). This can also be achieved by the amplification of the sensor signal, in this case as close as possible to the sensor such that the noise from the transmission is not amplified as well. Filtering can be achieved with an operational amplifier with a resistive-capacitive feedback, as shown in Fig. 4.15.

The transfer function $U(s)/U^{(c)}(s)$ can be obtained from the model of the filter shown in Fig. 4.15 and can be used for the characterization of this low-pass filter.

The following four equations with six variables ($i^{(c)}$, i_{FC}, i_F, V_n, $U^{(c)}$, and U) represent the analytical model of the active low-pass filter with a very high value of the input resistance R_{iA} (i.e., $i_i = 0$):

$$
\begin{aligned}
i^{(c)} &= i_{FC} + i_F; \\
i^{(c)} &= (U^{(c)} - V_n)/R_1; \\
i_F &= (V_n - U)/R_F; \\
i_{FC} &= C_F d(V_n - U)/dt.
\end{aligned}
$$

After eliminating the currents $i^{(c)}$, i_{FC}, and i_F, the remaining equation contains three variables (V_n, $U^{(c)}$, and U):

$$(U^{(c)} - V_n)/R_1 = C_F d(V_n - U)/dt + (V_n - U)/R_F.$$

A very high value of R_{iA} results not only in $i_i = 0$, but also gives $V_p = V_n = 0$, and the preceding equation becomes

$$U^{(c)}/R_1 = -C_F dU/dt - U/R_F.$$

The Laplace transform for zero initial conditions gives

$$U^{(c)}(s)/R_1 = -(sC_F + 1/R_F)U(s)$$

or

$$\frac{U(s)}{U^{(c)}(s)} = -\frac{R_F/R_1}{1 + sR_FC_F}.$$

This transfer function, with cut-off frequency $1/2\pi C_F R_F$, shows that the low-pass filter of Fig. 4.15 can filter out noise with frequency over $1/\ 2\pi C_F R_F$. For signals with frequencies over $1/2\pi C_F R_F$, this single-stage active filter reduces the magnitude of the output–input ratio by more than -3 dB.

Example: Low-Pass Active Filter Design

For the active filter shown in Fig. 4.15, calculate the value of the capacitance C_F for $R_I = R_F = 1\text{k}\Omega$ such that cut-off frequency is 100 Hz.

***Solution*:** The cut-off frequency equation,

$$f = 1/2\pi C_F R_F,$$

gives

$$C_F = 1/2\pi f R_F.$$

For $R_F = 1000\ \Omega$ and $f = 100$ Hz,

$$C_F = 1/2\pi(100)(1000) = 1.6 \cdot 10^{-6}\ F = 1.6\ \mu\text{F}.$$

4.4 DIGITAL-TO-ANALOG CONVERSION

Digital-to-analog converters (DACs) are required to convert digital output from computers to analog signals compatible with actuators.

A weighted resistors DAC is an example of an inverting operational amplifier used for computing a weighted sum (i.e., the output is a weighted addition of inputs). A typical eight-bit integrated circuit DAC based on weighted resistors, manufactured as a 16-pin dual-in-line package, is shown in Fig. 4.16. Only pins 1–8, 11–13, and 15–16 are identified, being the pins to which further reference is made.

The equivalent circuit of the weighted resistors DAC is shown in Fig. 4.17.

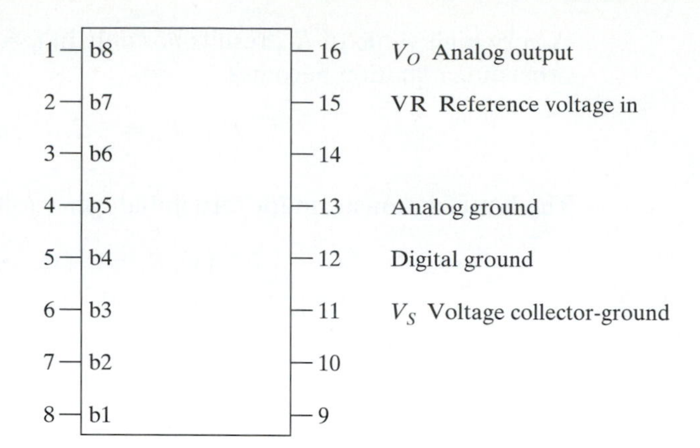

Figure 4.16

Pin assignment for a weighted resistors DAC

Pins 1–8 transmit the digital input with eight parallel bits to the DAC and determine the position of the eight switches identified by the same symbols. For the value $b_i = 1$ ($i = 1$ to 8), the corresponding (solid-state) switch is connected to V_R, while for $b_i = 0$ ($i = 1$ to 8), the corresponding switch is connected by a thick line to the common digital ground, pin 12. In Fig. 4.17, the positions of four switches realizing the digits $b_1 = 1, b_2 = 0, b_7 = 1$, and $b_8 = 1$ are shown. In series with the switch corresponding to $b_i = 1$ ($i = 1$ to 8), there is a resistance $(2^{i-1})R$. For example, between the reference voltage in V_R and the input V_n to the operational amplifier, for $b_1 = 1$, the resistance is $2^0 R = R$, while for $b_2 = 0$, the resistance is infinite (given that the switch is connected to the digital ground and not to V_n), for $b_7 = 1$, the resistance is $2^6 R = 64 R$, and for $b_8 = 1$, the resistance is $2^7 R = 128 R$.

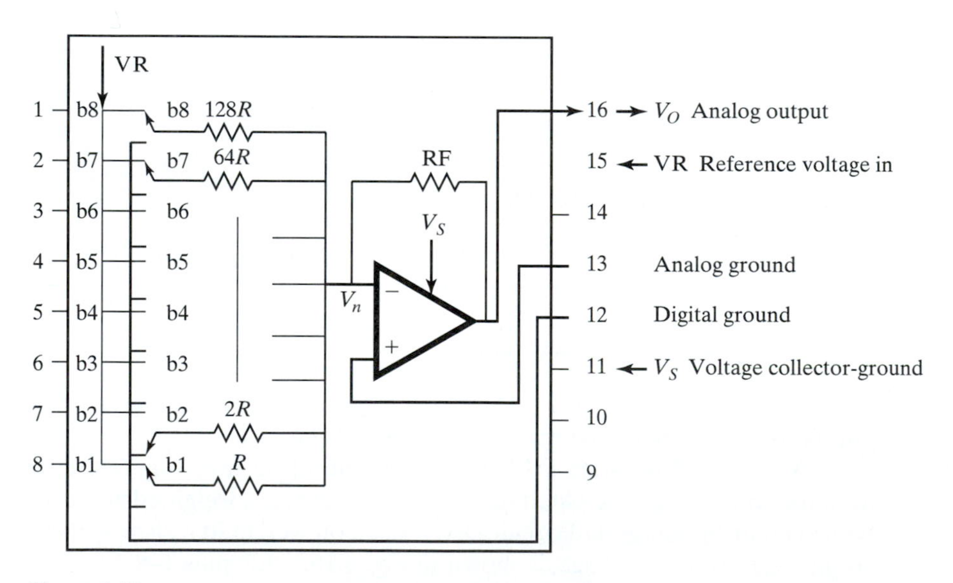

Figure 4.17

Equivalent circuit of the weighted resistors DAC

The relationship between the input V_R and the output V_o of the inverting amplifier was obtained earlier as

$$V_o = -(R_F/R_1)V_R,$$

where, in this case, R_I is the equivalent resistance resulting from all parallel resistances associated to switches in the position corresponding to $b_i = 1$ (i = 1 to 8), such that

$$1/R_1 = b_1/R + b_2/2R + \ldots. + b_8/128R = (1/(128R))(128b_1 + 64b_2 + \ldots. + b_8) = (2/(2^8R))(2^7b_1 + 2^6b_2 + \ldots. + 2^0b_8).$$

The foregoing input–output relationship becomes

$$V_o = -(2R_FV_R)(2^7b_1 + 2^6b_2 + \ldots. + 2^0b_8)/(2^8R) = -(2R_FV_R/R)(b_1/2 + b_2/2^2 + \ldots + b_8/2^8).$$

A binary number $b_1\, b_2 \ldots b_8$ has the decimal equivalent $2^7b_1 + 2^6b_2 + \ldots. + 2^0b_8$.

The bit b_1 is the most significant bit (MSB) and the bit b_8 is the least significant bit (LSB). The resolution has the same value as the V_o as for the LSB, given the LSB that corresponds to the smallest change in output voltage V_o.

For an n-bit DAC,

$$V_o = -(2R_FV_R)(2^{n-1}b_1 + 2^{n-2}b_2 + \ldots. + 2^0b_n)/(2^nR) = -(2R_FV_R/R)(b_1/2 + b_2/2^2 + \ldots. + b_n/2^n).$$

The full-scale range voltage is defined as

$$V_{fsc} = 2R_FV_R/R.$$

Example: 12-bit Weighted Resistors DAC

A 12-bit weighted resistors DAC has $R_F = 5\ \text{k}\Omega$, $V_R = 5$ V, and $R = 5\ \text{k}\Omega$. Calculate the value of V_o for

(a) the least significant bit;
(b) the most significant bit;
(c) the maximum value of the digital input;
(d) the binary input 100010010001; and
(e) the full-scale range voltage V_{fsr}.

Solution:

(a) For LSB, the binary input is 000000000001 such that

$$V_{o\,\text{LSB}} = -2(5)(1/2_{12}) = -0.00244\ \text{V} = -2.44\ \text{mV}.$$

The resolution is also 2.44 mV.

(b) For MSB, the binary input is 100000000000 such that

$$V_{o\,\mathrm{MSB}} = -2(5)(1/2^1) = -5 \text{ V}.$$

(c) The maximum value of the digital input is 111111111111 such that

$$V_o = -2(5)(1/2^1 + 1/2^2 + \dots + 1/2^{11} + 1/2^{12})$$
$$= -5(1 + 1/2^1 + 1/2^2 + \dots + 1/2^{11}) = -9.99756 \text{ V}.$$

(d) For the the binary input 100010010001,

$$V_o = -2(5)(1(1/2^1) + 0(1/2^2) + 0(1/2^3) + 0(1/2^4) + 1(1/2^5) + 0(1/2^6) + 0(1/2^7) + 1(1/2^8) + 0(1/2^9) + 0(1/2^{10}) + 0(1/2^{11}) + 1(1/2^{12})) = 5.35 \text{ V}.$$

(e) $V_{\mathrm{fsr}} = 2R_F V_R / R = 10$ V.

It can be observed that

$$V_{\mathrm{fsr}} - V_{o\,\mathrm{MSB}} = 10 - 9.99756 = 0.00244 \text{ V}$$

(i.e., the same value as the resolution).

A 12-bit weighted resistor DAC has resistances between R and $2^{11}R = 2048R$. Such a large range of value resistance limits of this type of DAC to a reduced value of the binary number length n. A better solution for higher values of n is the ladder resistances DAC [11].

The DAC can be modeled with passive components, represented by resistances, only at low frequencies. At higher frequencies, capacitancies and inductancies, however small, start affecting the output and have to be taken into account in modeling the conversion of the sequence of digital inputs. A DAC is often characterized by manufacturers by its 2% settling time (i.e., the time required for the output to reach a range of 2% above the steady-state value). The sequence of binary inputs has to be separated by a time longer than settling time to avoid signal corruption. For example, a 2% settling time of 100 μs limits the frequency of binary input signals to less than $1/100(10^{-6}) = 10{,}000$ Hz.

4.5 ANALOG-TO-DIGITAL CONVERSION

Analog-to-digital converters (ADC) are required to convert analog signals from sensors to a digital form compatible to PC digital input.

A typical integrated circuit eight-bit ADC, manufactured as a 22-pin dual-in-line package, is shown in Fig. 4.18. Only pins 3–8, 11, and 13–20 are identified, being the pins to which further reference is made. ADCs are available for 8-, 10-, 16-, 22-, and 24-bit digital conversion.

The equivalent circuit of the eight-bit ADC from Fig. 4.18 is shown in Fig. 4.19.

In Fig. 4.19, an internal DAC can be identified as the component with the analog output V_o and the inputs as the digital parallel bits b_1 to b_8, the voltage reference V_R, the voltage collector-ground V_s, the analog ground, and the digital ground.

The main operation of the ADC is concentrated around the loop Comparator-Registers-DAC.

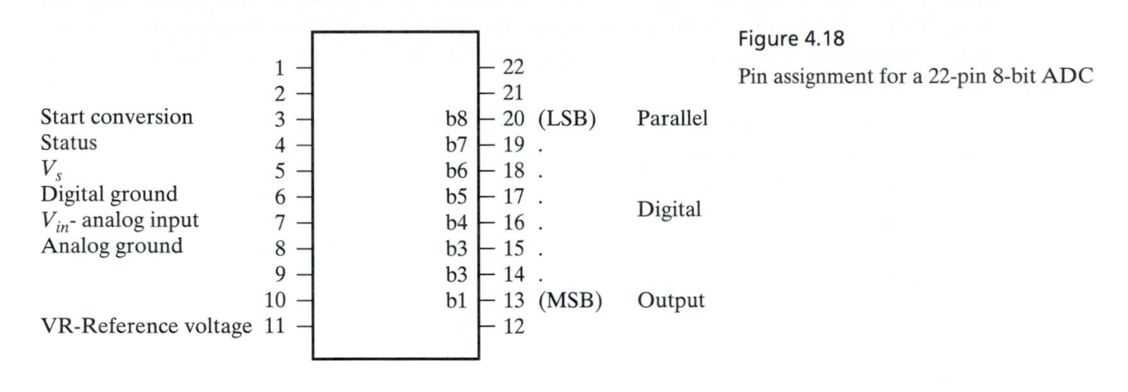

Figure 4.18

Pin assignment for a 22-pin 8-bit ADC

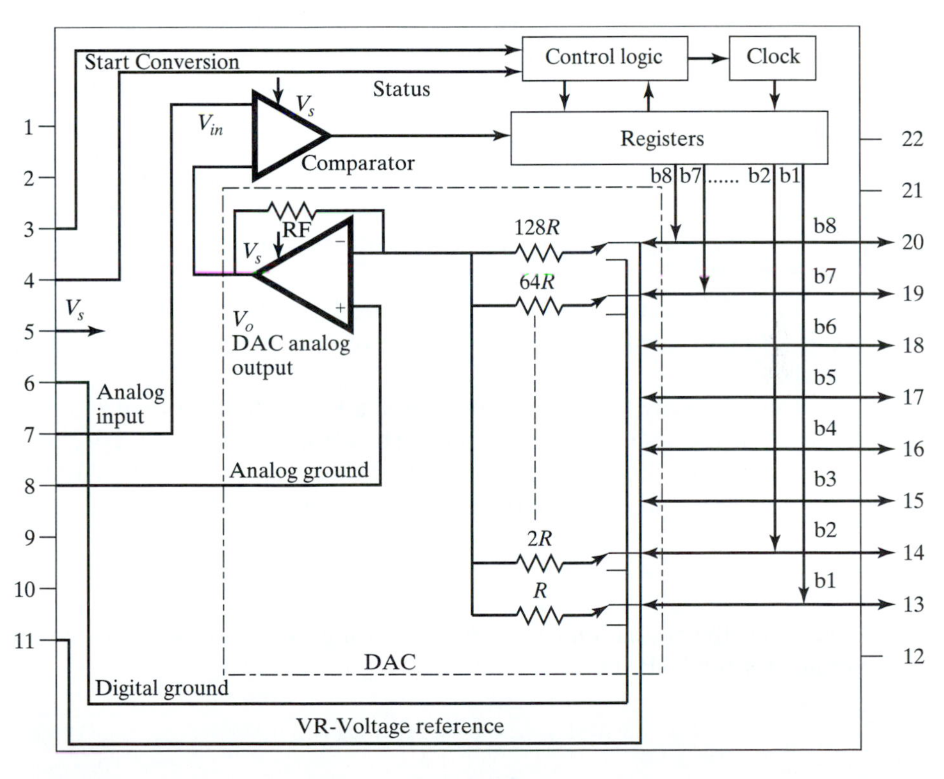

Figure 4.19

Equivalent circuit of the eight-bit ADC

The analog input V_{in} is compared to the DAC analog output V_o, and if

$$V_o < V_{in}, \text{ the output of the comparator goes high;}$$
$$V_o > V_{in}, \text{ the output of the comparator goes low.}$$

The digital content of registers b_1 to b_8 start at 00000000 and are modified in each clock cycle to induce a trial-and-error process. At the end of the trial-and-error process, the digital output of the registers b_1 to b_8 are sent to the pins 13–20 as parallel digital output of the ADC.

The conversion starts at the reception of start conversion high bit at the pin 3. The status of the ADC is available at the pin 4 as a busy high bit. The clock frequency is often from 0.25–80 MHz such that clock cycle time, the inverse of the clock frequency, is 0.0125 to 4 μs.

The trial-and-error process can be a simple upcount ADC, a successive approximations ADC [17], or a delta sigma ADC [2].

In this section, the issue of quantization, sampling, the upcount ADC, and the successive approximations ADC are presented.

Quantization The conversion of an analog signal into an n-bit digital number leads to a representation that has 2^n distinct digital values. The quantization error or the resolution corresponds to the smallest change of the analog input voltage that causes a bit change in the digital conversion.

In case of input voltages V_{in} larger than V_N ($V_{in} > V_N$), the conversion will be saturated by V_N and these input voltages will all be represented by the largest digital n-bit number 11–1. This error, due to saturation, can be reduced by signal conditioning to match the range of input voltages with one of the selectable scales, such that $\max\{V_{in}\} < V_N$.

In case of input voltages V_{in} with $\max\{V_{in}\} < V_N$, the part of the scale between $\max\{V_{in}\}$ and V_N will not be used and the resolution of the digital conversion will be reduced.

Consequently, the analog input voltage V_{in} has to be normalized on a 0–V_N scale, where V_N is the upper limit of the scale. The ideal selection is $\max\{V_{in}\} \cong V_N$.

Often, ADC boards have selectable scales with V_N of 0.1, 1, or 10 V.

V_N is determined by the full-scale range voltage given by

$$V_{fsr} = 2R_F V_R / R.$$

Given that the resolution r for a nominal range V_N is the same as the analog output voltage for the LSB, it follows that

$$r = 2R_F V_R / (R2^N).$$

For V_N chosen equal to $V_{fsr} = 2R_F V_R / R$, r becomes

$$r = V_N / 2^N.$$

Example: A 3-bit ADC

For an ADC with $n = 3$, $R_F = R = 5\ \text{k}\Omega$, 0–10 V input voltage scale, and $V_R = 5$ V

(a) calculate the full-scale range and nominal scale;
(b) calculate the resolution;
(c) calculate the amplification gain using an inverting amplifier for signal conditioning, assuming that the voltage output from a piezoceramic transducer can be up to 500 V;
(d) represent graphically the conversion of the 3-bit digital inputs to the internal DAC shown in Fig. 4.19 into voltage output V_o.

***Solution*:**

(a) $V_{\text{fsr}} = 2R_F V_R / R = 10\text{V} = V_N$.
(b) $r = V_N / 2^n = 10/8 = 1.25$ V.
(c) A gain of 10/500 = 0.02 reduces the maximum voltage 500 V to 10 V, achieving an ideal matching of the domain of input voltages, from 0 to 500 V, with the 0-to-10 V ADC scale.
(d) For a $n = 3$ bit $(b_1\, b_2\, b_3)$ DAC, there are $2^3 = 8$ distinct digital values, 000, 001, 010, 011, 100, 101, 110, and 111. The analog output voltage V_o signal is converted from eight discrete distinct steps. For $V_R = 5$ V and the resolution of 1.25 V, the eight values for V_o are

0.00, 1.25, 2.50, 3.75, 5.00, 6.25, 7.50, and 8.75 V.

Figure 4.20 displays the DAC conversion of the 3-bit digital inputs into voltage output V_o.

As a result of this internal DAC conversion, analog input voltages V_{in} of 9.00 and 9.60 V are found by the comparator the same as the V_o produced by DAC for 111. For the range of analog input voltage $V_{\text{in}} = 0$–10 V, the digital conversion gives

V_{in}	$b_1\, b_2\, b_3$	V_{in}	$b_1\, b_2\, b_3$
$0 \le V_{\text{in}} < 1.25$	0 0 0	$5.00 \le V_{\text{in}} < 6.25$	1 0 0
$1.25 \le V_{\text{in}} < 2.50$	0 0 1	$6.25 \le V_{\text{in}} < 7.50$	1 0 1
$2.5 \le V_{\text{in}} < 3.75$	0 1 0	$7.50 \le V_{\text{in}} < 8.75$	1 1 0
$3.75 \le V_{\text{in}} < 5.00$	0 1 1	$8.75 \le V_{\text{in}} < 10.0$	1 1 1

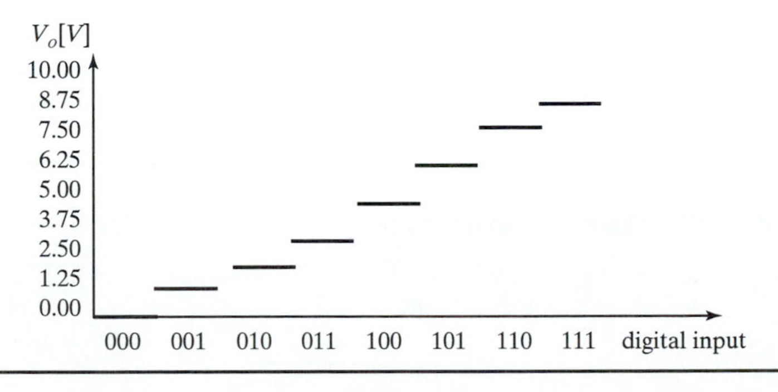

Figure 4.20

The DAC conversion of the 3-bit digital inputs into voltage output V_o

Sampling and Aliasing Analog signals normally vary in time, and the digital conversion is supposed to retain the information regarding this time variation. Empirical signals can contain a large variety of components of various frequencies, but only a limited range of frequencies, for example up to f_{max}, can be retained in the sampled signal. Digital conversion of the time-varying analog signals is obtained by sampling the signal at a sampling frequency f_s [samples/sec] or, its reciprocal, sampling period [sec/sample].

Sampling theorem states that the frequency content up to f_{max} of the analog signal can be reconstructed after sampling only if $f_s > 2 f_{max}$.

The frequency $f_N = f_s / 2$ is called Nyquist frequency. Components of the signal that have frequencies over f_N will result in aliasing (i.e., in false frequencies present in the reconstructed analog signal, but not existent in the original analog signal). To avoid alias frequencies, the analog signal frequencies $f > f_N$ have to be filtered before sampling with a low-pass filter called antialiasing filter [17].

Upcount ADC In the Upcount ADC, the registers start the trial and error approach with the lowest digital number and continue with one bit up increase, which also result in the increase of the output voltage of the internal DAC shown in Fig. 4.19. This upcounting continues until, for the first time, $V_o > V_{in}$ when the output of the comparator goes low, triggered by the release of the content of the registers, which is sent out as the parallel digital output equivalent to the analog input. The same trial-and-error process takes place at each sampling time and is repeated $(1 + 2^n)/2$ times for n-bit conversion.

Example: 3-Bit Upcount ADC

An analog voltage $V_{in} = 6.5$ V is converted by a $n = 3$-bit upcount ADC as follows:

Trial Digital Number	V_o	$V_o < V_{in}$
000	0	Yes
001	1.25	Yes
010	2.5	Yes
010	3.75	Yes
011	5.00	Yes
101	6.25	Yes
110	7.5	No

This conversion was finalized after six trials. Depending on the value of V_{in}, the conversion requires between 1 and $2^3 = 8$ trials, with an average of $(1 + 2^3)/2 = 4.5$ trials [17].

Successive Approximations ADC An n-bit binary number

$$b_1 \quad b_2 \quad \ldots \quad b_{n-1} \quad b_n$$

can also be denoted as

$$\mathrm{MSB}_1 \quad \mathrm{MSB}_2 \quad \ldots \quad \mathrm{MSB}_{n-1} \quad \mathrm{MSB}_n$$

such that the second most significant bit MSB_2 is b_2 and the least significant bit is

$$\mathrm{LSB} = \mathrm{MSB}_n = b_n.$$

The trial-and-error process starts with $\mathrm{MSB}_1 = 1$ and all other lower bits equal 0. If the analog voltage output of the internal DAC is

$V_o < V_{in}$, the output of the comparator goes high and the MSB_1 is maintained at 1;

$V_o > V_{in}$, the output of the comparator goes low and the MSB_1 is reset to 0.

At the next clock cycle, the process starts with $\mathrm{MSB}_2 = 1$ and all other lower bits equal 0. If the analog voltage output of the internal DAC is

$V_o < V_{in}$, the output of the comparator goes high and the MSB_2 is maintained at 1;

$V_o > V_{in}$, the output of the comparator goes low and the MSB_2 is reset to 0.

The process continues until all bits up to the LSB are scanned. At the end of the n clock cycles, the output of the registers is sent out as the parallel digital output of the ADC.

Example: A 3-Bit Successive Approximation ADC

The analog voltage input of $V_{in} = 6.5$ V is converted as follows:

	$\mathbf{MSB_1}$	$\mathbf{MSB_2}$	$\mathbf{MSB_3}$
Trial MSB_i	1	1	1
Trial number	100	110	101
V_o (Fig. 4.20)	5	7.5	6.25
$V_o < V_{in}$	Yes	No	Yes
Final MSB_i	1	0	1

Figure 4.21 shows the sequence of trials. The resulting digital output is 101, corresponding to $V_o = 6.25$ V. The conversion error, $V_o - V_{in} = 6.5 - 6.25 = 0.25$ V, is smaller than the resolution $r = 1.25$ V. The number of trials for 3-bit successive approximations ADC was 3, lower than six trials requires for the 3-bit upcount ADC for the same conversion. In general, an n-bit ADC conversion requires n trials of successive approximations, fewer than the average $(1 + 2^n)/2$ for upcounting.

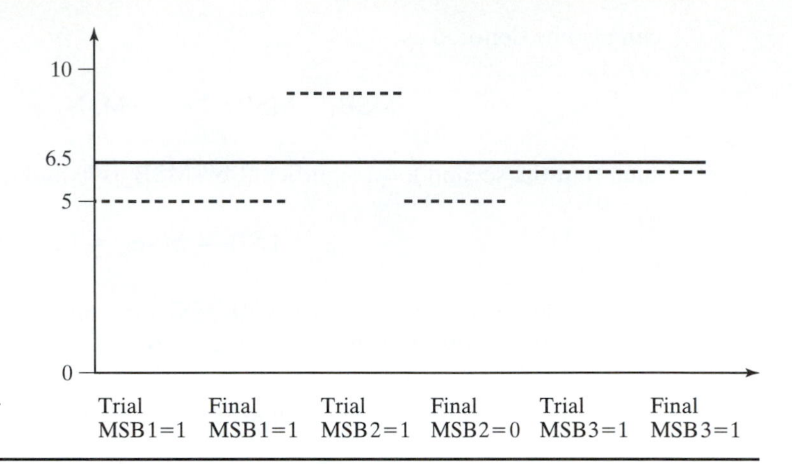

Figure 4.21

The sequence of trials in 3-bit successive approximations ADC for $V_{in} = 6.5$ V

4.6 POWER AMPLIFIERS AND ACTUATOR DRIVES

4.6.1 Types of Drives

Drives for electric motors can be classified as follows:

thyristors drives;
triac drives;
linear amplifier drives;
nonlinear amplifier drives.

This section will focus on drives for DC motors. Thyristor and triac drives use AC power supply and, besides rectifying it in DC, serve to control the DC motor velocity by modifying the duration of the rectification during an AC cycle. They are used for a variety of appliances and small tools, which have DC power supply and, in this case, benefit from avoiding a separate DC supply or a separate rectifier. They require, however, electrical or mechanical filtering of oscillatory components present in the rectified output voltage, and they are not convenient for torque control.

Linear and nonlinear amplifier drives require DC supply. Linear amplifiers provide an ideal solution for driving motors less than 500 W, but have high heat loss. Nonlinear amplifiers are most frequently of pulse width modulation (PWM) type and have lower performance than linear amplifiers in velocity, current of torque control of motors, but also have lower heat losses.

Only linear amplifiers drives for open-loop velocity control of unidirectional angular motion will be presented in this section. Bidirectional velocity motion and servomotor drives require more detailed presentation of operational and power amplifiers and are presented in specialized literature [23, 25, 28].

4.6.2 Linear Amplifier Drives for DC Motors

Out of the variety of linear amplifier drives presented here are

(a) power transistor drives;
(b) power operational amplifier drives;

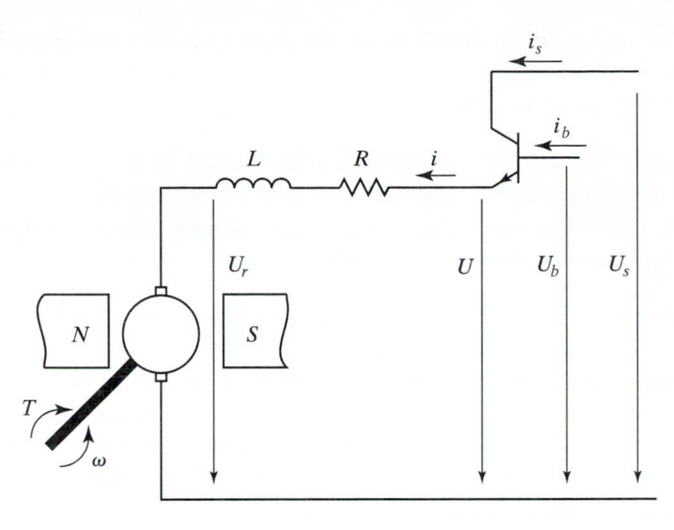

Figure 4.22

Unidirectional velocity control of a PM–DC motor with a power transistor drive

(c) power transistor-based amplifier drive for unidirectional velocity control (shown in Fig 4.22).

The subscripts for the transistor currents were presented in Section 4.2.1.

The power transistor is supplied by a constant DC voltage U_s of less than 35 V. A simplified linear model of the changes ΔI and ΔU for a bipolar junction transistor is given by [9]

$$\Delta i_s = \beta \, \Delta i_b$$

and

$$\Delta U_b = z_i \, \Delta i_b,$$

where β is the forward current transfer ratio $\Delta i_s/\Delta i_b$ and z_i is the input impedance $\Delta U_b/\Delta i_b$. Typically, Ω takes values between 20 and 150, while z_i could be between 500 Ω and 20 kΩ. For a current $i_s = 2$A, required for the supply of the rotor, the current i_b needed for controlling the angular velocity of the DC motor is in the range of 13 to 100 mA. This controlling current i_b is normally too high to be supplied directly by a digital-to-analog converter and a preamplifier stage has to be inserted between the ADC and the transistor base *b*.

This driver has several drawbacks, the most important being the high power loss. The power loss is given by

$$\Delta P = (U_s - U)i_s.$$

The power loss is significant and the resulting heat has to be removed to avoid overheating.

Another deficiency, mentioned in Section 4.2.1, is the dependence between forward current transfer ratio $\beta = \Delta i_s / \Delta i_b$ and the current i_s. The dependence limits the use of power transistors as linear amplifiers when precise control is desired.

A more advanced solution for DC motor drives is the power operational amplifier shown in Fig. 4.23. Power operational amplifiers resemble operational amplifiers, presented in Section 4.2.2, except for the following features:

The values for the current is can be a thousand times higher (e.g., 30 A for a power operational amplifier versus 30 mA for an operational amplifier);

The voltage Us can be 10 times higher (e.g., 200 V for a power operational amplifier versus 10–20 V for an operational amplifier).

Consequently, the power output of a power operational amplifier is over 1000 times higher than the power output of a typical operational amplifier.

Open-loop voltage gain of a power operational amplifier is similar to the open-loop gain A of an operational amplifier (i.e., $A = 10^4$ to 10^6). As a matter of fact, the first stage of amplification of the input voltage in a power operational amplifier is normally an operational amplifier followed by a power transistor-based amplifier.

The transfer of power is described by the balance equation

$$Ui = \eta\,(U_s i_s + U_b i_b),$$

where $\eta < 1$ is the power amplifier efficiency, which accounts for the power loss and is dependent on power amplifier operating current and voltage. Most of the power transfer is from $U_s i_s$ to Ui, while $U_b i_b$ is negligible, such that

$$Ui \approx \eta U_s i_s$$

and

$$i_s \approx Ui/(\eta U_s),$$

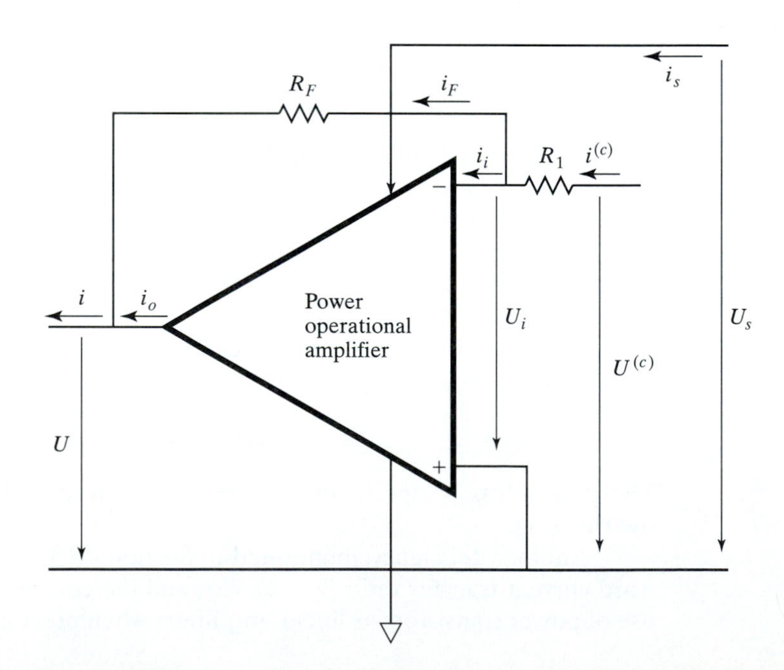

Figure 4.23

Unidirectional velocity control of a PM–DC motor with a power operational amplifier drive

where U_s, the supply voltage, is assumed constant for normal values of i_s, while U, i, and i_s are variable with regard to time.

The dependence between the gain β of the power transistor and the current i_s and the variability of the operational amplifier gain A are avoided by the use of passive resistances R_1 and R_F that determine the gain U/U_s.

The ideal power operational amplifier shares some of the properties of an ideal operational amplifier presented in Section 4.2 as, for example,

$$R_i = U_b/i_b = \infty \quad \text{and} \quad i_b = 0$$

such that, for the system for the system shown in Fig. 4.23, the following equations can be written:

$$i^{(c)} = (1/R_1)U^{(c)}$$

and

$$U = (-R_F/R_1)U^{(c)} - R_o i.$$

In the latter equation,

$$R_o = R_{oA}(R_1 + R_F)/(R_1 A)$$

and R_{oA} is the output resistance of the power operational amplifier [33]. For a value of R_{oA} of less than hundreds of ohms and the very high value of A, $R_o \cong 0$.

The equivalent circuit is shown in Fig. 4.24.

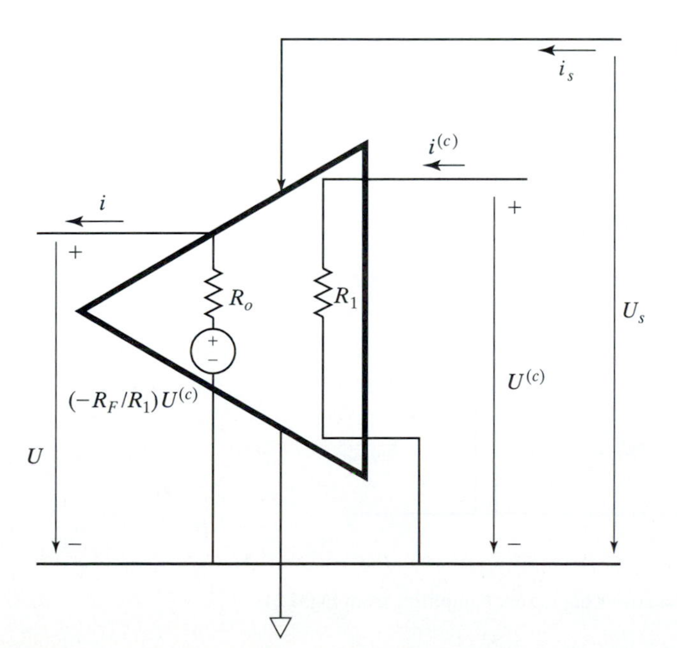

Figure 4.24

Equivalent circuit for diagram of Fig. 4.23

The input resistance of the inverting amplifier,

$$R_i = R_1 = U^{(c)}/i^{(c)},$$

is much smaller than the input resistance R_{oA} of the open-loop operational amplifier (10^{10} to 10^{14}).

A Simulink model of the equivalent diagram of Fig. 4.24 is shown in Fig. 4.25. This model represents a three-port element. Replacing the input *i* and output *U* for the DC power supply for the PM–DC motor, from Fig. 3.5, by the power operational amplifier of Fig. 4.25, the Simulink model of Fig. 4.26 is obtained.

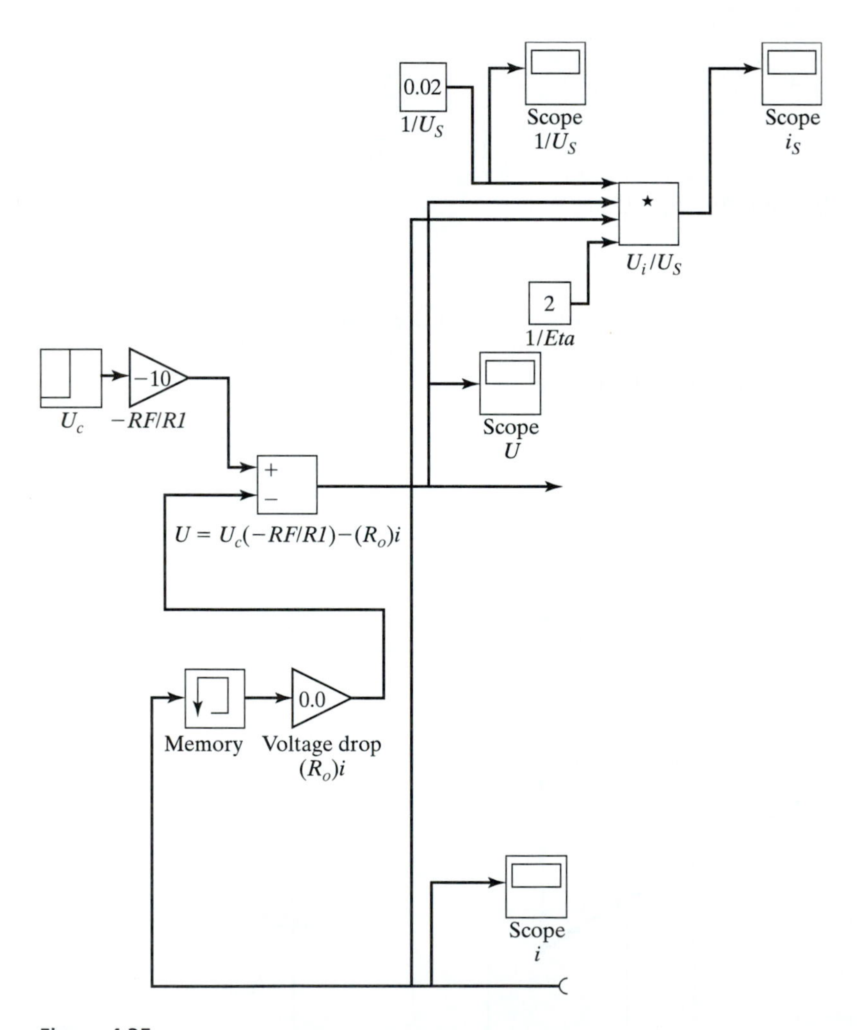

Figure 4.25

Simulink model of the power operational amplifier from Fig. 4.24

For the model in Fig. 4.26, the data are the same as for Fig. 3.5, except for $\eta = 0.5$ (denoted Eta in Fig. 4.26 such that 1/Eta = 2), $U_s = 50$ V (i.e., $1/U_s = 0.02$ [V^{-1}, A = 10^4, R1 = 1 kΩ, $R_F = 10$ kΩ, $R_{oA} = 100\Omega$, $R_o = R_{oA}(R_1 + R_F)/(R_1 A) = 100 + 10^4)/(10^3 10^4)$ = 0.01 Ω, $U^{(c)} = -1$ V, such that $U^{(c)}(-R_F/R_1) = 10$ V). The results of the simulation are shown in Fig. 4.27.

The results in Fig. 4.27(a) have the same shape as those shown in Fig. 3.6. The current i reaches a steady-state value of about 0.875 A. The voltage drop, on a small resistance $R_o = 0.01$ Ω, is very small and, consequently, the voltage U differs insignificantly from the step $U_s = 10$ V. The steady-state power output U_i of the power operational amplifier is $10 \cdot 0.875 = 8.75$ W. The angular velocity reaches a steady-state value of about 170 rad/s and the torque reaches a steady-state value of about 0.0170 Nm.

The results in Fig. 4.27(b) show a constant $1/U_s = 0.02$ [V^{-1}] and a current i_s with a steady-state value of 0.35 A. The steady-state power supply is $U_s i_s = 50 \cdot 0.35 = 17.5$ W. Taking into account power amplifier efficiency, $\eta = 0.5$, this confirms the preceding result for the power output of the power amplifier of 8.75 W.

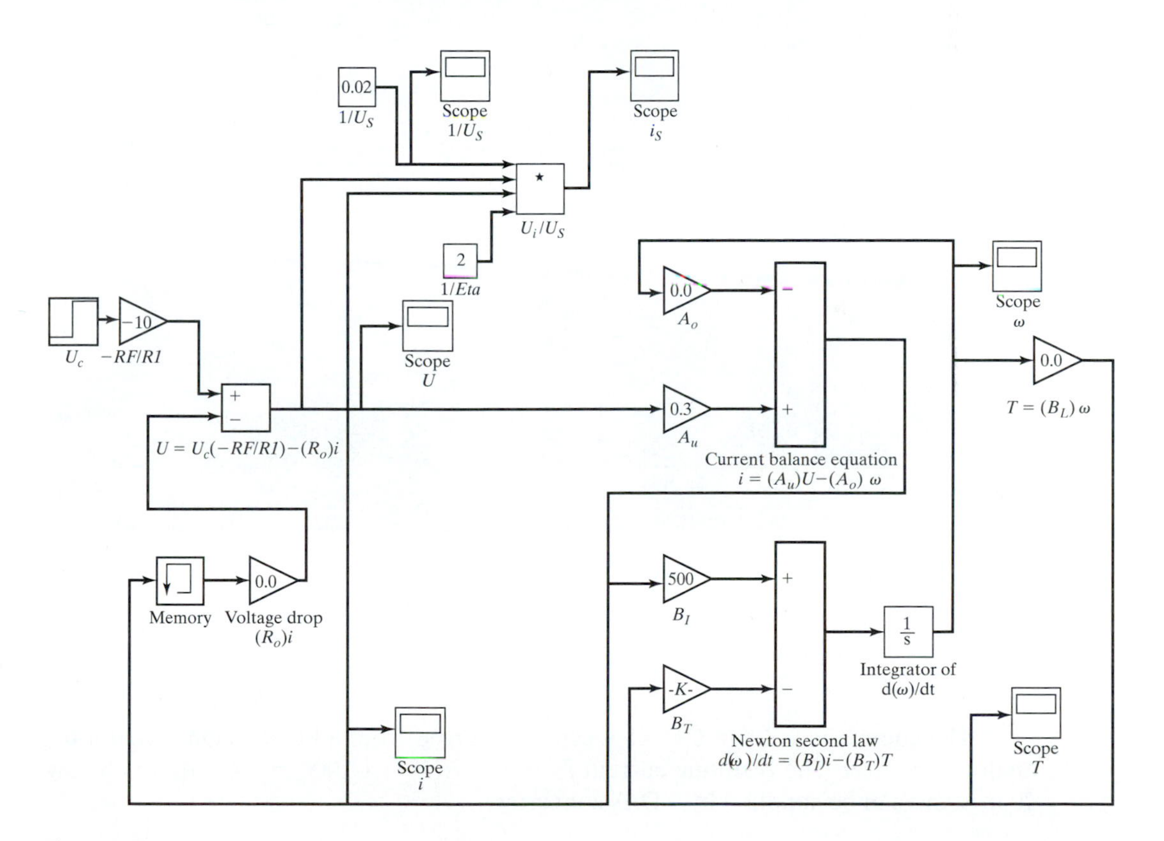

Figure 4.26

Simulink model of the PM-DC motor with a power operational amplifier drive

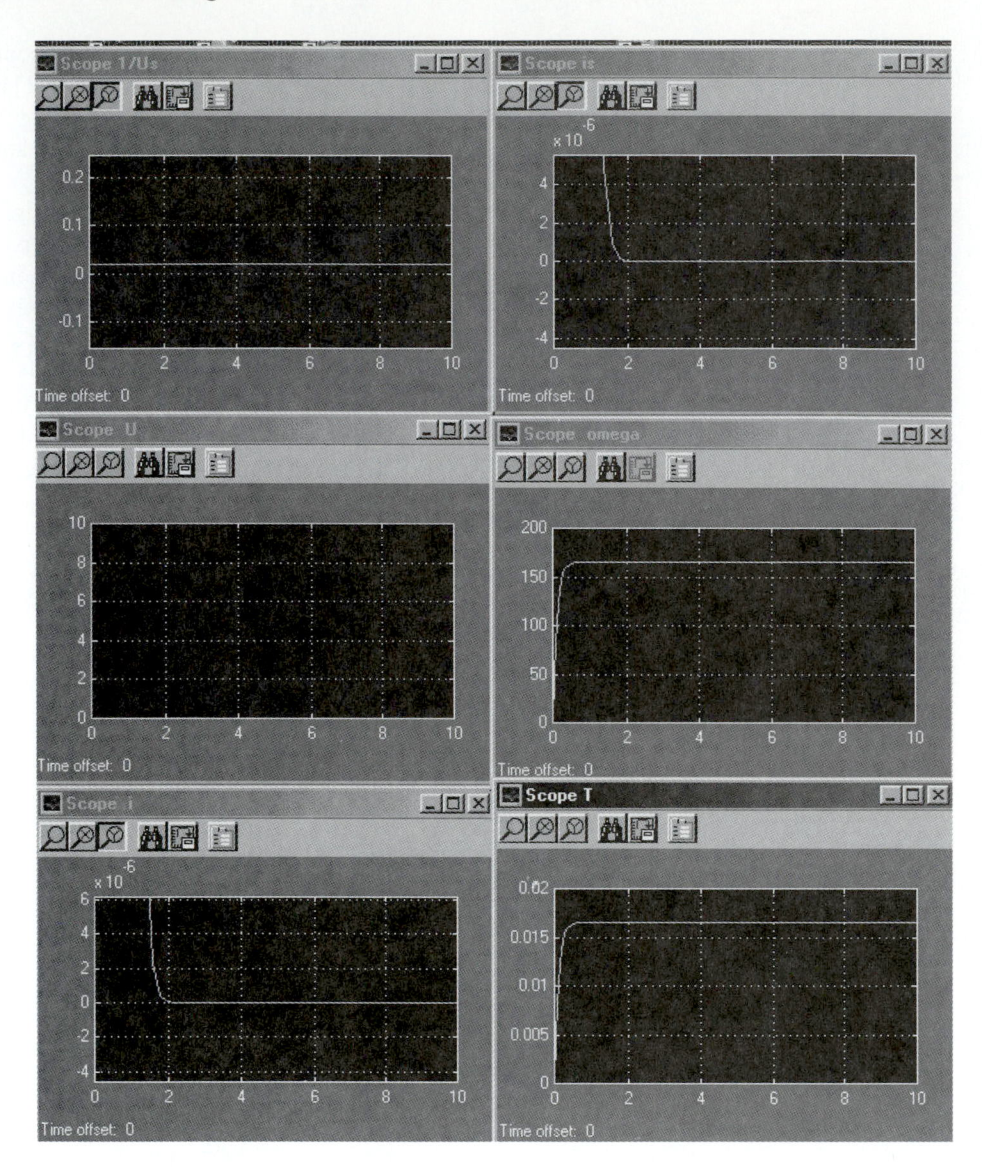

Figure 4.27

Simulation results for the PM–DC motor modeled in Fig. 4.26

The command voltage $U^{(c)} = 1$ V can be received from a PC through a digital-to-analog converter. The resulting current $i^{(c)} = U^{(c)}/R_1 = -1/1000 = -0.001$ A is below 20 mA and can be supplied by a D/A converter.

PROBLEMS

4.1. A weighted resistor digital-to-analog converter has $n = 8$ bits, the reference voltage $V_R = 10$ V, the most significant bit resistance $R = 12$ kΩ, and the feedback resistance of the operational amplifier $R_F = 6$ kΩ.

Calculate

(a) the output voltage V_o corresponding to the least significant bit;
(b) the output voltage V_o corresponding to the most significant bit;
(c) the maximum value of the output voltage V_o;
(d) the nominal full-scale output voltage V_o;
(e) the resolution;
(f) the output voltage V_o corresponding to the binary input 10101100.

4.2. A frequency sensor is connected to a computer through a digital-to-analog converter that has a 2% settling time of 130 μsec. Calculate the maximum frequency that can be measured.

4.3. The gain–bandwidth product (GBP) of an operational amplifier has the value 12 MhCz. The operational amplifier is an amplifier with a gain of 120 dB.

(a) Calculate the bandwidth of the amplifier.
(b) Can this amplifier be used for amplifying signals of 120 Hz? If not, calculate the gain of the amplifier having the same GB= 12 MHz.

4.4. Assume a weighted resistor digital-to-analog converter with $n = 4$ bit with $R = 1$ kΩ and $V_R = 5$ V. Calculate R_F such that analog output voltage V_o will take values from 0 to 10 V.

4.5. For the weighted resistor digital-to-analog converter described in problem 4.4, build a Simulink model and simulate the following cases:

(a) the output voltage V_o corresponding to the least significant bit;
(b) the output voltage V_o corresponding to the most significant bit;
(c) the maximum value of the output voltage V_o;
(d) the nominal full-scale output voltage V_o;
(e) the output voltage V_o corresponding to the binary input 1010.

4.6. Assume an analog-to-digital converter with $n = 3$ bit, $R = 1$ kΩ, $R_F = 2$ kΩ, and $V_R = 10$ V. Calculate the full scale range and nominal scale range as well as the resolution.

4.7. For the analog-to-digital converter described in problem 4.6, build a Simulink model for upcount method and simulate the following cases:

(a) the output for full-scale upper limit;
(b) the output for 1 V analog input;
(c) the outputs for inputs from 0 V, in steps equal to the resolution, up to the upper limit of the full scale.

4.8. For the analog-to-digital converter described in problem 4.6, build a Simulink model for the successive approximations method and simulate the computation of the digital output for 1 V and 3 V analog inputs.

4.9. Use the Simulink model developed in problem 3.4 and replace the battery model by a power operational amplifier model. For the same data as in problem 3.4 from Chapter 3 and $R_I = 1$ kΩ, $R_F = 10$ kΩ, and $U^{(c)} = -1$ V, compare the simulation results for the case of battery supply and power operational amplifier supply.

CHAPTER 5

Mixed Dynamic Systems Modeling and Simulation

5.1 OVERVIEW OF SYSTEM MODELING

5.1.1 Models, Simulations, and Experiments

The images and the descriptions we operate with are pictorial and textual models of the reality in conceptual-qualitative form. With such models, however, we cannot perform virtual experiments to test in advance our planned actions [41].

Structured models (which can be communicated to others in an unambiguous form) permit testing (simulating) planned actions to obtain a preview of the response. Structured models refer to a portion of the reality that can be separated from the rest and is referred to as a system or process.

In this chapter, structured modeling and simulation of processes will be presented. While a model is a simplified form of a process, a simulation is an experiment that gives the response of the model to a given input (actions on the system). Experimenting on the model is in fact a virtual experimentation, to be distinguished from experiments on the (real) process.

5.1.2 Inputs, Outputs, States, and Events

A simple model is shown in Fig. 5.1. It has only one input and one output and the system is represented by a single block. This simple model is suitable for representing a static system having the current output dependent only on the current input. An exam-

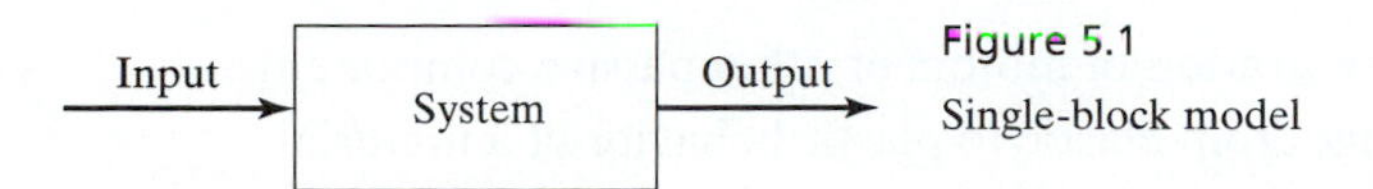

Figure 5.1
Single-block model

ple is a switch that can turn on and off a light bulb. The input is the switch position (on or off) and the output is the state of the bulb (lit, non-lit). This is the case of a static system given that the current output of the bulb does not depend on the history of the states of the bulb, but only on the current input.

Most systems require dynamic modeling, given that often the output depends not only on the input, but also on the (internal) state of the system. For example, the current flow rate of items processed by a server depends not only on the current input (incoming rate of items arriving for processing), but also on the current state of a buffer (accumulated items, waiting for service). Even in the case of a zero input, the output can be still significant as a result of processing items stored in the buffer. In case of no buffer, the system is a static system having the current output dependent strictly on the current input.

In general, the description of a system requires several variables. Besides input and output, state variables are often used. In fact, a system might have numerous variables, but not all are needed to have a complete description of the system at any time. The minimum set of such variables form the state vector of the system. The states form a necessary and sufficient set for this description, but the selection of the states from all variables is not unique. Normally, there are more variables than the minimum number needed to describe the system and many variables are dependent. Examples will be used to clarify the concept of state.

In a waiting line case, the number of items in the waiting can be used as state variable.

In a reliability model, the number of operating or failed components at a particular time characterize system state.

In a mechanical system, position, velocity, acceleration, force, power, and momentum are variables. Often position and velocity are chosen state variables, while force is considered as input and some position(s), velocity(ies), acceleration(s), or force(s) are chosen as outputs. The choice is application-dependent.

In an electrical system, electric charge, current, potential, voltage difference, and electric power are variables, and often currents and their derivatives and/or voltages and their derivatives are chosen as states. Some specific components, which interface the given system with other systems, sources, or loads, have their voltages or currents chosen as inputs or outputs.

The mathematical model uses the defined states, inputs, and outputs. When the model is a set of first order ordinary differential equations, the number of state variables equals the number of equations.

The concept of event is used in a confusing way in various books. Sometimes the event is considered as the instantaneous value of a state. In other books, event is considered as a significant change in input, state, or output [41].

In this book, event will be considered a significant change in continuous systems or a change of state in discrete events systems. Events can occur as a result of unusual inputs.

Examples are

the saturation of motors or other passive components,

change from elastic to plastic behavior of a material,

shutdown of a production or service facility as a result of a power failure, and
change in the structure of the system as a result of a collision in a mechanical system.

5.1.3 Types of Mathematical Models

Assume the state vector

$$x(t) = [x_1(t), x_2(t), \ldots, x_i(t), \ldots, x_n(t)],$$

where $x_i(t)$ is the ith state variable, a dependent variable, while the time is an independent variable (given that no input modifies time variable).

Various types of systems can be defined in accordance to the type of variable x and t.

The time variable can be a continuous variable or a discrete variable. This distinguishes between continuous-time and discrete-time systems.

The state variable x can be continuous or discontinuous. In the discontinuous case, when a finite number of changes of the state variable x take place over a randomly spaced time between changes, the model is called a discrete-event model. In a discrete-event system, both x and t are real numbers, but change of the value of x occurs only at specific moments in time.

Continuous Time Models Continuous time models use differential equations with solutions $x(t)$ of the type shown in Fig. 5.2. Most models of mechanical or electrical systems are of this kind.

Discrete Time Models In the case of discrete-time models, the time is equidistantly spaced (i.e., the time steps are of the same duration) and can be modeled with difference equations.

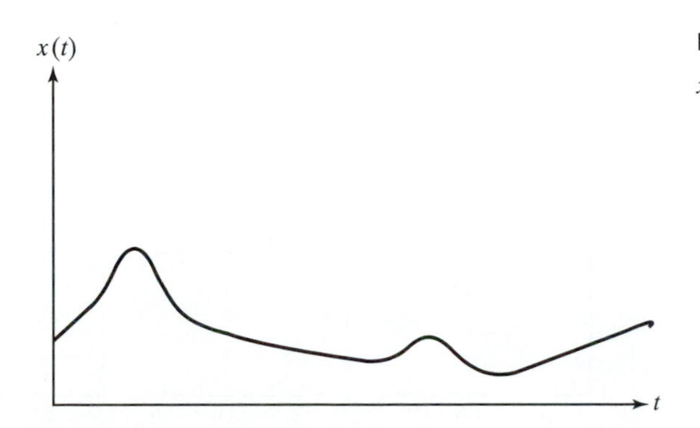

Figure 5.2
$x(t)$ for a continuous time model

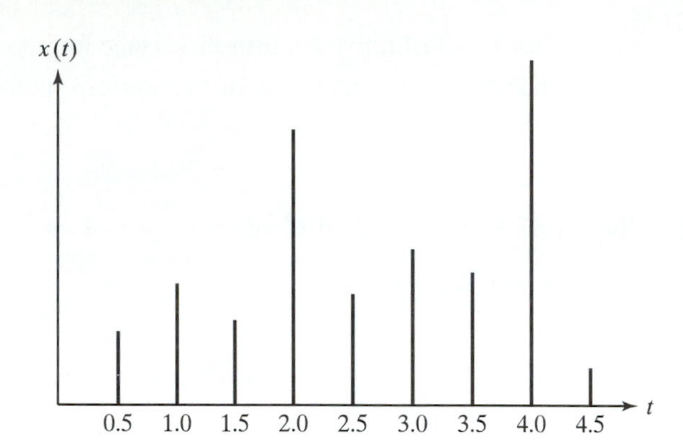

Figure 5.3
$x(t)$ for a discrete time model

Figure 5.3 illustrates discrete time variation of the state of a system. Sampled data models of continuous variables are also discrete time models, even if the system itself is a continuous time system.

Time t is a variable, which take discrete values equally spaced but, as shown in Fig. 5.3, not necessarily defined integer numbers. In a discrete time system, state variable x can be a real or an integer number and is defined only at discrete times t = 0.0, 0.5, 1.0, etc.

Discrete Events Models Discrete event modeling refers to the case of using events for representing state transitions that occur at nonequidistantly spaced moments in time.

A typical illustration is a waiting line model. The number of items in the waiting line changes discontinuously at random times, dependent of new arrivals and of end of services. Figure 5.4 shows $x(t)$ in a discrete event model.

In a discrete event system, the state variable x is an integer number and is defined continuously in time, even if the changes occur discontinuously. (The number of people in a waiting line is a good example of this model.)

The choice of continuous or discrete time model or discrete event model in a particular application is not subject to fixed or unique rules. For example, in computer animation for game programming, at an earlier time only discrete events models were

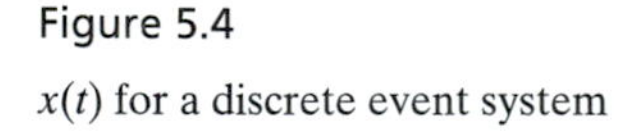

Figure 5.4
$x(t)$ for a discrete event system

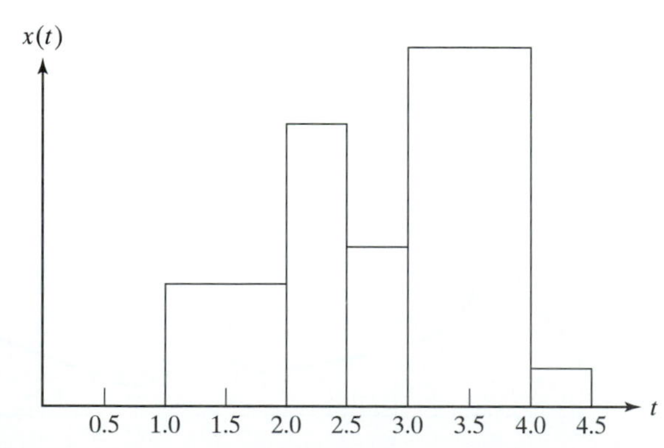

used, while more recently, analytical models reproducing physical behavior (collisions, explosion waves, etc.) are employed to add realistic feeling to the computer games.

5.1.4 System Representations Used in Modeling

A number of basic representations used in system modeling will be analyzed. The selection of these representations is based on the popularity of the models at the present time as well as on the potential for applicability in the near future.

The representations presented are used in commercial simulation packages and consequently the list is pragmatically rather than theoretically motivated. These representations are named differently by various authors [41].

Conceptual Models The use of natural language and of conceptual drawings are the means used in engineering design at conceptual design stage and represent typical conceptual models needed for communications with clients, upper management, and specialists. Knowledge representation in artificial intelligence also refers to mental models using pictorial form or language codification. These models are mostly qualitative and cannot be applied to obtain executable forms for computer simulation.

Analytical Models for Continuous Time Systems Analytical models use equations from physics (Newton equations of motion for mechanical systems, Kirchhoff equations for electric circuits, etc.) to derive a mathematical model of system dynamics. In the case of analytical modeling, the model and the simulation code are obtained directly from the differential equations, and the solution for simulations is obtained numerically. Simulation packages based on analytical modeling use high-level languages—like Fortran or C—or general-purpose simulation languages.

State Event Models A finite state automaton model represents all the states of the system (a finite number of discrete states) as nodes linked by branches with arrows, which represent events (feasible transitions from one state to another) [41]. Examples of models in this category are finite state automata, Markov models, finite event automata, and Petri nets.

The example of a finite state model shown in Fig. 5.5 is the model for the elevator in a building. In this example, the elevator is assumed used by five people—denoted A, B, C, D, E—to move as a group from a top floor to ground floor, taking into account that the elevator has a maximum capacity of two people. An input arrow indicates the start point, when all people are on the top floor, and an output arrow indicates the final point when all people arrived on the ground floor. In the figure each circle contains, on the upper line, the people on the top floor (the current state for the top floor) and, on the bottom line, the people on the ground floor. A zero denotes the state of no people on that floor. Each directed branch has attached the name of the people who move from one floor to another.

The transition diagram shown in the Fig. 5.5 displays two possible solutions from Start to Stop. In total there are $5 \cdot 4/2 = 10$ different possible solutions. If various other situations are considered, more solutions can be generated. For example, if useless

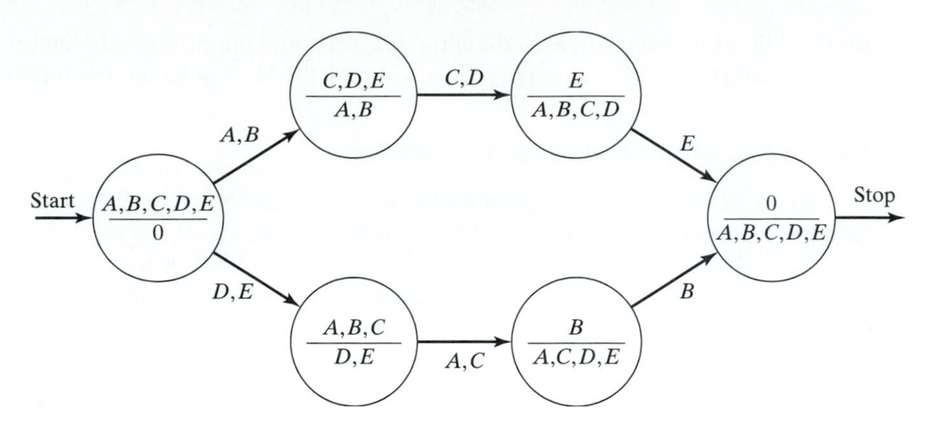

Figure 5.5

A state transition diagram

cases are considered (i.e., unnecessary transport, up and down, of some people), an infinite number of solutions can be generated.

Block Diagrams Block diagram components resemble to finite state automata components (nodes linked by directed branches), but they have different meanings. In the block diagram, the branches represent variables (inputs, outputs, states), while the nodes are blocks that transform the block input (in accordance to a block assigned function) into the block output.

In Fig. 5.6, the schematic diagram of a water tank system is shown. The following notations were used:

$$Q(t) = \text{incoming water flow rate [kg/s]},$$
$$Q_1(t) = \text{outgoing water flow rate [kg/s]},$$
$$H_1(t) = \text{head [m]},$$
$$A_1 = \text{cross-sectional area [m}^2\text{]},$$

and

$$R_1 = \text{resistance to flow from tank [m s/kg]}.$$

During a small time interval Δt, the conservation of liquid mass for the tank gives

$$[Q(t) - Q_1(t)]\Delta t = A_1 \cdot \Delta H_1(t).$$

Figure 5.6

Schematic diagram of a water tank system

For $\Delta t \rightarrow 0$,

$$d/dt \cdot H_1(t) = (1/A_1)[Q(t) - Q_1(t)].$$

Taking into account that a linearized model gives

$$\text{(outgoing flow rate from tank)} = \text{(head)/(resistance to flow from tank)},$$

the following equation results:

$$Q_1(t) = H_1(t)/R_1.$$

The former first-order differential equation and the latter algebraic equation represent the analytical model of the tank. The two equations have three variables—Q(t), $H_1(t)$, and $Q_1(t)$. $Q_1(t)$ can be chosen as state variable, $Q(t)$ is the obvious input, and the variable $H_1(t)$ can be chosen as output. Another choice could be $Q_1(t)$ as the output variable and $H_1(t)$ as the state variable. The algebraic constraint can be used to eliminate the dependent variable $Q_1(t)$ such that the mathematical model is reduced to a simple equation:

$$(d/dt)\, H_1(t) = -(1/R_1\, A_1)H_1(t) - (1/A_1)\, Q(t).$$

The block diagram realization of the tank model can be achieved with the blocks shown in Fig. 5.7 [26].

The block diagram shown in Fig. 5.8 is the model of the tank system. Systems do not have a unique block diagram model. For this tank, if $Q_1(t)$ is chosen as output, a different block diagram will result.

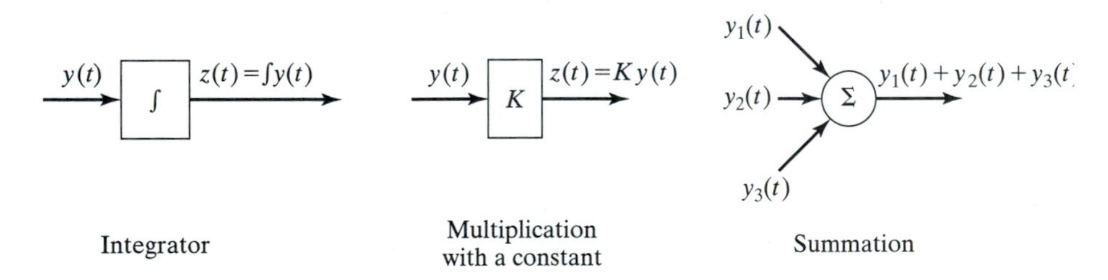

Figure 5.7

Blocks for block diagram realization

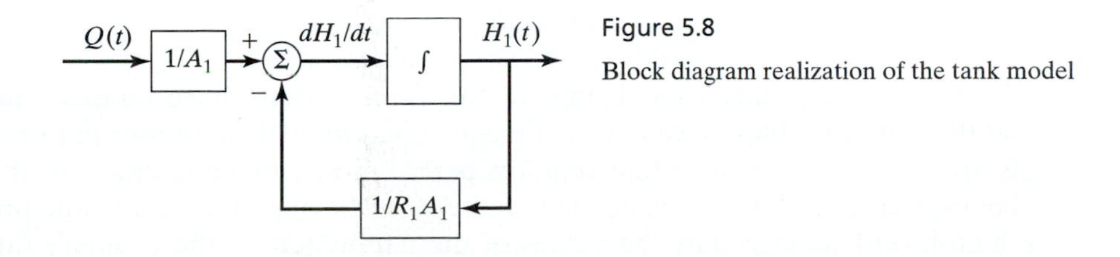

Figure 5.8

Block diagram realization of the tank model

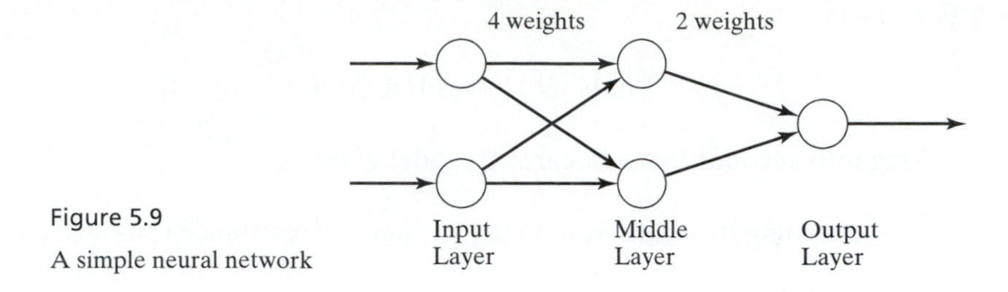

Figure 5.9
A simple neural network

Neural Network Models A neural network is also represented by nodes and directed branches (same as state event networks or block diagrams). These branches have attached weights w, while the nodes perform a nonlinear transformation of the sum of inputs using a saturation function as, for example, the sigmoid function: $1/(1 + \exp\{-x\})$. A simple neural network is shown in Fig. 5.9. Four weights for the branches going to the middle layer and the two weights of the branches going to output layer are initially unknown. These weights can be obtained for a particular application by "learning" (e.g., using back propagation method). Learning uses given sets of collected sets of input and output values reflecting the dynamics of the system to be modeled.

In dynamic systems modeling, neural networks can be seen as approximate solutions of the nonlinear differential equations that could describe the system. This approximate solution is a linear algebraic combination of sigmoid functions, which (after training) incorporates the information contained in the input and output values used for training. Given this procedure of obtaining the solution, the differential equations do not have to be derived and solved, as the solution is obtained from a trained neural network for a given set of values.

5.1.5 Textual and Graphical Programming: Object-Oriented Approach

High-Level Languages Modeling and simulation are used to obtain results of actions in a virtual environment with the intention of evaluating the results of similar actions in the real environment. High-level computer languages (e.g., Fortran or C), used to develop models and simulation codes, are text-oriented and approach the system under analysis as a whole. The textual program consists in a sequence of instructions.

General-Purpose Simulation Languages The effort of having to rewrite separate codes for each application has been alleviated by the introduction of general-purpose simulation languages (e.g., ACSL, SIMNON, DESIRE, etc.). Languages for continuous system simulation provide a predefined structure for writing and solving first-order ordinary differential equations with a nonlinear right-hand side—for example,

$$d\mathbf{x}(t)/dt = f(\mathbf{x}(t),t)$$

where $x(t)$ is an n-dimensional state vector. These systems have lumped parameters, and the state variables as well as the time are continuous. Continuous parameter models are a more special case that requires partial differential equations. In the case of general-purpose simulation languages, rather than having to write a whole program in a high-level language, only the values of the parameters of the ordinary differential equations have to be specified. The simulation is carried out by numerically solving all equations simultaneously. While programming is greatly simplified, the identification of the system components in the program is not easy.

Originally, general-purpose simulation languages were textual languages. Recently, graphical programming (e.g., the block diagram front-end ACSL) was added.

Graphical Programming Languages Textual programming requires an analytical model of the system that has to be available in state space format (i.e., as a set of first-order differential equations with derivatives on the left-hand side). Obtaining the model in this format requires extensive and often advanced engineering work. Also, in some cases, an analytical model contains algebraic loops (i.e., equations having the same variable on the left-hand as well as on the right-hand side) and cannot be immediately cast into state space format [35, 42].

New programming languages for simulation were developed for various engineering fields to alleviate these difficulties. A solution is textual programming that starts with specifying circuit topology. Spice is an example for electric and electronic circuits [65].

More recently graphical programming languages were developed that permit the design of models without having first to develop a state space model. Analytical modeling, state space formatting of the equations, and algebraic loop handling is hidden from the model developer and is achieved algorithmically. Working Model is an example of a mechanical systems simulator such that the model is built graphically using lumped parameter components and defining joining constraints [31, 72]. LabVIEW is a virtual instrumentation software for monitoring and control of processes, which can also model monitoring and system dynamics using an icon-based editor [73]. Spice, Working Model, and LabVIEW are domain-specific software and use icons of electric, mechanic, and monitoring components, respectively, for graphical editor-based programming.

Graphical programming languages for generic continuous or discrete time simulation use block diagram editors for abstract block functions. Simulink and SYSTEM BUILD are examples of such simulation packages [35, 64]. They require specification of the blocks in the form of state space equations, transfer function, etc., and analytical modeling has to be carried out before graphically developing the model.

The object-oriented approach was adopted by several simulation packages. Simulink is based on MATLAB, which—starting with version 4—was rewritten using object-oriented programming. LabVIEW was also designed using object-based technologies. In order to clarify the specificity of the object-oriented approach, this topic will be covered in a separate section.

Object-Oriented Programming for Simulation A clear mapping between the components of the simulation code and the corresponding real counterparts appears only in object-oriented programming (OOP) languages like Smalltalk, C++, or Java. This mapping is very useful in model development and model updating and, for this reason, will be presented here as part of the object-oriented approach to modeling (OOM) and to simulation code development [41, 42].

In OOM, the states can be associated to the variable attributes, but supplementary definitions are needed to distinguish the variables from states. While inputs and outputs are linking a system with other systems or with the environment, the states (like the attributes in OOM) are not accessible directly from outside the system and can be changed only by modifying the input to the system (message passing in OOP).

Given the specificity of the object-oriented approach, which is very different from the traditional engineering analytical modeling, the basic concepts of the OO approach will be presented. In presenting OOM, the keywords, syntax, and structure of OOP will be used. The examples for modeling, even if written similarly to an OOP, are only pseudocodes and could not be compiled and executed. The purpose is not to learn one or another OOP language, but to use OO approach as a unifying framework for modeling, in a structure readily formatted, which can eventually be used for OOP. The examples will only be illustrations of modeling and, in general, might be too simple or too specific to justify further efforts to transform them in OOP. These pseudocodes of OOM have the purpose of helping to build abilities to read OOP, often needed to a user of commercial simulation packages. Writing OOP simulation packages requires good knowledge of a current version of an OOP language (C++, Java, Smalltalk, etc.) [27, 55, 57]. The users of commercial simulation packages, however, do not need programming knowledge for using simulators.

5.1.6 Simulation Packages

Simulation software was, in the beginning, developed specifically either for continuous systems or for discrete event systems. At the present time, many established simulation software packages contain tools for simulating a combination of both systems, but the stress on one kind of system is often obvious.

SIMULA, SLAM II, and GPSS are system simulation packages mainly for discrete events, but they also can model simple continuous systems. SIMNON, DYMOLA, and ACSL are mainly continuous systems simulation packages, but the last two can simulate simple discrete events systems [41, 50, 60, 69].

Simulink and SYSTEM BUILD were developed for both continuous and discrete systems modeling and simulation [35], [64]. Often, discrete event simulation packages (SLAM II, GPSS, etc.) contain random number generators, Monte-Carlo simulation facilities, variance analysis tools, sensitivity analysis routines, and other statistical analysis tools [41].

At the present time, the trend of combining discrete events and continuous systems approaches in one simulation package is further enhanced by the OO approach, which facilitates the unification of these approaches. In OOP, the use of objects permits hiding an internal model by encapsulation, such that only objects linked by message passing is visible to the user [41]. Examples of domain-specific simulation software using an object-based approach (already mentioned earlier) are Working Model and LabVIEW [72, 73]. Examples of generic simulation software (also mentioned earlier) are Simulink and SYSTEM BUILD [35, 64]. These simulators use signal or data flow along the links only in the assigned direction from a given block output to another block input. Power flow representation, however, is often needed in engineering simulations and, in these cases effort–flow variables replace simple signals. DYMOLA and OMOLA are examples of effort–flow based representation of systems in computer modeling packages [42, 70, 74].

5.2 BLOCK DIAGRAMS AND STATE SPACE MODELING

5.2.1 Block Diagrams

In the block diagrams, the branches represent variables (inputs, outputs, states), while the nodes are blocks that transform the input into the block according to a block assigned function. In Section 5.1, the example of a block diagram of a tank system was presented. In this chapter, a two-tank system will be analyzed [26]. A first example is shown in Fig. 5.10.

In Fig. 5.10, the following notations are used:

$$
\begin{aligned}
Q(t) &= \text{incoming fluid flow rate [kg/s]},\\
Q_1(t) &= \text{tank 1 outgoing flow rate [kg/s]},\\
Q_2(t) &= \text{tank 2 outgoing flow rate [kg/s]},\\
H_1(t) &= \text{tank 1 head [m]},\\
H_2(t) &= \text{tank 2 head [m]},\\
A_1 &= \text{cross-sectional area of tank 1 [m}^2\text{]},\\
A_2 &= \text{cross-sectional area of tank 2 [m}^2\text{]},\\
R_1 &= \text{resistance to flow from tank 1 [m} \cdot \text{s/kg], and}\\
R_2 &= \text{resistance to flow from tank 2 [m} \cdot \text{s/kg]}.
\end{aligned}
$$

The following equations can be obtained as the analytical model of the system (Section 5.1):

$$
\begin{aligned}
(d/dt)\, H_1(t) &= (1/A_1)[Q(t) - Q_1(t)] = (1/A_1)Q(t) - (1/A_1)Q_1(t);\\
(d/dt)\, H_2(t) &= (1/A_2)[Q_1(t) - Q_2(t)] = (1/A_2)Q_1(t) - (1/A_2)Q_2(t);\\
Q_1(t) &= H_1(t)/R_1;\\
Q_2(t) &= H_2(t)/R_2.
\end{aligned}
$$

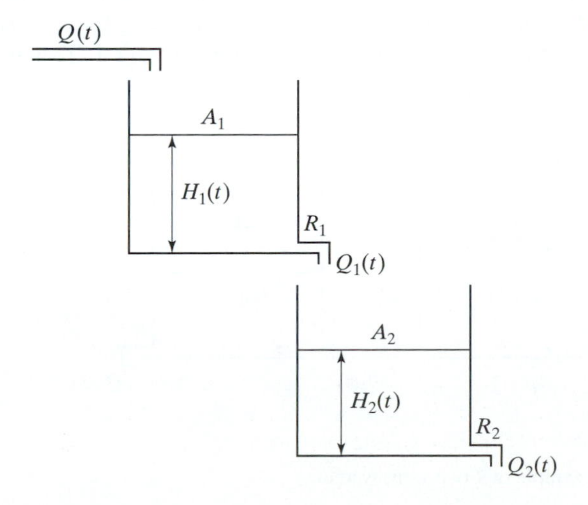

Figure 5.10

A first example of a two-tank system

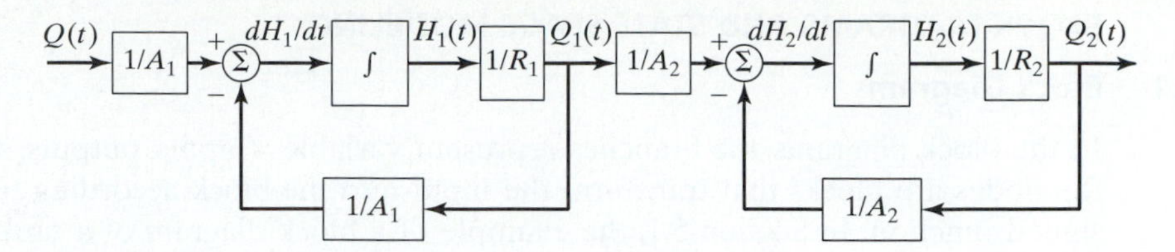

Figure 5.11

Block diagram of the first example of a two-tank system

The two first-order differential equations and the two algebraic equations contain five variables that are functions of t: $H_1(t)$, $H_2(t)$, $Q(t)$, $Q_1(t)$, and $Q_2(t)$. Two variables, $H_1(t)$ and $H_2(t)$, can be chosen as state variables; $Q(t)$ is the obvious input and $Q_2(t)$ can be chosen as output. (Other choices can be made, depending on the application.)

The block diagram, shown in Fig. 5.11, represents a two-tank system. In this block diagram, there is only forward interaction from tank 1 to tank 2 (and no feedback interaction from tank 2 to tank 1). This system is called two noninteracting tanks.

Each tank can be identified as a distinct part of the block diagram, and these two parts are identical except for the subscript 1 for tank 1 and 2 for tank 2. This symmetry is obviously a feature that is useful in transforming the four previous equations and the block diagram in an OO model, as it will be shown later in this chapter.

A second example is the two-tank system shown in Fig. 5.12. The following equations can be obtained as the mathematical model:

$$(d/dt)\,H_1(t) = (1/A_1)[Q(t) - Q_1(t)] = (1/A_1)Q(t) - (1/A_1)Q_1(t);$$
$$(d/dt)\,H_2(t) = (1/A_2)[Q_1(t) - Q_2(t)] = (1/A_2)Q_1(t) - (1/A_2)Q_2(t);$$
$$Q_1(t) = [H_1(t) - H_2(t)]/R_1;$$
$$Q_2(t) = H_2(t)/R_2.$$

Compared to the analytical model of the first example of a two-tank system, the obvious change for the second example is the replacement of $H_1(t)$ by $H_1(t) - H_2(t)$ in the third equation to account for the tank 2 direct connection to tank 1. This makes $Q_1(t)$

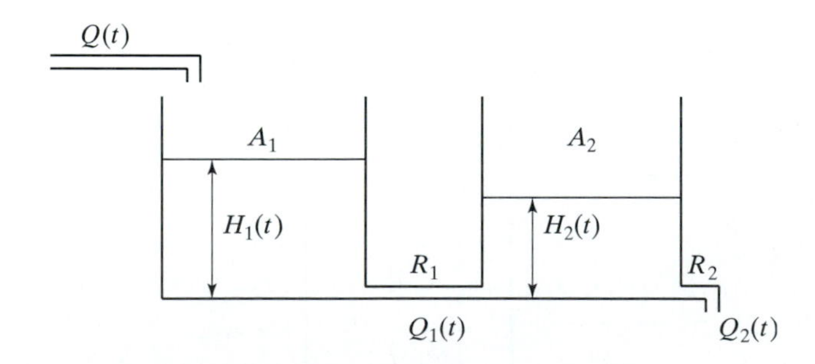

Figure 5.12

A second example of a two-tank system

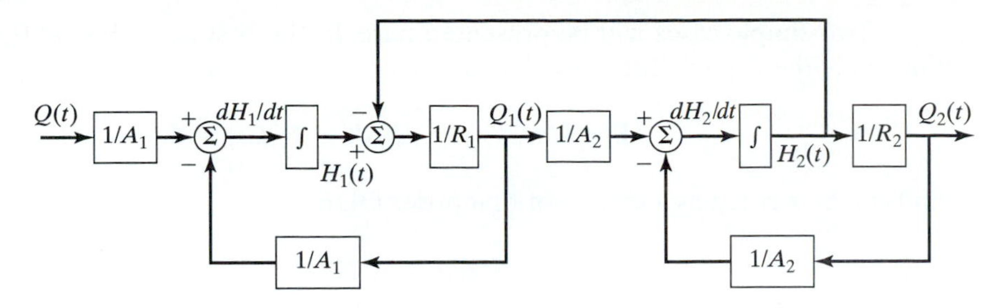

Figure 5.13

Block diagram of the second example of a two-tank system

dependent on $H_1(t) - H_2(t)$ (obviously, if $H_1(t) = H_2(t)$, then $Q_1(t) = 0$) and the system can be called two interacting tanks. As in the previous case of noninteracting tanks, in this case of interacting tanks, the two first-order differential equations and the two algebraic equations contain five variables that are functions of t: $H_1(t)$, $H_2(t)$, $Q(t)$, $Q_1(t)$, and $Q_2(t)$. Again, two variables, $H_1(t)$ and $H_2(t)$, can be chosen as state variables, $Q(t)$ is the obvious input, and $Q_2(t)$ can be chosen as output.

The block diagram shown in Fig. 5.13 represents a two-interacting-tank system. In the present case of the two-interacting-tank-system, the identification of the two tanks in the block diagram is maintained, but a feedback interaction is added. As the number of interactions between the subsystems of a system increases, the block diagram model becomes more difficult to formulate, and the direct solution to the set of differential or difference equations or algebraic equations might become more efficient.

5.2.2 State Space Models

Differential Algebraic Equations (DAE) and Ordinary Differential Equations (ODE) Graphical programming using block diagram models requires only icons manipulation and wiring block. The graphical model is afterwards automatically converted into a high-level language program containing the analytical model, which can remain hidden to the user.

Simulations can be performed directly by solving the analytical model, which, in general, is a system of N differential algebraic equations (DAE) [75]

$$\mathbf{F}(d\mathbf{X}/dt, \mathbf{X}, t) = 0,$$

where $\boldsymbol{F}$ is a vector of N differential or algebraic equations and $\boldsymbol{X}$ is a vector of N variables.

Many models of physical systems consist in a mixture of algebraic and differential equations. In particular, lumped parameters models for electric networks and multibody mechanical systems are primarily modeled by a system of DAE.

Computer codes were developed for solving systems of DAE using traditional high-level languages (Fortran, C, etc.), OO languages (C++, Java, J++, etc.), or general-purpose simulation languages (ACSL). No general solution for a system of DAE exists. Numerical solvers for DAE were recently extensively discussed [42, 68, 69, 70, 74].

Two simple cases will be presented here. In the first case, the DAE system contains only the explicit differential equations

$$\mathbf{F}(d\mathbf{X}/dt,\mathbf{X},t) = d\mathbf{X}/dt - \mathbf{f}(\mathbf{X},t) = 0$$

and can be written as a system of first order ODEs:

$$d\mathbf{X}/dt = \mathbf{f}(\mathbf{X},t).$$

In a second case, part of the equations of $\mathbf{F}(d\mathbf{X}/dt, \mathbf{X}, t) = 0$ are assumed to be algebraic equations. The original system of N DAE can be written in this case as the system of n implicit first-order differential equations

$$F_i(d\mathbf{X}/dt,\mathbf{X},t) = 0, \qquad i = 1 \text{ to } n,$$

and a system of m implicit algebraic equations

$$F_j(\mathbf{X}) = 0, \qquad j = 1 \text{ to } m,$$

where $n + m = N$. This is actually the general form of the system of DAE.

The vector $\mathbf{X}$ of N variables can be partitioned into a vector $\mathbf{z}$ of m variables, which appear only in algebraic form (i.e., do not appear in differential form) and a vector $\mathbf{x}$ of $n = N - m$ variables that appear also in differential form. In this case,

$$F_i(d\mathbf{x}/dt,\mathbf{x},\mathbf{z},t) = 0, \text{ and } i = 1 \text{ to } n$$
$$F_j(\mathbf{x},\mathbf{z}) = 0, \text{ and } j = 1 \text{ to } m.$$

In many cases, these m algebraic equations can be analytically solved and an $mx1$ vector $\mathbf{z}$, the explicit solution to the algebraic equations, can be obtained:

$$\mathbf{z} = \mathbf{g}(\mathbf{x}).$$

The variables of the vector $\mathbf{z}$ can now be eliminated from the differential equations, yielding

$$G_i(d\mathbf{x}/dt,\mathbf{x},t) = 0, \qquad i = 1 \text{ to } n.$$

Assuming that this system of implicit differential equations can be analytically converted into a system of explicit first-order n ordinary differential equations (ODEs) $d\mathbf{x}/dt = \mathrm{f}(\mathbf{x}, t)$, the following system of $n + m = N$ equations is obtained:

$$d\mathbf{x}/dt = \mathbf{f}(\mathbf{x},t);$$
$$\mathbf{z} = g(\mathbf{x}).$$

The system of n ODE contains n unknowns, and readily available numerical solvers can be used for obtaining the solution $\mathbf{x}(t)$. The system of m algebraic equations immediately gives the solutions for m variables of the vector $\mathbf{z}$, using the explicit equations $g(\mathrm{x})$ for known $\mathbf{x}$.

For a system of DAE in the form

$$F_i(d\mathbf{x}/dt,\mathbf{x},\mathbf{z},t) = 0, \qquad i = 1 \text{ to } n;$$
$$F_j(\mathbf{x},\mathbf{z}) = 0 \qquad j = 1 \text{ to } m,$$

no analytical solutions for obtaining **z** and $d\mathbf{x}/dt$ are available. Numerical solvers for some forms of the implicit DAE can be found in the specialized literature [74, 75].

Example: Conversion of DAE into ODE Model for Two Noninteracting Tanks

Considering the case of the two noninteracting tanks shown in Fig. 5.10, the analytical model is given by the following $n = 2$ differential equations and $m = 2$ algebraic equations:

$$(d/dt)\, H_1(t) = (1/A_1)[Q(t) - Q_1(t)],$$
$$(d/dt)\, H_2(t) = (1/A_2)[Q_1(t) - Q_2(t)],$$
$$Q_1(t) = H_1(t)/R_1,$$

and

$$Q_2(t) = H_2(t)/R_2.$$

Using the notations from the previous section yields

$$\mathbf{x} = [H_1 H_2]_t$$

and

$$\mathbf{z} = [Q_1 Q_2]_t.$$

The two algebraic equations in matrix form are cast in the format $\mathbf{z} = g(\mathbf{x})$:

$$\begin{bmatrix} Q_1 \\ Q_2 \end{bmatrix} = \begin{bmatrix} 1/R_1 0 \\ 0 1/R_2 \end{bmatrix} \begin{bmatrix} H_1 \\ H_2 \end{bmatrix}.$$

After eliminating $\mathbf{z} = [Q_1 Q_2]_t$ from the two differential equations the result, given in the format $d\mathbf{x}/dt = \mathbf{f}(\mathbf{x}, \mathrm{t})$, is

$$(d/dt)\, H_1(t) = (1/A_1)[Q(t) - H_1(t)/R_1]$$

and

$$(d/dt)\, H_2(t) = (1/A_2)[H_1(t)/R_1 - H_2(t)/R_2]$$

or, in matrix form,

$$\begin{bmatrix} dH_1/dt \\ dH_2/dt \end{bmatrix} = \begin{bmatrix} -1/A_1R_1 0 \\ 1/A_2R_1 -1/A_2R_2 \end{bmatrix} \begin{bmatrix} H_1 \\ H_2 \end{bmatrix} + \begin{bmatrix} 1/A_1 \\ 0 \end{bmatrix} Q.$$

The input into the system Q appears as an extra term on the right-hand side of the system of equations. A state space format with explicit outputs is presented in the next section.

Linear ODE in State Space Format The state space matrix form is given by

$$(d/dt)\,\mathbf{x}(t) = \mathbf{A}\mathbf{x}(t) + \mathbf{B}\mathbf{u}(t)$$

and

$$\mathbf{y}(t) = \mathbf{C}\mathbf{x}(t) + \mathbf{D}\mathbf{u}(t),$$

where $\mathbf{x}(t)$ is the vector of state variables, $\mathbf{u}(t)$ is the vector of input variables, and $\mathbf{y}(t)$ is the vector of output variables [26].

This is a standard notation used in computer-aided engineering software (e.g., Simulink). The steady state solution can be obtained directly, as in the previous paragraph, by solving the algebraic equations for $\mathbf{x}$:

$$\mathbf{A}\mathbf{x} + \mathbf{B}\mathbf{u} = \mathbf{0}.$$

The solution, for a nonsingular matrix A, is

$$\mathbf{x} = \mathbf{A}^{-1}\mathbf{B}\mathbf{u}$$

Example: The Model of Two Noninteracting Tanks in State Space Format

In matrix form, the model for the two noninteracting tanks was

$$\begin{bmatrix} dH_1/dt \\ dH_2/dt \end{bmatrix} = \begin{bmatrix} -1/R_1\, 0 \\ 1/R_1\, -1/R_2 \end{bmatrix} \begin{bmatrix} H_1 \\ H_2 \end{bmatrix} + \begin{bmatrix} 1/A_1 \\ 0 \end{bmatrix} Q.$$

Assuming the output from the system is chosen to be $Q_2(t)$, the following equation links the output to the variables present in the preceding differential equations:

$$Q_2(t) = H_2(t)/R_2 = [0\,1\,/\,R_2][H_1\; H_2]_t.$$

A linear ODE in state space format can be obtained using the following notation:

$$\mathbf{x}(t) = [H_1\; H_2]_t;$$
$$\mathbf{u}(t) = Q;$$
$$\mathbf{y}(t) = Q_2(t);$$
$$\mathbf{A} = \begin{bmatrix} -1/R_1 & 0 \\ 1/R_1 & -1/R_2 \end{bmatrix};$$

$$\mathbf{B} = \begin{bmatrix} 1/A_1 \\ 0 \end{bmatrix};$$

$$\mathbf{C} = [0 \quad 1/R_2];$$

$$\mathbf{D} = 0.$$

Example: A Higher Order ODE Model

In Fig. 5.14 is shown a spring balance using the deflection $Y(t)$, due to an extra mass M, from a steady state absolute position $Y(0)$ of the unloaded spring.

Newton's second law gives the second-order differential equation

$$M(d^2Y/dt^2) + B(dY/dt) + KY = Mg,$$

where M is the mass, B is the damping coefficient, and K is the spring coefficient. Damping coefficient B accounts for the structural damping of the spring material. In the steady state, all derivatives are zero, so that

$$KY = Mg,$$

or $M = (K/g)Y$, which confirms that spring deflection is proportional to the mass M.

The second-order differential equation can be transformed in state space form. To obtain two first-order ODE, we define the new variables

$$x_1 = Y$$

and

$$x_2 = dY/dt = d\,x_1/dt,$$

so that $d^2Y/dt^2 = dx_2/dt$.

The second-order differential equation can be rewritten in the new notations as

$$M(dx_2/dt) + B(x_2) + Kx_1 = Mg,$$

or

$$dx_2/dt = -(K/M)x_1 - (B/M)x_2 + g.$$

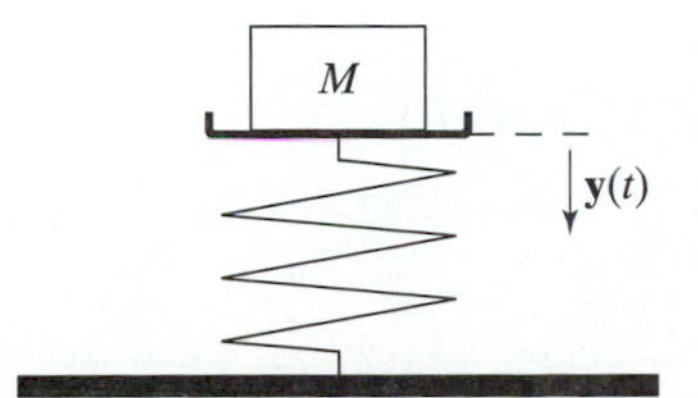

Figure 5.14
A spring balance

Taking into account that $dx_1/dt = x_2$, the last two ODE equations represent the state space form of the analytical model.

Difference Equations Models Difference equations are discrete time counterparts of differential equations. Numerical solutions of differential equations are formulated for their difference translation. The earlier second-order equation,

$$M(d^2Y/dt^2) + B(dY/dt) + KY = Mg,$$

can serve as an illustration for this translation. Backward differences for a small time step T are used as approximations for differentials at $t = kT$:

$$dY/dt = [-Y((k-1)T) + Y(kT)]/T;$$
$$d^2Y/dt^2 = [Y((k-2)T) - 2Y((k-1)T) + Y(kT)]/T^2.$$

The second-order differential equation can be approximated at kT, for a small T, by

$$M[Y(k-2) - 2Y(k-1) + Y(k)]/T^2 + B[-Y(k-1) + Y(k)]/T + KY(k) = Mg$$

or

$$M[Y(k-2) - 2Y(k-1) + Y(k)] + B[-Y(k-1) + Y(k)]T + KY(k)T^2 = MgT^2.$$

Rearranging terms yields

$$Y(k)[M + BT + KT^2] + Y(k-1)[-2M - BT] + Y(k-2)M = MgT^2.$$

Solving for $Y(k)$ gives

$$Y(k) = \{Y(k-1)[2M + BT] - Y(k-2)M + MgT^2\}/[M + BT + KT^2],$$

a second-order difference equation in $Y(k-2)$, $Y(k-1)$, and $Y(k)$.

The conversion into state space form (i.e., into a set of two first-order difference equations) is also needed for solving using computer-aided engineering packages and is performed, similarly to the differential equations case, by defining two new variables, namely,

$$x_1(k) = Y(k-1)$$

and

$$x_2(k) = Y(k)$$

as well as the delay operator D, yielding

$$DY(k) = Y(k-1) = x_1(k) = x_2(k-1) = Dx_2(k)$$

and

$$DY(k-1) = Y(k-2) = Dx_1(k),$$

so that

$$Y(k-2) = Dx_1(k),$$
$$Y(k-1) = Dx_2(k),$$

and

$$Y(k) = x_2(k).$$

The preceding second-order difference equation becomes

$$x_2(k) = \{[2M + BT]/[M + BT + KT^2]\}Dx_2(k) = \{M/[M + BT + KT^2]\}Dx_1(k) + MgT/[M + BT + KT^2].$$

Taking into account that the operator D gives $x_1(k) = Dx_2(k)$, the last two first-order difference equations represent the state space model.

5.2.3 Simulation Example with Simulink

In Fig. 5.11, an example of a block diagram for a two-tank system is shown. Simulink graphical model of this block diagram is presented in Fig. 5.15 using the following parameters: $A_1 = A_2 = 1\ \text{m}^2$, and $R_1 = R_2 = 100\ \text{s/m}^2$.

The flow rates Q are given in this example in m^3/s and the resistances to flow $R = Q/H$ in s/m^2. The initial states are assumed $H_1(0) = H_2(0) = 1$ m. The input flow rate is a step input with the amplitude 0.001 m^3/s:

$$Q(t) = 0 \quad \text{for } t < 0$$
$$0.001\ m^3/s \quad \text{for } t \geq 0$$

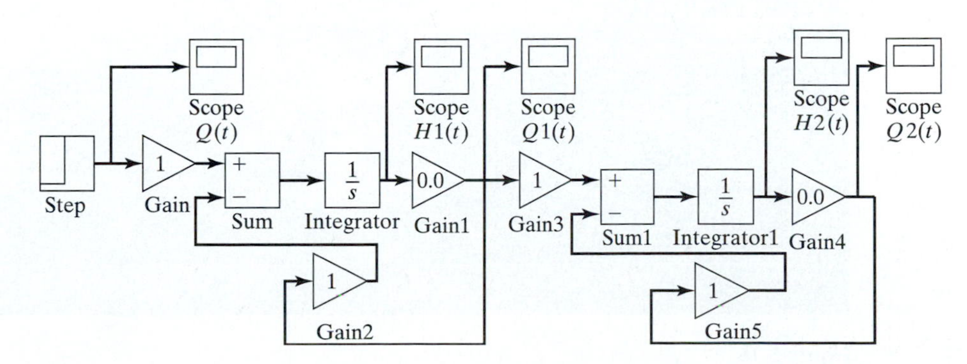

Figure 5.15

Simulink model of the two-tank system shown in Fig. 5.10

Figure 5.16 shows the simulation results. For the assumed step input with amplitude $Q(t) = 0.001$ m^3/s, the flow rate $Q_1(t)$ and the head $H_1(t)$ reach steady state after 400 s while the flow rate $Q_2(t)$ and the head $H_2(t)$ reach steady state after 600 s. At steady state, as expected, $Q(t) = Q_1(t) = Q_2(t) = 0.001$ m^3/s. The results for the transient part are not representing accurately the real dynamics given that the linear equations used

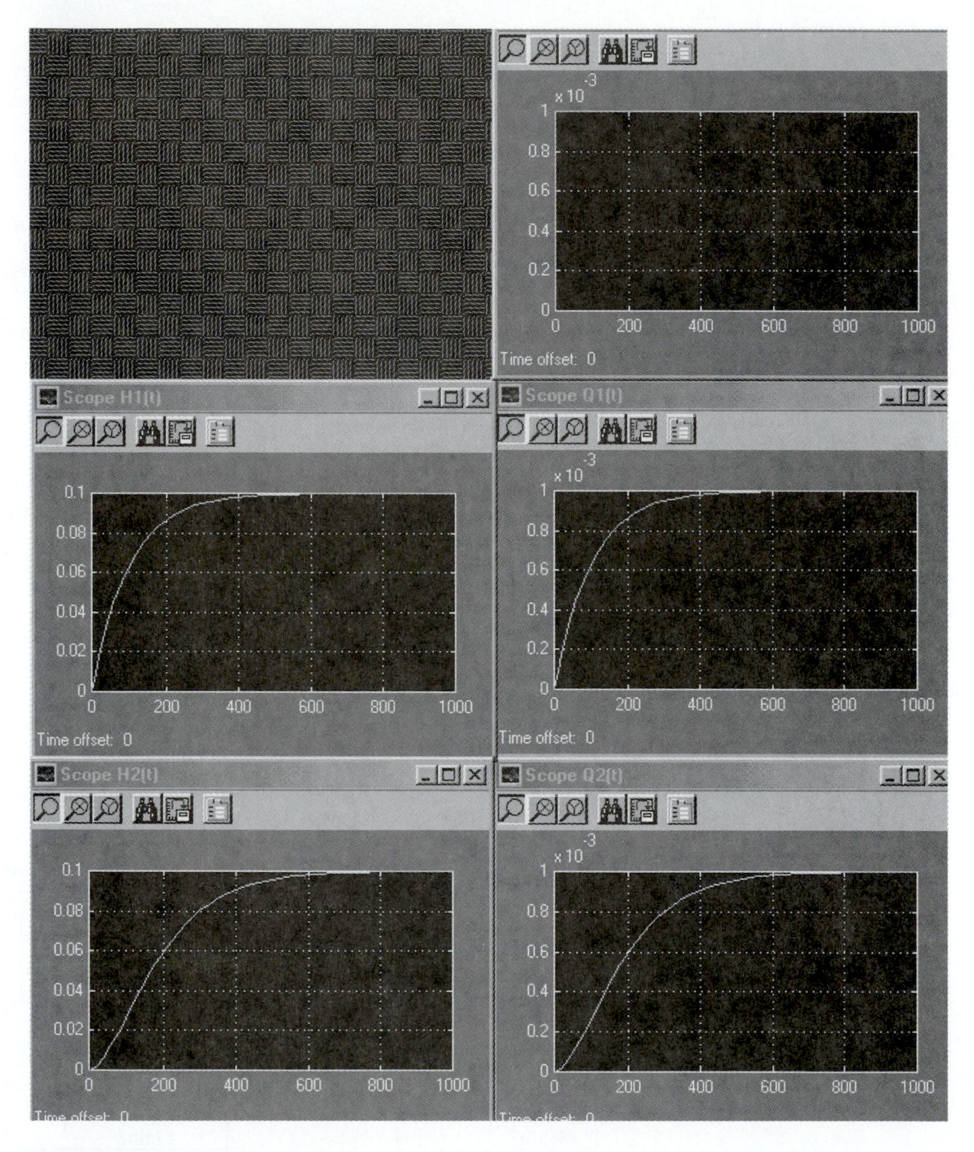

Figure 5.16

Simulink simulation results of two-tank model from Fig. 5.15

for the block diagram of Fig. 5.10 are valid only for small variations about a steady state regime.

5.3 OBJECT-ORIENTED MODELING: SIGNAL AND POWER TRANSMISSION

5.3.1 Object-Oriented Approach

Background Decomposition of systems in interconnected subsystems is an approach widely used in industry for modular design and interchangeability of components. Upgrading and repairs are facilitated by this approach. Computer hardware followed this trend while software lagged behind. Programming languages reflected this situation [82].

High-level languages (e.g., Fortran) are designed around a main program and subroutines. The programs are coded using a flow chart and are executed in a sequence of instructions interrupted by logic conditions for feedback and feedforward loops. The user could not interact with the programs executed in a batch. Today, this approach is justified only in the case of modeling large systems with a very large number of complex internal interactions, which cannot be decomposed by computationally efficient approach in interconnected subsystems.

The extensive use of feedforward and feedback loops in high-level languages was restricted by the structured programming approach (e.g., Modula-2 or Ada). This approach uses top-down design (stepwise design stating first the goal and ending with a detailed structure), defines modules with extensive inner links and minimal external interconnections, and uses private data limited to instructions belonging to a particular unit. These features were later incorporated in the object-oriented approach.

The object-oriented (OO) approach enhanced the features of the structured programming approach and several other features were added. In the OO approach, modules became objects, private data were limited to instructions of a module, and encapsulation procedure was made available. This approach is bottom up, starting with defining objects, classes and couplings.

Objects in the object-oriented approach are characterized by internal state values, which remain unchanged unless an input makes an inner procedure (method) of the object change them [81]. Objects communicate by message passing. Objects are defined as instances of classes such that objects of a class share the same methods and attributes. Most of the concepts used in object-oriented programming have names with redefined meaning, often significantly different from common usage. Moreover, the relationships of these concepts with common descriptions of the same real system become more transparent after analyzing a variety of examples of object-oriented pseudocodes. It is helpful, for this reason, to provide sufficient comments to clarify the examples of pseudocodes.

Classes In this presentation, pseudocodes based on the features of C++ language will be used to exemplify OOP. These pseudocodes are not full programs and cannot be executed [37, 41]. Moreover, C++ language is not a prerequisite for reading these

C++ illustrations; comments (text from // to the end of the line) were added to help explain the examples. OOP defines (or declares or creates) classes as follows:

```
#include <iostream.h>        //Inclusion of a C++ library containing display() function
                             //used subsequently
class Office                 //The keyword class with the name Office, a user-defined
                             //type
{                            //An opening brace
public:                      //A keyword describing the level of access to data members;
                             //in this case the access is not restricted
  Office(int fl, int no);    //Constructor (i.e., a function or a method), which creates
                             //an object belonging to the class Office
  void display();            //A function, contained in the iostream.h library, to print
                             //data
  ~Office();                 //Destructor of the object created earlier
private:                     //A keyword stating the level of access to member
                             //functions; in this case it hides from the user the
                             //variables used internally in the class
  int floor, number;         //Data members declared as private
};                           //A closing brace
```

This example contains three functions or methods, viz.,

```
Office()
display()
~Office()
```

also called "member functions" of the class, and two instances of Office, namely, the integers (int)

```
fl
no
```

also called "data members" of the class.

Encapsulation This example permits us to introduce an essential characteristic of OOP: the encapsulation or packaging together of data "fl", "no" and behavior, methods or functions, "Office()", "display()", "~Office()". The keywords public and private allow the user to see only the public variables "fl" and "no" and not the private variables "floor" and "number". (In general, public attributes could be characters, which are easier to understand, while attributes could be numbers, which are easier to use in computer representation.)

Data types that are part of OOP are called primitive data types and are almost the same in all high-level languages (Fortran, C, C++, Java, etc.) For example, we have the following:

```
int    //Integer
char   //Character
float  //Floating point or real number
```

Objects OOP permits the creation of new data types (NDT) such as "Office" with objects, such as "MyOffice" or "ChiefOffice," as instances of a class type "Office."

The construction (creation) of these objects can be done with special operators such as "new," which construct and initialize objects and allocate memory.

The following is an example of construction of the two objects of the class "Office":

```
Office * MyOffice;           //declaration of "Office" class as data type for
                             //the object "MyOffice"
Office * ChiefOffice;

MyOffice = new Office;       //creation of the object "MyOffice" using the
                             //operator new
ChiefOffice = new Office;
```

These two objects, "MyOffice" and "ChiefOffice," have the same attributes (or data members) as the class "Office" (i.e., "fl" and "no"). Specific values can be assigned to "MyOffice":

```
fl = 5;
no = 53;
```

The attributes (data members) can be static (constants, like the values assigned to "fl' and "no" in the previous example), or they can be variable (states, a new attribute for the class "Office" or the temperature "temp"). The particular values assigned to attributes are called attribute values.

To de-allocate the memory when these objects are no longer needed, the operator "delete" can be used for object destruction (e.g., ~MyOffice).

Inheritance An important feature of OOP is the inheritance, which permits us to reuse the code for a class of general types (a base class) for the programming of a class of a more particular type (a derived class). For the "Office" example, a base class could be "room" (for which the attributes are "fl" and "no") and the derived class "room_office" (for which the attributes are "fl", "no", "temp", and "noise"). The derived class "room_office" inherits the attributes "fl" and "no" from the base class "room".

The inheritance is illustrated in the following example:

```
#include <iostream.h>
enum location {fl, no};
enum conditions {temp, noise};

class room
{
public:
 room (int fl, int no);
 void print();
};

class room_office: public room  //The keyword "public" in the header means that the
                                //public members of the base class "room" are
                                //inherited by the derived class "room_office"
```

```
{
public:
 room_office (int fl, int no, float temp, char* noise); //"temp" is a real number
                                                         //(float); "noise"
                                                    // is a character datum (for ex.,
                                                    // high, medium, low)
 void print();
};
```

5.3.2 State Space Representation and OOP

OOP concepts can be seen as a generalization of basic system theory models.

A linear state space model for a continuous time system is given by ordinary differential equations (ODE). In matrix form, the state space model is

$$(d/dt)\,\mathbf{X} = \mathbf{AX} + \mathbf{BU},$$

where **X** is a state (variable attributes) vector, **A** and **B** are constant attributes matrices, and **U** is a input (actions) vector.

For a nonlinear discrete time system, difference equations can be used. In matrix form, the nonlinear discrete time model is

$$\mathbf{X}(k + 1) = f(\mathbf{X}(k),\mathbf{U}(k)).$$

The same restriction to changing the states **X** applies.

Message Passing In the case of the OO approach, objects cannot have the states (variable attributes) directly changed by another object due to encapsulation, which forbids the access from outside that object to the internal variable attributes (states). The change of state can be achieved only indirectly, using the inputs applied to the object. Consequently, the communication between objects takes place by message passing from one object to another as inputs to the methods. The message is either a signal (a piece of information) or, in the context of Section 5.3.4, an energy flow.

5.3.3 Signal Transmission in Block Diagram Models

Block diagrams, as for example those shown in Figs. 5.11 and 5.13, contain variables associated to the directed links between blocks. These variables can be seen as signals containing the information transmitted from the output of one block to the input of another block. In control engineering, signal flow graphs are sometimes used as an equivalent alternative form of block diagrams.

In communication systems, only the information contained in the signals and not the power transmitted by the carrier of this information is important. In this case, blocks represent transformations applied to the signal transmitted (e.g., delays, attenuation, or filtering). On a communication link, signals can be transmitted bi-directionally. Block diagram models represent only unidirectional transmission, from the designated output of one block to the designated input of another block. The model associated with a block corresponds only to the transfer from the input to the output.

In other engineering systems, the power transmitted by the carrier becomes important, and the equations describing their dynamics are written for variables like force and velocity in mechanical systems and voltage and current in electrical circuits. Equations using these variables can be used for block diagram modeling, as exemplified in Section 5.2. Again, while power can flow bidirectionally on a transmission line, a block diagram model represents only a single direction of the transmission. The conclusion is that state space models, transfer function, and block diagram representations require the assignment of the direction of the signal from one component of the model to another.

Example: A Simple Mechanical System, Mass-Spring-Damper

Consider first a simple mechanical system example, shown in Fig. 5.17 composed of a mass m, a spring k, and a damper b, subject to a force F. The velocity v is assumed the output.

Newton's second law gives

$$F = m(d^2x/dt^2) + b(dx/dt) + kx.$$

Given that

$$y = dx/dt,$$

the foregoing differential equation can be written, using v as a variable, as

$$F = m(dv/dt) + bv + k\int vdt.$$

The corresponding block diagram can be obtained from this time-domain equation using elementary blocks of the type shown in Fig. 5.7, or from the s-domain equation, obtained from the Laplace transform of the previous equation. The latter block diagram is more compact and is typical to control system modeling.

The Laplace transform for zero initial conditions gives

$$F(s) = msv(s) + bv(s) + (k/s)v(s),$$

or

$$v(s) = \frac{1}{ms + b + k/s}F(s).$$

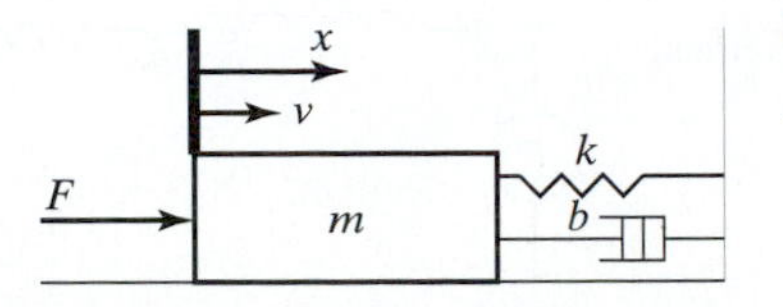

Figure 5.17

A mass-spring-damper example

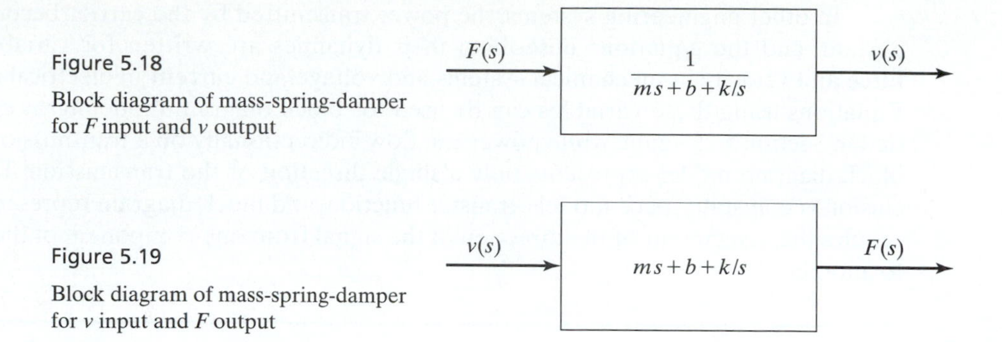

Figure 5.18
Block diagram of mass-spring-damper for F input and v output

Figure 5.19
Block diagram of mass-spring-damper for v input and F output

The block diagram shown in Fig. 5.18 represents the mass-spring-damper system for F input and v output. The function in the block is called the transfer function of the system [26].

The same mass m, spring k, and damper b system, subject to a velocity v input and with the force output F, is described by the equation

$$F(s) = (ms + b + k/s)v(s)$$

and has the block diagram shown in Fig. 5.19.

Due to the input and output assignment, the same system is modeled differently when the variables F and v change designation. This restricts modularity and interchangeability to modules with identical input and output assignment. Moreover, the power transfer cannot be verified directly from this model.

5.3.4 Power Transmission Models

Effort–Flow Variables and Two-Port Models Two-port models were introduced for representing components of electric networks using two terminals for each port. Alternative names for two-port components of a network are four-terminal network and quadripole [77]. The two-pole port has associated two variables, a current I and a voltage V, which permit the calculation of the power $P = VI$ transferred through the port.

Example: Resistance-Inductance Circuit

A circuit, with an inductance L and a resistance R, supplied by an ideal voltage E source (i.e., with zero internal impedance), is shown in Fig. 5.20.

As a result of three cuts, Fig. 5.21 is obtained.

Figure 5.20
A resistance-inductance R-L circuit

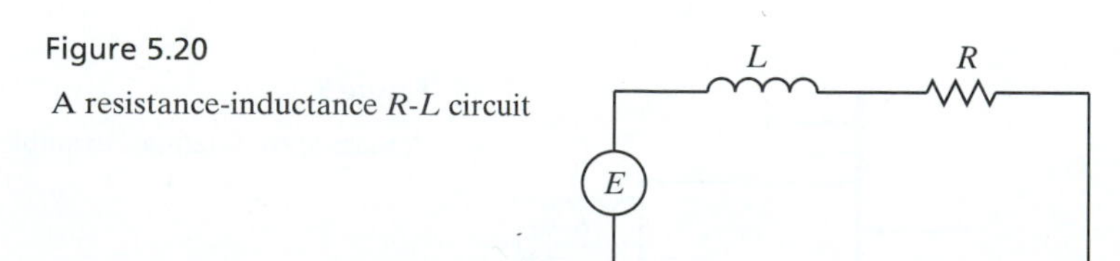

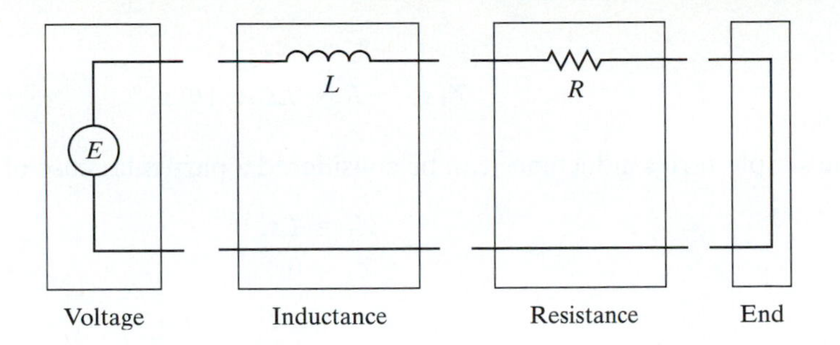

Figure 5.21

Three cuts in R-L circuit

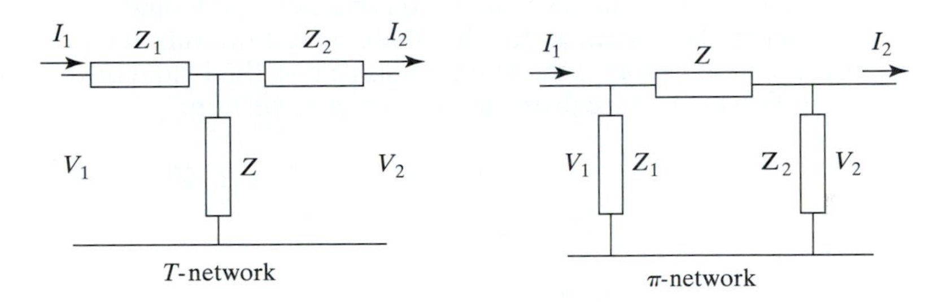

Figure 5.22

T-network and π-network

In linear network analysis, generic two-port elements are a T-network or a π-network, shown in Fig. 5.22. These networks are equivalent and can replace any passive network between a voltage source and a passive load.

A resistance–inductance–capacitance (R–L–C) series circuit, subject to voltage V, is shown in Fig. 5.23. The following voltage drop equation can be written:

$$V(t) = Ri(t) + Ldi(t)/dt + C\int i(t)dt.$$

The Laplace transform of this equation, for zero initial conditions, gives

$$V(s) = Ri(s) + Lsi(s) + C\,i(s)/s.$$

The impedance of the resistance–inductance–capacitance series circuit is given by

$$Z(s) = V(s)/i(s),$$

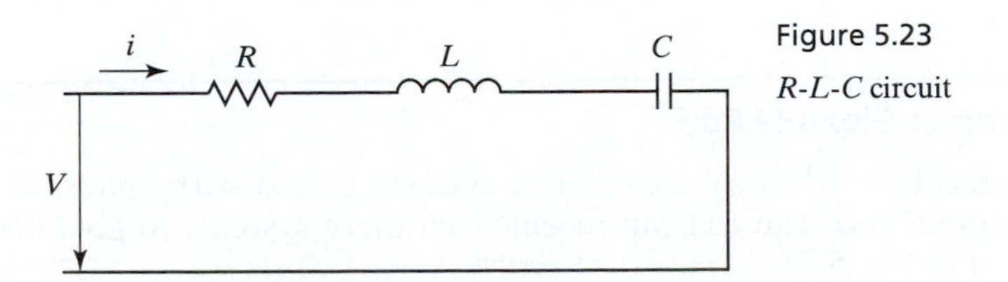

Figure 5.23

R-L-C circuit

or

$$Z(s) = R + Ls + 1/Cs.$$

The simple series inductance can be considered a particular case of a T-network with

$$Z_1 = Ls,$$
$$Z_2 = 0,$$

and

$$Z = \infty.$$

In Fig. 5.20, the four variables V_1 and I_1 on one side and V_2 and I_2 on the other side, permit us to write the equations for the model based on only the inner structure of the two-port element, ignoring the voltage source and the load. For the T-network of Fig. 5.22, the following voltage drop equations can be written:

$$V_1 - I_1 Z_1 = (I_1 - I_2)Z;$$
$$(I_1 - I_2)Z - I_2 Z_2 = V_2.$$

Solving for (V_2, I_2) as a function of (V_1, I_1) gives

$$I_2 = -V_1(1/Z) + I_1(Z_1 + Z)/Z$$

and

$$V_2 = I_1 Z - (Z + Z_2)I_2 = I_1 Z - (Z + Z_2)(-V_1(1/Z) + I_1(Z_1 + Z)/Z),$$

or

$$V_2 = V_1(Z + Z_2)/Z + I_1(-(Z_1 + Z)(Z + Z_2)/Z + Z).$$

The preceding equations for I_2 and V_2 depend only on V_1 and I_1 and the impedances Z_1, Z_2, and Z, which are composed of R, L, or C, in series. Any lumped parameter electric system can be sectioned by cuts into subsystems interfaced by only voltage and current in the cuts.

In the case of solid-body mechanics, free-body diagrams represent components of a multibody system by cutting each body from the system and by representing boundary effects using force f and velocity v, which also permit the calculation of the power $P = fv$.

Example: Flexible Rod

Assume two arbitrary mechanical systems linked with spherical joints by a flexible horizontal rod. The rod can be cut from these systems to give the free-body diagram shown in Fig. 5.24.

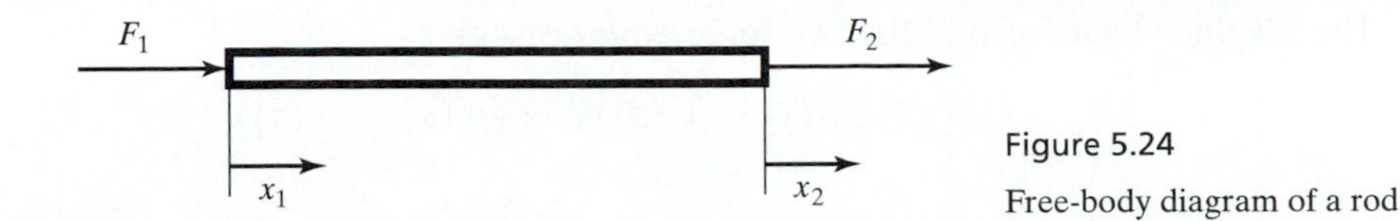

Figure 5.24
Free-body diagram of a rod

In Fig 5.24, for each cut, the internal force F and the absolute velocity v are identified. Assuming that the flexible rod is represented by a lumped parameter model, shown in Fig. 5.25, the following equations can be written:

$$F_1(t) = (x_1(t) - x_2(t))k + (v_1(t) - v_2(t))b;$$
$$F_2(t) = -[(x_1(t) - x_2(t))k + (v_1(t) - v_2(t))b] = -F_1(t).$$

Even if there is a spring and a damper between the two forces, the equality $F_2(t) = -F_1(t)$ reflects the fact that in this model the force change is transmitted instantaneously. Lumped parameter models do not account for propagation delay.

Obviously,

$$v_1(t) = dx_1(t)/dt$$

and

$$v_2(t) = dx_2(t)/dt.$$

The Laplace transform of these two equations gives, for zero initial conditions,

$$v_1(s) = sx_1(s)$$

and

$$v_2(s) = sx_2(s)$$

or

$$x_1(s) = v_1(s)/s$$

and

$$x_2(s) = v_2(s)/s.$$

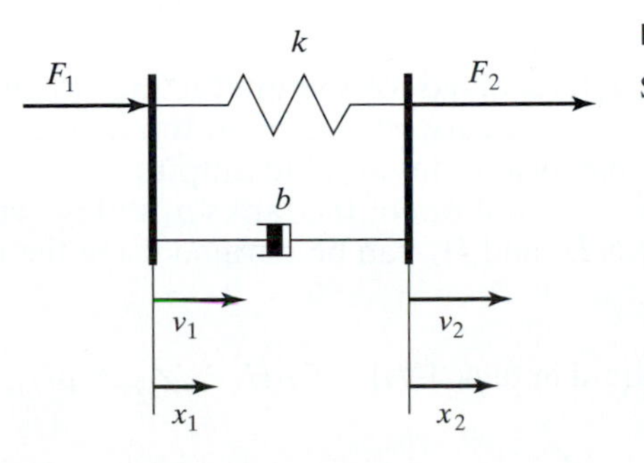

Figure 5.25
Spring damper model of the rod

The Laplace transform of the two force equations gives

$$F_1(s) = (x_1(s) - x_2(s))k + (v_1(s) - v_2(s))b$$

and

$$F_2(s) = -F_1(s).$$

Using $x_1(s) - x_2(s) = (v_1(s) - v_2(s))/s$ in the equation for F_1 yields

$$F_1(s) = (v_1(s) - v_2(s))k/s + (v_1(s) - v_2(s))b = (v_1(s) - v_2(s))(k/s + b)$$

and

$$F_2(s) = -F_1(s),$$

and the following two equations result:

$$v_1(s) = -[1/(k/s + b)]F_2(s) + v_2(s);$$
$$F_1(s) = -F_2(s).$$

These two equations contain the cut variables F and v function only of internal variables b and k. Lumped parameter mechanical systems can be sectioned by cuts into subsystems interfaced by only force and velocities defined regarding the cuts. For a flexible shaft, with cut parameters torque T and angular velocity ω, the model is similar:

$$\omega_1(s) = -[1/(k/s + b)]T_2(s) + \omega_2(s);$$
$$T_1(s) = -T_2(s).$$

A generalization to a variety of engineering systems can based on the two port components that have associated a flow or through variable f and an across or effort variable e, giving the power as the product (flow)(effort) [77]. This is the power crossing the junction of two components connected through a particular port.

Example: Two-Tank System, Effort–Flow Model

In the case of the two interacting tanks system (Fig. 5.12), the two tanks could still be identified in the block diagram, but a feedback interaction was added (Fig. 5.13). The output Q_1 of the first tank depends, however, not only on the state H_1 of the first tank, but also on the state H_2 of the second system, and the modularity is not any more obvious.

An OO model can, however, be derived for the interacting system of tanks. The approach is to define the connection by cuts, which, given the input, have local variables that completely determine the local states and the output.

For example, the pressures at the bottom of the tanks p_1 and p_2, which are determined by the corresponding heads H_1 and H_2, can be defined using the mass density ρ of the liquid. We have

$$p_1 = (\text{weight of liquid in tank 1})/A_1 = \rho H_1 A_1/A_1 = \rho H_1,$$
$$H_1 = p_1/\rho,$$

and

$$H_2 = p_2/\rho,$$

so that the flow rate is

$$Q_1(t) = [H_1(t) - H_2(t)]/R_1 = [p_1(t) - p_2(t)]/(\rho R_1).$$

The four equations of the model of interacting tanks become

$$(d/dt)\, p_1(t) = (\rho/A_1)[Q(t) - Q_1(t)] = (\rho/A_1)Q(t) - (\rho/A_1)Q_1(t),$$
$$(d/dt)\, p_2(t) = (\rho/A_2)[Q_1(t) - Q_2(t)] = (\rho/A_2)Q_1(t) - (\rho/A_2)Q_2(t),$$
$$Q_1(t) = [p_1(t) - p_2(t)]/(\rho R_1),$$

and

$$Q_2(t) = p_2(t)/(\rho R_2).$$

Two new classes of objects can be defined:

tank with valve, with

inputs

inputs Q_i the first terminal flow rate,
p_i, the first terminal pressure

outputs

Q_o, the second terminal flow rate,
p_o, the second terminal pressure

state p, the pressure at the bottom of the tank (variable attribute)
methods

$$(d/dt)\, p_o(t) = (1/A)Q_i(t) - (1/A)Q_o(t)$$
$$Q_o(t) = [p_i(t) - p_o(t)]/(R).$$

Pipe with
inputs

Q_i, the first terminal flow rate,
p_i, the first terminal pressure

outputs

Q_o, the second terminal flow rate
p_o, the second terminal pressure,

state none
methods

$$p_o = p_i$$
$$Q_o = Q_i.$$

In order to obtain an object-oriented model for the two interacting tanks system, effort (pressure) variable and a flow (flow rate) variable at each terminal are first defined. This solution can be also applied for other lumped parameter systems. In the case of distributed parameter systems, the interactions might become too complex to be reducible to an equivalent noninteractive multi-input multi-output system, as in the previous case of two interactive tanks. When the number of interactions between the subsystems of a system increases, the block diagram and OO models become more difficult to formulate and the direct solution to the set of differential or difference equations algebraic equations might become more attractive. In such cases, a non-object-oriented approach might be justified. Examples of such distributed parameter systems are the acoustic field transmitting audible vibrations to flexible structures, heat transmitted by heat radiation and convection to the components located in the vicinity, electromagnetic field influence, pressure wave during storms, human interactions between parallel servers, etc. In these cases, any object under the field influence has to be modeled not only with the effort–flow transmitted at the connected terminals, but also the interactions through the field, difficult to model with object-oriented approach.

Paynter called the effort–flow modeling a noncausal description because the direction of the power flow in the junction is bilateral. This differs from the block diagram description in which signals have assigned unidirectional flow. Paynter developed bond graph modeling based on two port descriptions. The same description, using effort–flow two–pole ports, is used in object-oriented modeling of mixed engineering systems. The theoretical background of this description can be found in Hamiltonian dynamics for obtaining power transfer equations [80].

5.3.5 Modeling Mechatronic Systems Using Object-Oriented Representation

In the previous section, a generic object-oriented approach is presented. For mechatronic systems, the object-oriented approach uses mechatronic objects, which model mechatronic components. Typical components for lumped parameter models are resistance, inductance, capacitance, and transformers for electrical systems and mass, spring damper, gears for mechanical systems as well as motors and generators for electomechanical components. Each mechatronic object incorporates the function or method (differential or algebraic equations) and communicates (can be connected) with other objects using only message passing (input and output variables).

Object-oriented modeling of mechatronic systems has specific features:

the mechatronic objects are connectable to each other at ports with effort–flow variables,

mechatronic system model is built from objects by imposing constraints on message passing such that power balance is respected at the connections of objects.

In bond graph models, power dissipation is modeled at the point where it occurs, while in objects only the power transfer at ports is explicit.

Power balance equations take specific forms in particular connections. For electric components connected at the same node, given that all links to the that node have the same voltage, the power balance equation is reduced to a current balance equation (i.e., Kirchhoff's first law). For mechanical components, in a free-body diagram, at a cut all links have the same velocity (and position) and the force balance equation replaces the power balance equation [78].

Mechatronic objects contain algebraic or differential equations as methods for linking the port variables.

Example: A Mixed System Effort Flow Model: Permanent Magnet DC Motor-Generator

Figure 5.26 reproduces here, for convenience, the PM–DC machine presented in Section 3.1, where T is the load torque, ω is the angular velocity, U is the DC supply voltage, i is the current, J is the rotor moment of inertia, and b the viscous friction coefficient.

The electric model of the rotor is given by the lumped parameters R and L, the resistance and inductance of the rotor winding circuit.

Power Conversion Electromechanical power conversion, assumed lossless, results in mechanical power equal to electrical power. (See Section 3.1.) That is,

$$T_r\omega = U_r i,$$

where T_r denotes the torque applied on the rotor and U_r is the back electromagnetic force.

For a DC motor,

(a) $T < T_r$, where T represents the load torque and

(b) $U > U_r$, where U represents the supply voltage. The current flows from the voltage supply source toward the rotor as shown in Fig. 5.26.

For a DC generator,

(a) $T > T_r$, where T represents the driving torque and

(b) $U < U_r$, where U represents the voltage supplied to the electric load. The current flows from the rotor toward the electric load cut (U, i), opposite to the current shown in Fig. 5.26.

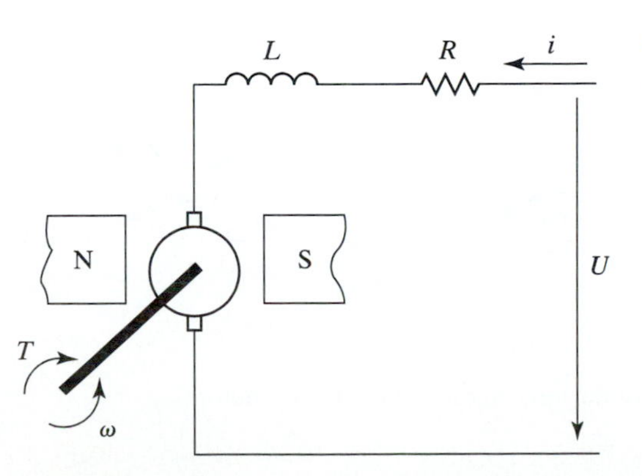

Figure 5.26

A permanent magnet DC machine diagram

For a motor, the power will flow from the electrical cut (U, i) towards the mechanical cut (T, ω). For a generator, the power will flow from the mechanical cut (T, ω) toward the electrical cut (U, i).

Assignment of the input and the output to correspond to the power transfer from the electrical to the mechanical side for a motor, or from the mechanical to the electrical side for a generator, is called direction of causality [77]. The assignment of power flow

$$(Ui) \rightarrow (T\omega)$$

corresponds to a unilateral signal flow for a DC motor, while the assignment of power flow

$$(T\omega) \rightarrow (Ui)$$

corresponds to a unilateral signal flow for a DC generator. The noncausal model (where no direction of causality was assigned), described by the last two equations, is equivalent to a bilateral signal flow [77]

$$(Ui) \leftrightarrow (T\omega).$$

Noncausal models can be formulated for a variety of other systems where bilateral power flow is possible, for example, systems based on thermoelectric or piezoelectric phenomena. These systems can also operate as motors (output cut temperature–heat flow or strain–stress output) or DC generators (voltage–current cut).

DC Motor Model The DC motor model, derived in Section 3.1 for the electrical and mechanical parts of the motor, is reproduced again for convenience in Figs. 5.27 and 5.28

The PM–DC motor model is given by

$$\omega(s) = -((Ls + R)/K)i(s) + (1/K)U(s)$$

and

$$T(s) = (K + (Js + b)(Ls + R)/K)i(s) - ((Js + b)/K)U(s).$$

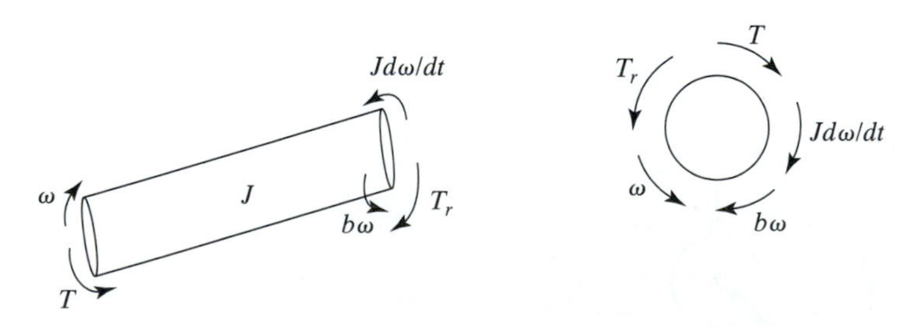

Figure 5.27

Free-body diagram for the mechanical part of the DC motor

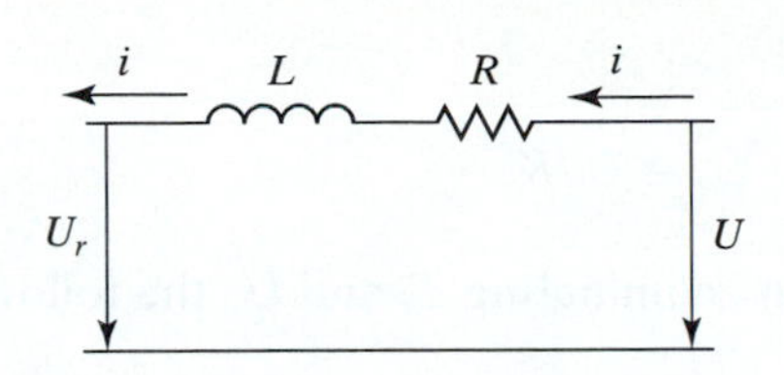

Figure 5.28

Two-port—(U_r,I) and (U,i)—circuit of the electrical part of the DC motor

DC Generator Model A DC generator model is presented as a basis for deriving later in this chapter the tachometer model. The mechanical and electrical subsystems are shown in Figs. 5.29 and 5.30.

Newton's second law for the free-body diagram of Fig. 5.29 gives

$$J d\omega_g/dt = T - b\omega_g - T_r.$$

The voltage drop equation for Fig. 5.30 yields

$$U_r = L di_g/dt + R i_g + U.$$

The model is completed by the two algebraic equations resulting from the lossless power conversion equation

$$T_r \omega_g = U_r i_g,$$

written as

$$T_r/i_g = U_r/\omega_g = K$$

or

$$T_r = K i_g$$

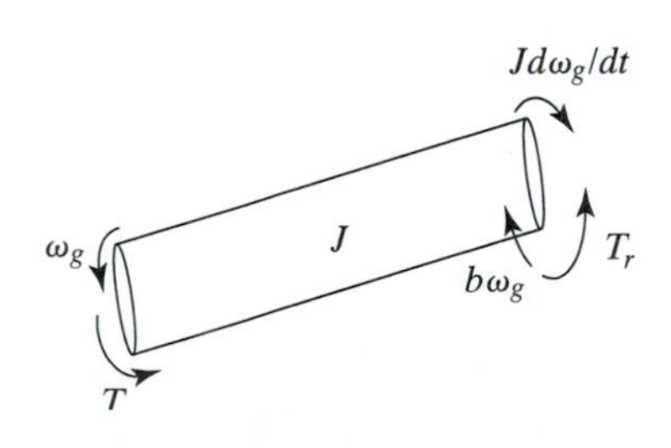

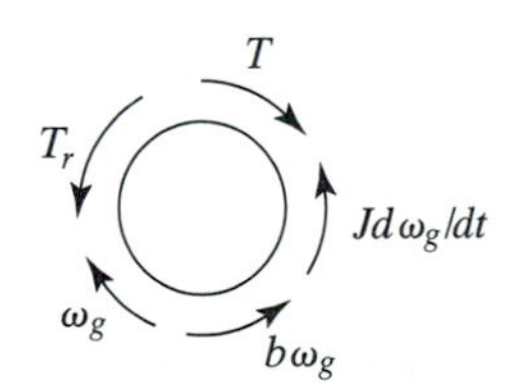

Figure 5.29

Free-body diagram for the mechanical part of the DC generator

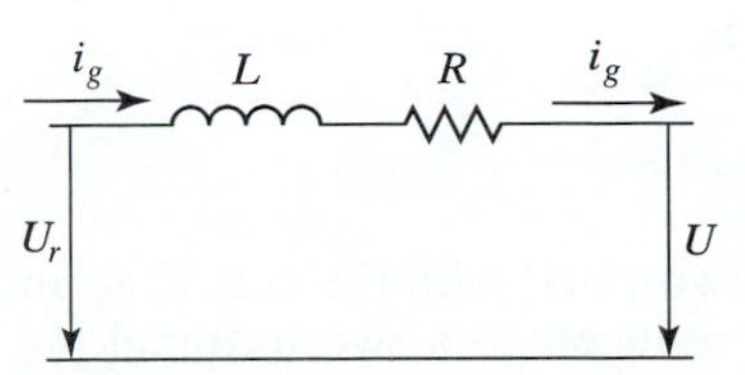

Figure 5.30

The circuit with two ports—(U_r,i_g) and (U,i_g)—of the electrical part of a DC generator

and

$$\omega_g = U_r/K.$$

After applying Laplace transform and eliminating T_r and U_r, the following model of a DC generator is obtained:

$$U(s) = -((Ls + R)/K)T(s) + (K + (Js + b)(Ls + R)/K)\omega_g(s);$$
$$i_g(s) = (1/K)T(s) - ((Js + b)/K)\omega_g(s).$$

Example: Reduction of Power to Signal Transmission: Tachometer Model

The tachometer is a transducer that converts angular velocity input ω into a voltage output U. It is basically a DC generator, with a very low mechanical power consumption, installed on the output shaft of a DC motor. Ideally, the tachometer functions in accordance with a linear algebraic model

$$U = k\omega_g$$

[i.e., with no nonlinearity and no dynamics (no differential term)]. Being a generator, the tachometer specificity can be outlined by deriving ideal tachometer model from the DC generator model. This requires a design with a relatively very low L (which is the case of most generators), low R, and large K; these conditions reduce the first equation of the DC generator to

$$U(s) \cong K\omega_g(s)$$

(i.e., exactly the same as the linear algebraic equation of the ideal tachometer) [1].

Another assumption could be that the output port (U,i) of the generator is connected to a voltmeter with a very high input resistance

$$R_M \cong \infty$$

such that

$$i_g = U/R_M \cong 0.$$

The second equation of the DC generator model, for $i_g = 0$, gives

$$T(s) = (Js + b)\omega_g(s).$$

In this case, the first equation is reduced again to

$$U = k\omega_g.$$

The DC generator model with four terminals (two-port) with the cuts T, ω_g and U, i_g is a power transmission model. This model was reduced to a two-terminal (ω_g and U)

model of a velocity transducer (tachometer). The two-terminal model is a signal transmission model, specific to transducers. In fact, for a tachometer only the signal transmission is important and power consumption is supposed to be negligible. The resulting $i_g \cong 0$ gives actually zero power consumption ($Ui = T\omega_g \cong 0$).

5.3.6 Simulation Software

Object-oriented models can be programmed as computer codes for execution using object-oriented programming languages (for example, C++). While the rigor of object-oriented programming languages is a plus, often the convenience of an already known non-object-oriented language, like C or Fortran, justifies their use even if they cannot reproduce all the rigor of object-oriented programming languages [38].

MATLAB, after version 4, and Simulink were developed using object-oriented programming languages. Simulink permits graphical programming for simulating systems based on signal transmission, and it uses block diagram modeling. Signal flow and power transmission modeling of mechatronic systems can be illustrated using Simulink, even if this cannot provide the rigor of a simulator using directly object-oriented programming for mixed engineering systems like DYMOLA or OMOLA [42, 76].

5.4 VIRTUAL PROTOTYPING AND HARDWARE-IN-THE-LOOP EXPERIMENTATION

5.4.1 Signal to Power Conversion in Mechatronic Systems

Virtual prototyping uses detailed models, simulation, and animation for arriving at an efficient design. This requires models, which not only can reproduce the dynamics of the system, but also can help test various design alternatives. Replacing components of the system model can be achieved by a modular modeling of signal and power transmission components. As long as input and output of signal transmission components are identified in the model, these components can be easily replaced by other components with the same input and output. Power transmission components are more difficult to replace due to the effort–flow cuts. The interface between a signal transmission component and a power transmission component is another fundamental characteristic of mechatronics systems.

The execution of the commands given to actuators represents a conversion of a signal, containing a power value command, into a power output of the actuator. For the case of electric motors this is achieved by motor drivers, which modulate available electric power from a source in accordance with the power value of a command signal. A solution, useful for precise DC servomotors, is to use power operational amplifiers as a controllable voltage source (for motion control) or as a controllable current source (for torque control) [1, 23–25].

Example: Controllable Voltage Source for a DC Motor Driving a Fan

A fan with moment of inertia J_f and an air drag coefficient B_f is assumed to be linked by a flexible shaft with stiffness coefficient k_f to a DC motor (Fig. 5.31). Determine the relationship between the output angular velocity of the fan ω_f and the voltage U of a controllable voltage source.

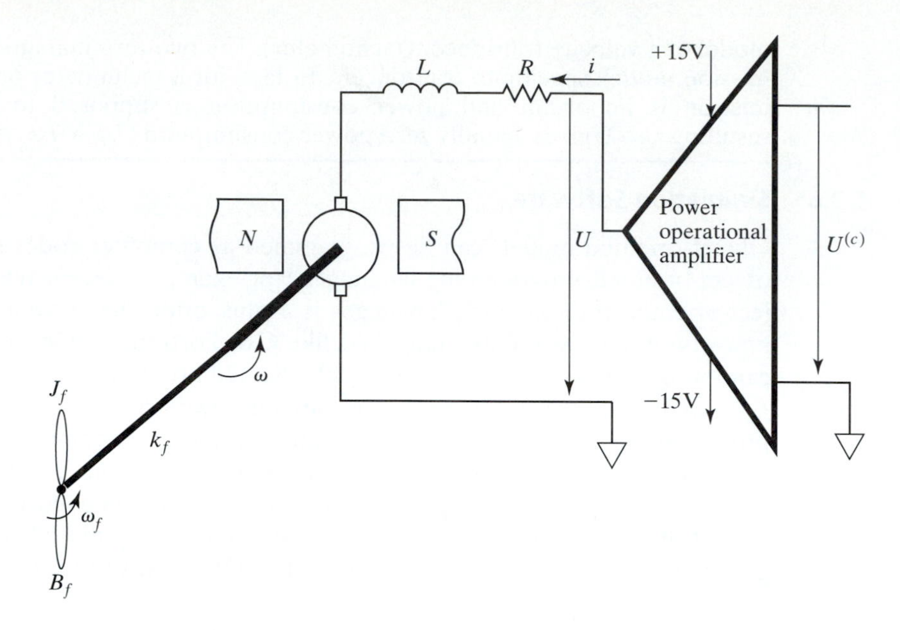

Figure 5.31
A DC motor for a fan

***Solution*:** The DC motor model, derived in Section 3.1, for the cut variables $T(s)$ and $\omega(s)$, is

$$T(s) = -((Js + b)/K)U(s) + (K + (Js + b)(Ls + R)/K)i(s)$$

and

$$\omega(s) = (1/K)U(s) - ((Ls + R)/K)i(s),$$

or, in matrix form,

$$\begin{bmatrix} T(s) \\ \omega(s) \end{bmatrix} = \begin{bmatrix} -(Js + b)/K & K + (Js + b)(Ls + s)/K \\ 1/K & -(Ls + R)/K \end{bmatrix} \begin{bmatrix} U(s) \\ i(s) \end{bmatrix}.$$

The free-body diagrams for the motor load is shown in Fig. 5.32.

Modular modeling of this system requires the derivation of motor load equations for the appropriate cut variables. For the flexible shaft, the cut variables are $T(s)$ and $\omega(s)$ toward the motor rotor and $T_f(s)$ and $\omega_f(s)$ toward the fan.

From the free-body diagrams of the two ends of the flexible shaft, shown in Fig. 5.32, we can write the torque balance equations

$$T = k_f(\theta - \theta_f)$$

and

$$T_f = k_f(\theta_f - \theta)$$

such that

$$T_f = -T.$$

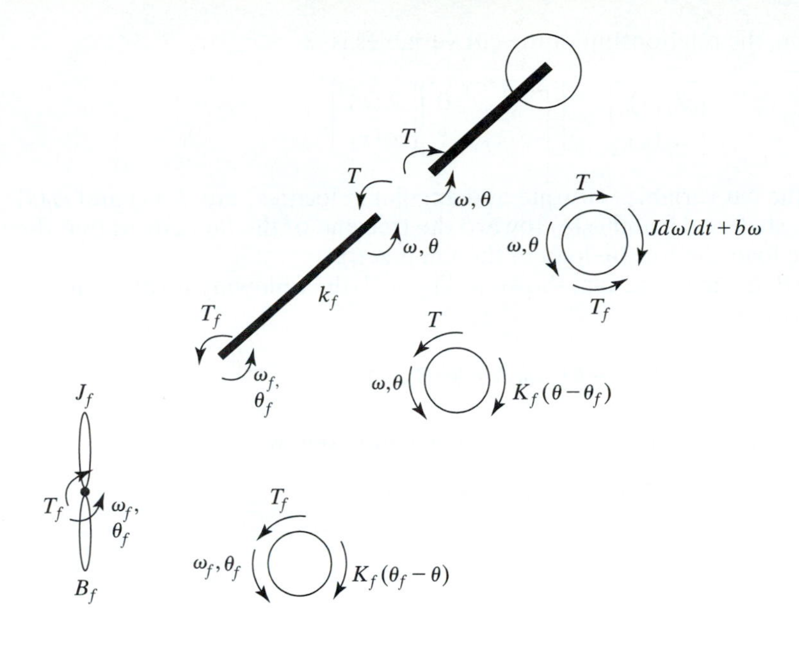

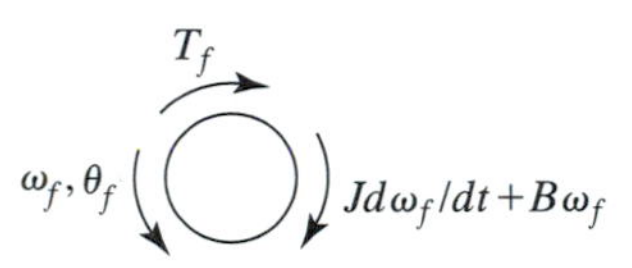

Figure 5.32

Free-body diagrams for the motor load

In order to use only angular velocities, variables θ and θ_f are replaced by ω and ω_f, taking into account the relationships

$$\omega(t) = d\theta(t)/dt$$

and

$$\omega_f(t) = d\theta_f(t)/dt.$$

After taking the Laplace transform for zero initial conditions, we obtain

$$\theta(s) = \omega(s)/s$$

and

$$\theta_f(s) = \omega_f(s)/s.$$

In this case, the equation for the torque becomes

$$T(s) = k_f(\omega(s) - \omega_f(s))/s,$$

or

$$\omega_f(s) = -(s/k_f)T(s) + \omega(s).$$

In matrix form, the relationship of the cut variables is

$$\begin{bmatrix} T_f(s) \\ \omega_f(s) \end{bmatrix} = \begin{bmatrix} -1 & 0 \\ -s/k_f & 1 \end{bmatrix} \begin{bmatrix} T(s) \\ \omega(s) \end{bmatrix}.$$

For the fan, the cut variables (torque and angular velocities) are $T_f(s)$ and $\omega_f(s)$ toward the flexible shaft and 0 and $\omega_f(s)$ toward the free end of the fan. Given that the fan is a passive free load, the torque load of the fan is zero.

For the body diagram of the fan, shown in Fig. 5.32, the following torque balance equation can be written:

$$T_f = -(J_f d\omega_f/dt + B_f\omega_f).$$

Or, in Laplace transform form, for zero initial conditions, we have

$$T_f(s) = -(J_f s + B_f)\omega_f(s),$$

or

$$T_f(s)/(J_f s + B_f) + \omega_f(s) = 0.$$

In matrix form, the relationship of these cut variables is

$$\begin{bmatrix} 0 \\ \omega_f(s) \end{bmatrix} = \begin{bmatrix} 1/(J_f s + B_f) & 1 \\ 0 & 1 \end{bmatrix} \begin{bmatrix} T_f(s) \\ \omega_f(s) \end{bmatrix}.$$

In order to obtain a relationship between $\omega_f(s)$, $U(s)$, and $i(s)$, the cut variables $T(s)$ and $\omega(s)$ have to be eliminated by using the matrix equation of the DC motor:

$$\begin{bmatrix} 0 \\ \omega_f(s) \end{bmatrix} = \begin{bmatrix} 1/(J_f s + B_f) & 1 \\ 0 & 1 \end{bmatrix} \begin{bmatrix} -1 & 0 \\ -s/k_f & 1 \end{bmatrix}$$
$$\begin{bmatrix} -(Js + b)/K & K + (Js + b)(Ls + R)/K \\ 1/K & -(Ls + R)/K \end{bmatrix} \begin{bmatrix} U(s) \\ i(s) \end{bmatrix}.$$

This matrix equation maintains the mapping between the model components and the physical system components; that is,

$$[0\ \omega_f(s)]_t = [\text{fan model}][\text{flexible shaft model}][\text{PM–DC motor model}][U(s)\quad i(s)]_t,$$

where the subscript t denotes vector transpose.

For the output of the system (the angular velocity ω_f of the fan), the right-hand side of the equation contains the 2 by 2 matrices corresponding to

the fan model (with parameters J_f and B_f),
the flexible shaft model (with parameter k_f),
the DC motor model (with parameters J, b, L, R, and K),

and the inputs vector [U(s) i(s)]t to the DC motor.

The conservation of the mapping represents a modularity feature of the model and can be used for changing various components of the model while conserving the same cut variables. This will be illustrated in the example of virtual prototyping that follows.

The reduced model is obtained by matrix multiplication. The multiplication of the first two matrices on the right-hand side gives

$$\begin{bmatrix} 0 \\ \omega_f(s) \end{bmatrix} = \begin{bmatrix} -1/(J_f s + B_f) - s/k_f & 1 \\ -s/k_f & 1 \end{bmatrix} \begin{bmatrix} -(Js + b)/K & K + (Js + b)(Ls + R)/K \\ 1/K & -(Ls + R)/K \end{bmatrix} \begin{bmatrix} U(s) \\ i(s) \end{bmatrix}.$$

Finally, multiplying the remaining two matrices on the right-hand side produces

$$\begin{bmatrix} 0 \\ \omega_f(s) \end{bmatrix} = \begin{bmatrix} (-1/(J_f s + B_f) - s/k_f)(-(Js + b)/K) + 1/K \\ (-s/k_f)(-(Js + b)/K) + 1/K \end{bmatrix.$$

$$\begin{matrix} (-1/(J_f s + B_f) - s/k_f)(K + (Js + b)(Ls + R)/K) - (Ls + R)/K \\ (-s/k_f)(K + (Js + b)(Ls + s)/K) - (Ls + R)/K \end{matrix}\Bigg] \begin{bmatrix} U(s) \\ i(s) \end{bmatrix}.$$

The two scalar equations are

$$((-1/(J_f s + B_f) - s/k_f)(-(Js + b)/K) + (1/K))U(s) +$$
$$((-1/(J_f s + B_f) - s/k_f)(K + (Js + b)(Ls + R)/K) - (Ls + R)/K)i(s) = 0$$

and

$$((-s/k_f)(-(Js + b)/K) + (1/K))U(s) +$$
$$((-s/k_f)(K + (Js + b)(Ls + R)/K) - (Ls + R)/K)i(s) = \omega_f(s).$$

In order to obtain a relationship between $\omega_f(s)$ and $U(s)$, the variable i (s) has to be eliminated. The first scalar equation gives

$$i(s) = -U(s)((-1)/J_f s + B_f) - s/k_f)(-(Js + b)/K) + (1/K))$$
$$/((-1/(J_f s + B_f) - s/k_f)(K + (Js + b)(Ls + R)/K) - (Ls + R)/K),$$

while the second equation, after eliminating $i(s)$, yields

$$\omega_f(s) = ((-s/k_f)(-(Js + b)/K) + (1/K)) + ((-s/k_f)(K + (Js + b)(Ls + R)/K)$$
$$- (Ls + R)/K)(-((-1/(J_f s + B_f) - s/k_f)(-(Js + b)K) + (1/K))$$
$$/((-1/(J_f s + B_f) - s/k_f)(K + (Js + b)(Ls + R)/K) - (Ls + R)/K))U(s).$$

This shows that, for known passive load J_f, B_f, and k_f, in case of a controllable U from a voltage source, the power transmission model gives a direct relationship of the output angular velocity $\omega_f(s)$ and the voltage U, decoupled from the current i:

$$\omega_f(s) = F_f(s)\,U(s).$$

Here, the complex function

$$F_f(s) = ((-s/k_f)(-(Js + b)/K) + (1/K)) + ((-s/k_f)(K + (Js + b)(Ls + R)/K) - (Ls + R)/K) (-((-1/(J_f s + B_f) - s/k_f)(-(Js + b)/K) + (1/K)) /((-1/(J_f s + B_f) - s/k_f)(K + (Js + b)(Ls + R)/K) - (Ls + R)/K))$$

is dependent only on the parameters of the system. This function $F_f(s)$ is the transfer function from the input $U(s)$ to the output $\omega_f(s)$.

This shows that the angular velocity ω_f can be open-loop controlled by U. The relationship resembles the signal transmission relationships, but has an associated current equation. The current i can be computed for the known U by means of the relationship

$$i(s) = F_i(s)\,U(s),$$

where

$$F_i(s) = -((-1/(J_f s + B_f) - s/k_f)(-(Js + b)/K) + (1/K)) /((-1/(J_f s + B_f) - s/k_f)(K + (Js + b)(Ls + R)/K) - (Ls + R)/K).$$

This decoupled result can be implemented by the modulation of electric power input with a power operational amplifier having practically zero internal impedance (ideal Thevenin equivalent of a linear circuit of the DC supply). In the example shown in Fig. 5.31, the signal $U^{(c)}$ is a voltage containing the command for the angular velocity ω_f. The power operational amplifier also provides the required current i that is dependent on the ω_f. In fact, the power operational amplifier achieves the command $U^{(c)}$ and provides at the same time the power Ui. This distinguishes this power transmission case from a simple signal transmission case.

The decoupling of ω_f from i justifies in such cases the use of signal transmission models and the corresponding block diagrams, rather than the complete effort–flow and power transmission models. Such cases occur for all servomotors and explain why they are often modeled in control engineering using block diagrams. Modeling and simulation of actuators, particularly when modularity is desired, requires, complete effort–flow and power transmission models. Virtual prototyping implies testing design alternatives on system models. A modular model permits the replacement of components having the same inputs and outputs, for signal transmission components, and the same effort–flow cuts, for the power transmission components. Only series systems, (e.g., motor drives) are investigated here. Electric networks and multibody dynamic systems often result in complex models described by differential algebraic equations. As mentioned in Section 5.1, these complex systems can be modeled using advanced simulation packages and are subject to specialized literature [42, 70, 74].

To illustrate the modular modeling of power transmission components, a previous model of the DC motor driving a fan is considered again for showing how the flexible shaft, modeled by a spring coefficient k_f, can be replaced by another flexible shaft, that has also significant viscous friction modeled by a coefficient b_f.

Example: Virtual Prototyping of the DC Motor Driving a Fan

The DC motor driving a fan, shown in Fig. 5.31, is reviewed and a new design is proposed. A new model is required for the case in which original flexible shaft if replaced, for a cheaper design, by a shaft with significant viscous friction in the bearings modeled by b_f. The new design is shown in Fig. 5.33.

In a modular model, this design change can be easily incorporated as long as the new component has the same cuts as the old one. Figure 5.34 shows the free-body diagrams for the new design.

In Fig. 5.34, the viscous friction coefficient is assumed to be affecting the fan end of the flexible shaft; only the free-body diagram at that end of the flexible shaft is modified by the inclusion of an extra viscous friction reaction torque $b_f\omega_f$ in the torque balance equation. We have

$$T_f(s) = k_f(\omega_f(s) - \omega(s))/s + b_f\omega_f,$$

or

$$\omega_f(s) = (s/(k_f + sb_f)T_f(s) + (k_f/(k_f + sb_f))\omega(s).$$

The motor end of the shaft maintains the previous torque balance equation:

$$T(s) = k_f(\omega(s) - \omega_f(s))/s.$$

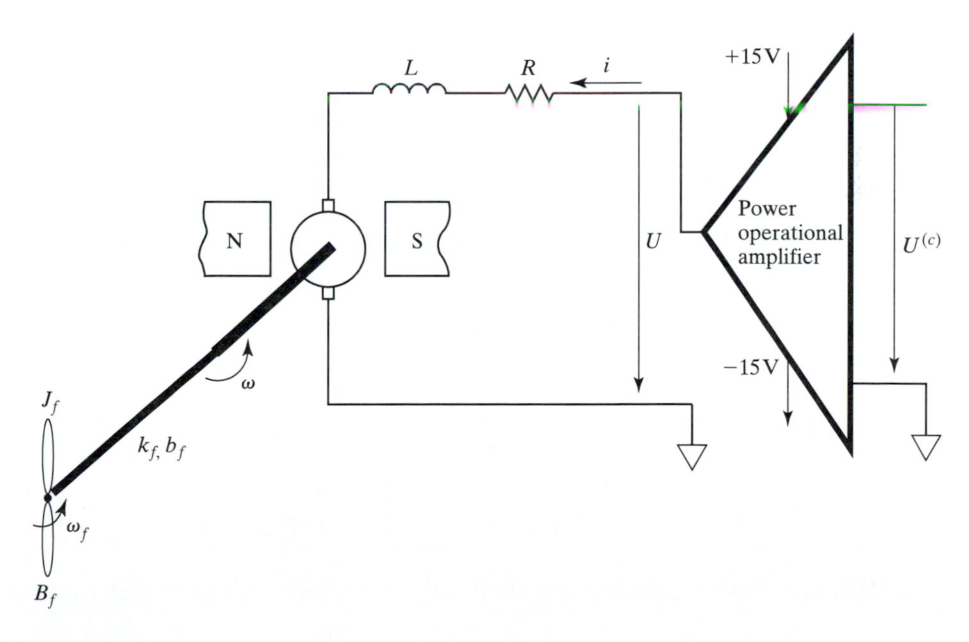

Figure 5.33

DC motor driving a fan with a flexible shaft with significant viscous friction

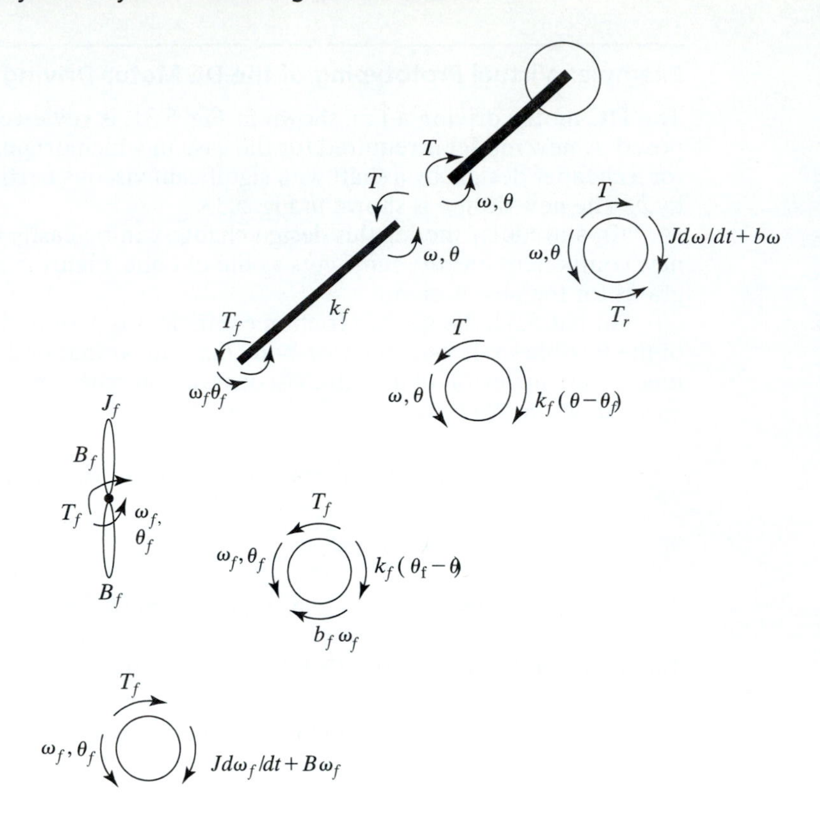

Figure 5.34

Free-body diagrams for the system shown in Fig. 5.33

In this case,

$$T_f(s) = -T(s) + b_f\omega_f$$

and

$$\omega_f(s) = (s/(k_f + sb_f)(-T(s) + b_f\omega_f) + (k_f/(k_f + sb_f))\omega(s),$$

or

$$\omega_f(s) = -T(s)((k_f + sb_f)/k_f)((s/(k_f + sb_f)) + \omega(s).$$

In matrix form,

$$\begin{bmatrix} T_f(s) \\ \omega_f(s) \end{bmatrix} = \begin{bmatrix} -1 & b_f \\ -s/k_f & 1 \end{bmatrix} \begin{bmatrix} T(s) \\ \omega(s) \end{bmatrix}.$$

This new model of the flexible shaft can now replace the old model; we have

$$\begin{bmatrix} 0 \\ \omega_f(s) \end{bmatrix} = \begin{bmatrix} 1/(J_f s + B_f) & 1 \\ 0 & 1 \end{bmatrix} \begin{bmatrix} -1 & b_f \\ -s/k_f & 1 \end{bmatrix} \begin{bmatrix} T(s) \\ \omega(s) \end{bmatrix},$$

or

$$\begin{bmatrix} 0 \\ \omega_f(s) \end{bmatrix} = \begin{bmatrix} -1/(J_f s + B_f) - s/k_f & b_f/(J_f s + B_f) + 1 \\ -s/k_f & 1 \end{bmatrix} \begin{bmatrix} T(s) \\ \omega(s) \end{bmatrix}.$$

In order to obtain a new relationship among $\omega_f(s)$, $U(s)$, and $i(s)$, the cut variables $T(s)$ and $\omega(s)$ have to be eliminated using the matrix equation of the DC motor:

$$\begin{bmatrix} 0 \\ \omega_f(s) \end{bmatrix} = \begin{bmatrix} -1/(J_f s + B_f) - s/k_f & b_f/(J_f s + B_f) + 1 \\ -s/k_f & 1 \end{bmatrix}$$

$$\begin{bmatrix} -(Js + b)/K & K + (Js + b)(Ls + s)/K \\ 1/K & -(Ls + R)/K \end{bmatrix} \begin{bmatrix} U(s) \\ i(s) \end{bmatrix}.$$

This example illustrates that a modular model, using effort–flow cuts for the power transmission components mimics the replacement of a component by another. This is achieved by the replacement of the model modules of the components, in a matrix multiplication, without affecting the models of the other components.

5.4.2 Transition from Virtual Prototype to Physical Prototype

Virtual prototyping uses detailed simulation of all components. In such a case, signal transmission is represented by the signal flow among components. Block diagram modeling is based on the representation of data flow from block inputs, modified by block diagram functions to produce block outputs. The effort–flow approach introduces modularity and permits object-oriented modeling. Power transmission is represented by the coupled equations of effort and flow signals. The interface between adjacent power transmission elements uses the same effort–flow cuts.

The transition from virtual to physical prototyping can take the intermediate step of hardware in the loop simulation. In this case, physical components, inserted in the simulator, are interfaced by signal to power transmission components using various power amplifying devices, based on either controllable effort or controllable flow. Power operational amplifiers can serve as controllable voltage (or, in other cases, controllable current) amplifiers with negligible output impedance and available-on-demand current (or voltage). Servovalves often serve to control fluid flow rate. In case of a large upstream fluid reservoir, a servovalve can achieve the required flow rate independent of the load. Similarly, carburetors modulate the flow rate of gasoline in accordance to the position of the acceleration pedal; obviously, the specific energy content of the gasoline remains constant, independent of the motor load.

5.4.3 Hardware-in-the-Loop Experimentation

Hardware-in-the-loop simulation (or experimentation) uses a combination of components, some consisting of simulation models and others consisting of actual hardware. This kind of simulation requires proper interfacing of the simulated and physical components. In the case of signal transmission components, interfacing might require signal conditioning, which in general, can be easily achieved.

Example: Hardware-in-the-Loop Simulation of a DC Motor Angular Velocity Measurement

For a DC motor fitted with an angular position sensor, the velocity $\omega(t)$ can be obtained as the derivative of the position $\theta(t)$:

$$\omega(t) = d\theta(t)/dt.$$

This numerically computed signal $\omega(t)$ gives the simulated output of an angular velocity sensor, for example, of a tachometer. In Fig. 5.35, the output ω of a velocity sensor is simulated by the computed derivative $\omega = d\theta/dt$.

Hardware-in-the-loop simulation can achieve the simulation of a new type of component or subsystem by model-based control of physical actuators.

There is significant resemblance between interfacing issues in mechatronic systems and hardware-in-the-loop simulation, due to the similarity of situations of conversion of signals containing power commands into actual modulated power transfer. While power operational amplifiers as controllable voltage source (or, in other cases, controllable current source) can be used in some applications of hardware-in-the-loop simulation, this is not, however, always applicable. In the conversion of energy from an external source into a modulated power transfer in a mechatronic system, ideal converters can be used to represent a variety of situations:

- a DC power supply that provides a controllable voltage independent from the current demand in normal conditions (i.e., excluding short circuits);
- a gas tank and a servovalve that provides a controllable flow rate of gas for a gas engine at a constant specific energy content of the gas;
- a tank, storing a fluid under pressure, with a servovalve that provides a controllable flow rate at a constant pressure.

In hardware-in-the-loop simulations, this decoupling of flow and effort variables of the power source might not be realistic and actual dependence might have to be simulat-

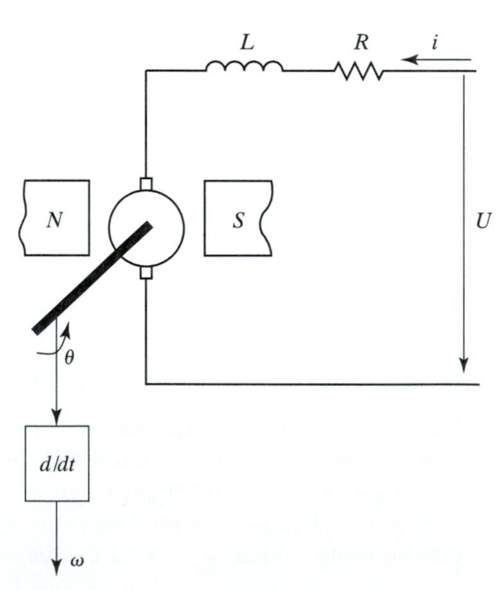

Figure 5.35
Numerical determination of angular velocity from position sensing

ed and converted in modulated power output. This example of a DC motor driving a fan can serve as an illustration of the effort–flow coupling of the DC power source.

5.4.4 Simulink Modular Modeling Using Encapsulation

Simulink procedure for "encapsulating a subsystem" facilitates modular modeling of mechatronic systems. The example that is presented for the encapsulation of the Simulink model is the PM–DC motor with DC supply and load represented in Fig. 3.5. The model in Fig. 3.5 contains the subsystem consisting in the model of the PM–DC motor shown in Fig. 3.4, plus the subsystem containing the DC supply and the subsystem containing the load added in Fig. 3.5. By encapsulating these three subsystems, the Simulink model is reduced to the three blocks shown in Fig. 5.36.

The encapsulation, as illustrated in Fig. 5.36, permits modular exchange of blocks as long as the designations of inputs and outputs are respected. Simulink uses a signal transmission approach in a block diagram representation and, as such, does not permit object-oriented modeling of power transmission due to causal assignment of the input

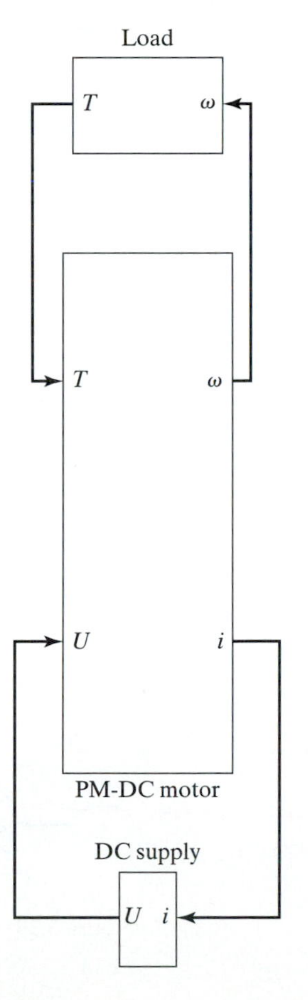

Figure 5.36

Encapsulated Simulink model of the PM–DC motor with DC supply and load

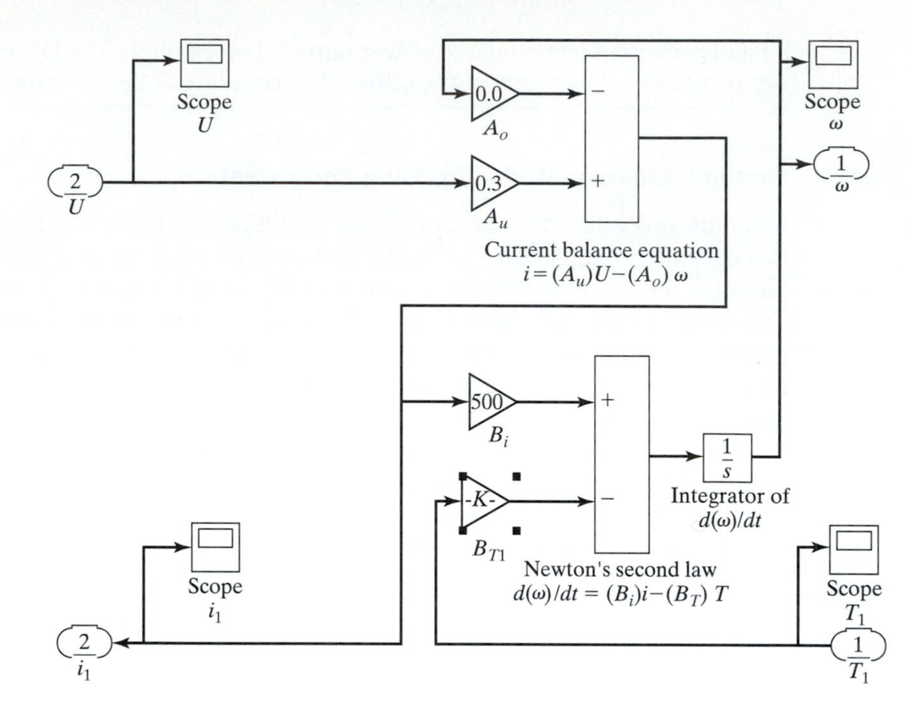

Figure 5.37

PM–DC motor block

and output variables. Simulink encapsulation is, however, useful for modular modeling. The encapsulation in Simulink model has the advantage of retaining the identity of the subsystem blocks not only during graphical programming of the model, but also during the execution of the simulation. During the execution, data flow approach leads to the solution of each block as soon as all its inputs are available. While this data flow approach slows down the simulation, it has advantages during program debugging. In order to speed up the execution, the Simulink model has to be compiled in a high-level language program that can be executed quickly.

The blocks from Fig. 5.35, for the PM–DC motor, DC supply, and load, are shown in Figs. 5.37, 5.38, and 5.39, respectively.

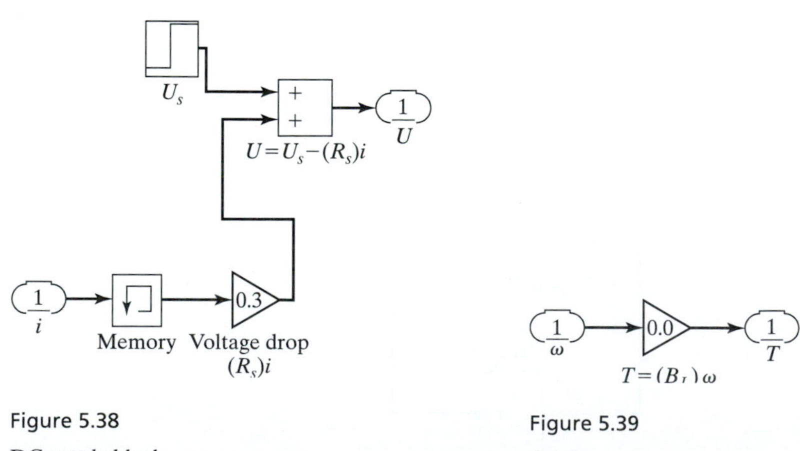

Figure 5.38

DC supply block

Figure 5.39

Load

5.5 NEURAL NETWORK MODELS

5.5.1 Types of Neural Networks

Artificial neural networks (referred to as NNs) consist of layers of linear and nonlinear algebraic functions that transform a given input x_i (i = 1 to n) in an output. The parameters w (called weights) of the linear functions for jth neuron, $\Sigma w_{ij}x_i$ (j =1 to m) are determined in an NN learning stage using an approach that process sets of given input and output sets of values. The nonlinear functions $f()$ of the NN can contain shape parameters that are usually chosen off-line based on practical considerations.

There are two broad categories of NN [71]:

nonrecurrent NN, in which no feedback is present and the directed branches transfer data only in the feedforward direction, from input toward output;

recurrent, in which feedback link connections are also present in layers (for example, in Hopfield NN).

The frequently used feedforward NN will be presented here.

5.5.2 Layers and Weights, Inputs and Outputs, Threshold Function

A three-layer feedforward NN (n neurons input layer, m neurons hidden layer, and s neurons output layer) is shown in Fig. 5.40, where

x_i are the inputs (for i = 1 to n),

y_l are the outputs (for l = 1 to s),

w_{ij} are the weights between the input and the hidden layer for i = 1 to n and j = 1 to m, and

W_{jk} are the weights between the hidden and the output layer for j = 1 to m and k = 1 to s.

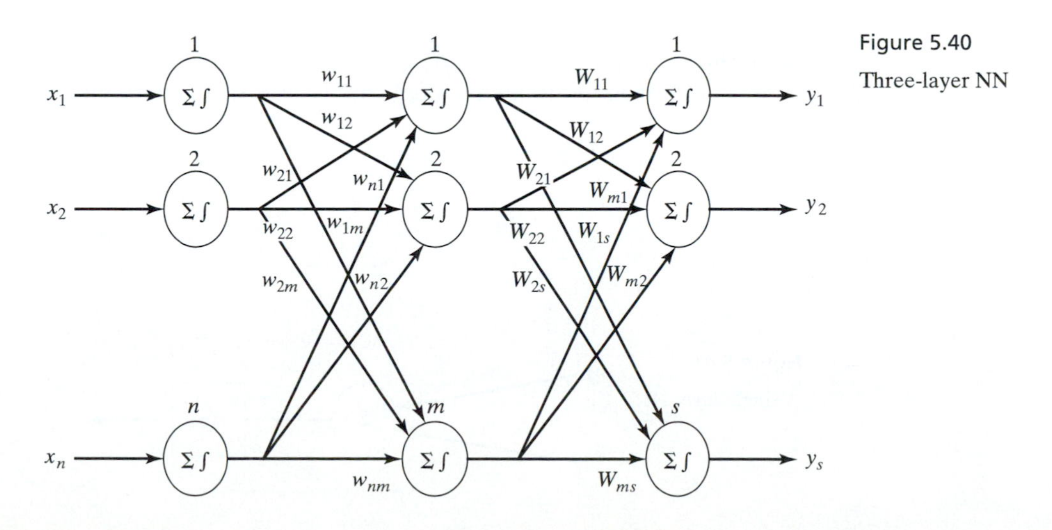

Figure 5.40
Three-layer NN

This is a block diagram with oriented branches (the input to the branch is multiplied by the branch weight to give the output of that branch) and blocks, in this case circles, representing nodes and containing summation of inputs followed by nonlinear rescaling (threshold function) of the result of summation.

For any neuron j of the three layers, the two symbols in the circles have the following meanings:

Σ—the weighted summation of inputs to neuron j, $\Sigma_{\text{for all } i}\, w_{ij}\, x_i = z_j$;

∫—the threshold function, a nonlinear nondecreasing function $f(z_j)$ for z_j varying between $-\infty$ and $+\infty$.

The function $f(z_j)$ takes values from 0 to 1 (or −1 to 1) and has the form of a saturation function for the summation z_j.

A simple threshold function is a unit step function

$$f(z) = 1 \quad \text{for } z \geq 0$$
$$= 0 \quad \text{elsewhere}$$

Often, as a threshold function, we use the sigmoid function

$$f(z) = 1/(1 + \exp\{-cz\}),$$

where c is a parameter that determines the shape of the saturation function.

Simple examples will illustrate the use of single-layer and multilayer NN. These examples are much simpler than real applications of NN, but were chosen for facilitating manual calculations.

Example: A Single-Layer NN Model of a Logical "or"

Consider, as an example, two binary inputs x_1 and x_2 and the binary output y shown in Fig. 5.41. The circle in Fig. 5.40 containing the symbols for the summation Σ as well as the threshold function, is separated in Fig. 5.41 into the summation block Σ, indicating that

$$z = w_{11}x_1 + w_{12}x_2,$$

and the unit step threshold function block $f(z)$.

This NN will be trained to model a logical "or" ($y = x_1$ or x_2) defined as follows:

x_1	x_2	y
0	0	0
0	1	1
1	0	1
1	1	1

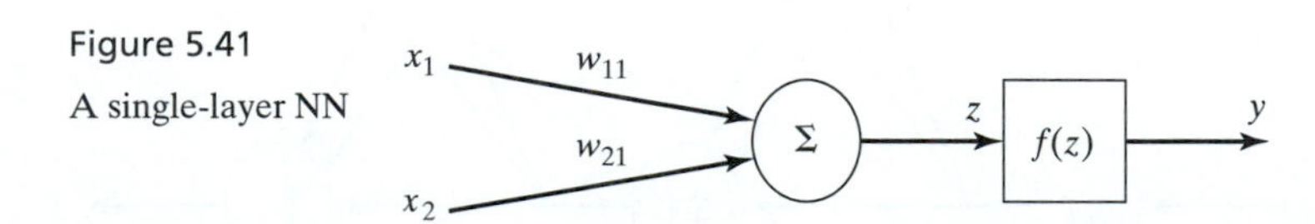

Figure 5.41
A single-layer NN

This logical "or" can be represented graphically using a linear function $x_2 = Ax_1 + B$, which, in an (X_1, X_2) plane, defines a straight line separating the point $y = 0$ from the points $y = 1$, as shown in Fig. 5.42.

The coordinates (X_1, X_2) of four points correspond to the following values of y as defined for logical "or":

$$(0, 0) \text{ for } y = 0;$$

$$(0, 1) \text{ for } y = 1;$$

$$(1, 0) \text{ for } y = 1;$$

$$(1, 1) \text{ for } y = 1.$$

The value $y = 0$ is represented as an empty circle and is shown in the origin. The values $y = 1$ are marked by x and are shown as three other points [42].

A straight line separates the points having $y = 0$, the origin, from the points having $y = 1$. Assuming that it intersects the axes at the points (0.5,0) and (0,0.5), the coefficients A and B can be obtained by making the equation $x_2 = Ax_1 + B$ verify these two points. We have

$$\text{for } (0.5, 0), \qquad 0 = A0.5 + B,$$

and

$$\text{for } (0, 0.5), \qquad 0.5 = A\,0 + B,$$

from which it follows that

$$B = 0.5$$

and

$$A = -1$$

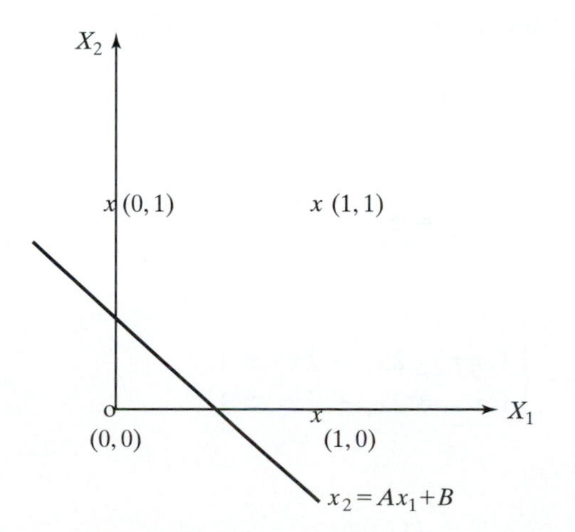

Figure 5.42

Geometric representation of the logical "or"

or

$$x_2 = -x_1 + 0.5$$

or

$$2x_1 + 2x_2 = 1.$$

This result permits us to interpret geometrically the single-layer NN shown on Fig. 5. 41.

The function

$$z = 2x_1 + 2x_2$$

corresponds to the summation block Σ; that is,

$$z = w_{11}x_1 + w_{12}x_2$$

with $w_{11} = 2$ and $w_{12} = 2$.

The straight line $z = 0$ separates the plane X_1,X_2 as follows:

$z > 1$—the plane above the line is defined.
$z < 1$—the plane below the line is defined.

Figure 5.42 shows only the origin (which has $z = 0$) below the line and all other three points (which have $z = 1$) above this line.

Based on the foregoing information, the threshold function $f(z)$, corresponding to a step function, can be defined as

$$y = 0 \text{ for } z \leq 1$$

and

$$y = 1 \text{ for } z > 1.$$

This equivalence permits us to define a single-layer NN with weights

$$w_1 = 2$$

and

$$w_2 = 2,$$

the step threshold function

$$f(z) = \begin{cases} \text{for } z = 2x_1 + 2x_2 \leq 1 \\ \text{for } z = 2x_1 + 2x_2 > 1\text{'} \end{cases}$$

and the output

$$y = f(z).$$

x_1 $w_{11} = 2$
x_2 $w_{21} = 2$
Σ z
$f(z) = 1$ for $z \geq 1$
$= 0$ for $z < 1$
$Y = f(z)$

Figure 5.43

Single-layer NN model for logical "or"

The resulting NN, shown in Fig. 5.43, can be verified for the four sets of values of the logical "or":

x_1	x_2	$z = 2x_1 + 2x_2$	$f(z) = y$
0	0	0	0
0	1	2	1
1	0	2	1
1	1	4	1

This NN has been obtained based on geometric considerations. In Section 5.5.3, a training algorithm will be used to obtain the weights w_1 and w_2.

A single-layer NN can serve as a model for systems with very reduced complexity. Minsky and Papert have demonstrated that the above single layer NN cannot model "exclusive or." A multilayer NN can, however, serve as a model in this case [71].

Example: A Multilayer NN Model of a Logical "exclusive or"

The "exclusive or" is defined as follows:

x_1	x_2	Y
0	0	0
0	1	1
1	0	1
1	1	0

The "exclusive or" differs from "or" only in the last row.

The corresponding (X_1, X_2) plane representation in Fig. 5.44 shows two empty circles for $y = 0$ and two x for $y = 1$. As can be seen, no single straight line can separate

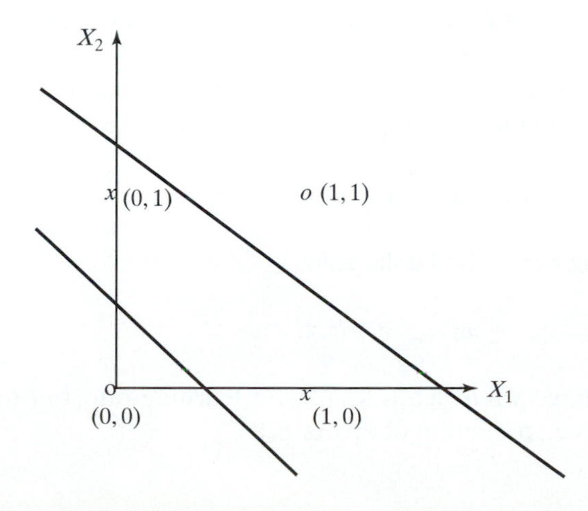

Figure 5.44

Geometric representation of the logical "exclusive or"

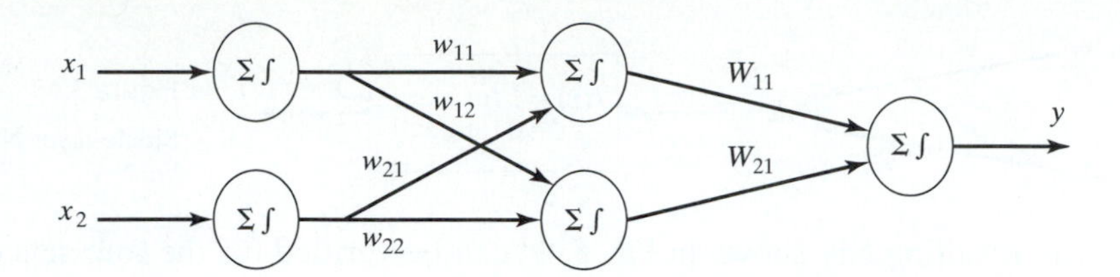

Figure 5.45

Multilayer NN model for logical "exclusive or"

the $y = 0$ from $y = 1$ points, such that a single-layer NN cannot be used. Two straight lines and a corresponding multilayer NN have to be used [42, 71].

A multilayer NN is shown in Fig. 5.45. Input layer has $n = 2$ neurons (required for the two inputs x_1 and x_2), hidden layer has $m = 2$ neurons, and output layer has $s = 1$ neuron (required for the single output y).

In the previous examples, the weights could be determined by geometric considerations and using simple algebraic equations for the binary values. In general, for NN modeling, numerous floating-point inputs and outputs have to be considered and iterative methods for determining the weights have to be used in a so-called NN training.

5.5.3 Learning and Training

Training a neural network is used to make a NN learn from a set of data and to be able to reproduce later, when used, a similar behavior to the one contained in that set of data. Supervised learning is based on calculating the weights w such that, for the given inputs, the outputs of the set of data used for teaching and the output of the learned NN are acceptable close. Unsupervised learning leads to grouping in clusters the input data subsets, which produce similar output without having output data available for teaching the NN.

This presentation will focus on delta rule for supervised leaning, a frequent method used in training NN.

Delta Rule for Supervised Learning of Feedforward NN Consider a single-layer NN with $i = 1$ to n inputs x_i and $l = 1$ to s outputs y_l [42, 71].

The output error is the difference of the desired output Y_l (i.e., the output of the data set for training NN output) and the NN output y_l to the same input:

$$E_l = Y_l - y_l.$$

The weights w_{lj} are updated using a so-called delta rule:

$$w_{1j,\text{ new}} = w_{1j,\text{ old}} + gE_1x_j$$

Here, g is a constant that, for a large value, leads to a faster learning rate, but to more uncertainty about convergence (i.e., reduction of errors E_l).

Example: Delta Rule Training of a Single-Layer NN Model

The example of the logical "or" is used again for illustrating the application of the delta rule. Consider the logical definition of the "or":

x_1	x_2	y
0	0	0
0	1	1
1	0	1
1	1	1

The NN, shown in Fig. 5.41, has

$$z = w_1 x_1 + w_2 x_2$$

and

$$y = f(z) = 0 \text{ for } z \leq 1$$
$$y = f(z) = 1 \text{ for } z > 1^{\cdot}$$

For this example, it is assumed that $g = 1$. Learning starts with arbitrary values for the weights (chosen at random or based of previous experience). The learning process is as follows:

(a) The initial values are chosen arbitrarily. We have

$$w_1 = 0$$

and

$$w_2 = 1$$

such that

$$z = w_{11} x_1 + w_{12} x_2 = x_2.$$

$E_1 = Y_1\text{-}y_1$ is calculated as follows:

x_1	x_2	$z = x_2$	Y	y	E
0	0	0	0	0	0
0	1	1	1	0	1
1	0	0	1	0	1
1	1	1	1	0	1

A row with nonzero error will be chosen. The third row is selected because it has x_1 nonzero, and a new nonzero weight $w_{1,new}$ can be obtained:

x_1	x_2	E
1	0	1

The new weights are

$$w_{1,\text{new}} = w_{1,\text{old}} + E\,x_1, \qquad \text{or} \qquad w_{1,\text{new}} = 0 + 1 \cdot 1 = 1$$

and

$$w_{2,\text{new}} = w_{2,\text{old}} + E\,x_2, \qquad \text{or} \qquad w_{2,\text{new}} = 1 + 1 \cdot 0 = 1.$$

(b) For the new weights, we have

w_1	w_2
1	1

E is calculated as follows:

x_1	x_2	$z = x_1 + x_2$	Y	y	E
0	0	0	0	0	0
0	1	1	1	0	1
1	0	1	1	0	1
1	1	2	1	1	0

A row with nonzero error will again be chosen. The second row is selected because it has x_2 nonzero, and a new nonzero weight $w_{2,\text{new}}$ can be obtained:

x_1	x_2	E
0	1	1

The new weights are

$$w_{1,\text{new}} = 1 + 1 \cdot 0 = 1$$

and

$$w_{2,\text{new}} = 1 + 1 \cdot 1 = 2.$$

(c) For the new weights, we have

w_1	w_2
1	2

E results as follows:

x_1	x_2	$z = x_1 + 2\,x_2$	Y	y	E
0	0	0	0	0	0
0	1	2	1	1	0
1	0	1	1	0	1
1	1	3	1	1	0

A row with a nonzero error will again be chosen. The third row is again selected because it is the only row with nonzero E:

x_1	x_2	E
1	0	1

The new weights are

$$w_{1,\text{new}} = 1 + 1 \cdot 1 = 2$$

and

$$w_{2,\text{new}} = 2 + 1 \cdot 0 = 2.$$

This set of weights, namely,

w_1	w_2
2	2

gives

x_1	x_2	$z = 2x_1 + 2x_2$	Y	y	E
0	0	0	0	0	0
0	1	2	1	1	0
1	0	2	1	1	0
1	1	4	1	1	0

At the end of the third updating, all E values are zero signaling that the NN has learned the input–output set of values of the logic "or."

This result obtained by the learning NN confirms the result of Section 5.5.2 for the example of a single-layer NN model of a logical "or,"

$$z = 2x_1 + 2x_2,$$

obtained on the basis of a geometric representation.

5.5.4 Applications

Traveling Salesperson The traveling salesperson problem is a binary linear programming problem. The problem is formulated as follows [71]:

notations

n cities to be visited by the salesperson in a nonrestricted order; all n cities have to be visited in a tour.

X_{ij} = 1 if the salesperson, currently in city i, visits the next city j;
= 0 if, from city i, the next city visited is not j.
d_{ij} is the distance from city i to city j

formulation

The linear function of the optimization criterion is

$$\text{Min} \sum_{i=1}^{n} \sum_{j=1}^{n} X_{ij}\, d_{ij} \quad \text{for } j = 1 \text{ to } n \quad /^{*}\text{minimization of total length of the tour}^{*}/$$

subject to the following linear constraints:

$$\sum_{i=1,\ i \text{ not } j}^{n} X_{ij} = 1 \qquad \text{for } j = 1 \text{ to } n \qquad /^{*}\text{for any } j \text{ there is only one } i \text{ to come from}^{*}/$$

$$\sum_{j=1,\ j \text{ not } i}^{n} X_{ij} = 1 \qquad \text{for } i = 1 \text{ to } n \qquad /^{*}\text{from any } i \text{ the move is to only one specific } j^{*}/$$

The solution is constrained to binary values of X_{ij} and represents a sequence of values designating the order of the n cities to be visited.

This problem does not have a general solution for all possible examples, except for the complete enumeration of all tours, followed by the total length evaluation for all tours, and the final choice of the tour having minimum total length. This solution requires the evaluation of $n!$ possible tours, which, for n=13, gives 6.2 billions possible tours [71]. A more efficient solution to this problem can be obtained using NN, but the minimum is not guaranteed. For NN, n^2 neurons are needed and the weight training can be performed using delta rule until the errors are acceptable. The minimization criterion is for this NN a quadratic function of X_{ij} [71]. The resulting floating-point values for X_{ij} are close to 0 and 1 and have to be rounded up to 0 or 1.

System Dynamics The system dynamics for linear time invariant systems is described by the matrix equations

$$(d/dt)\, X(t) = AX(t) + BU(t)$$

and

$$Y(t) = CX(t) + DU(t).$$

This system of equations can be solved by numerical integration but, if the system is nonlinear or discontinuous, the numerical solution can be difficult to obtain if not impossible in real-time applications.

NN have been proposed and tested for simulating system dynamics [42, 71]. The NN is trained for sets of values for $d/dt\ X(t)$, $X(t)$, and $U(t)$ until the errors between NN results and system dynamics results are acceptably low.

Commercial packages for neural networks are available. An example is neural network toolbox using MATLAB [37].

PROBLEMS

5.1. State the main distinguishing features of inputs, outputs, states, and events.

5.2. State the main distinguishing features of continuous time, discrete time, and discrete event models.

5.3. Develop a state transition diagram for seven people passing through a revolving door that allows a maximum of two people at a time.

5.4. Draw a block diagram for a water tank system with two incoming flow rates and one outgoing flow rate.

5.5. Consider the following data for the input (x_1,x_2) and output (Y):

x_1	x_2	Y
0	0	1
0	1	0
1	0	0
1	1	0

(a) Is a single-layer neural network or a two-layer neural network needed for modeling?

(b) Use the delta rule for supervised learning of the neural network based on the given data.

5.6. State the main distinguishing features of high-level languages, general-purpose simulation languages, graphical programming languages, and object-oriented programming languages.

5.7. Draw a block diagram for a noninteracting two-tank system with two incoming flow rates and one outgoing flow rate for each tank.

5.8. Draw a block diagram for an interacting two-tank system with two incoming flow rates and one outgoing flow rate for each tank.

5.9. Build a Simulink model of the interacting two-tank system shown in Fig. 5.13, using the data from chapter 5.2.3.

5.10. State the main distinguishing features of an object-oriented model.

5.11. Develop a pseudocode for a pool of four different cars.

5.12. Develop the matrix model using effort–flow cuts for the DC motor driving a fan, shown in Figure 5.31, by including viscous friction coefficients on the shaft bearings on the rotor side b_r and on the load side b_f.

5.13. Build the Simulink model for the DC motor driving a fan described in problem 5.12, for the same data used for Figs. 5.37 to 5.39 and for $b_r = b_f = 0.001$ [Nms/rad].

5.14. Use Simulink encapsulation procedure to reduce the Simulink model developed for problem 5.13 to the encapsulated form shown in Fig. 5.36.

C H A P T E R 6

Data Acquisition and Virtual Instrumentation

6.1 COMPUTER-BASED MONITORING AND CONTROL

Data acquisition was traditionally carried out manually by writing down the readings of various instruments at specified moments in time. Later, recorders automated the process of retaining the data acquired from instruments on paper plots. Further development of data loggers permitted us to maintain the values of the readings from instruments versus time printed in a numerical form.

Traditionally, instrumentation manufacturers provided specific functions, a given architecture, and fixed interfaces for measuring devices, which limited the domain of applications for these devices. Moreover, traditional instruments required extensive time and specific skills for adjusting the measuring range and for saving and documenting the results.

The advent of microprocessors in the measurement and instrumentation field produced rapid modifications and, soon, instrumentation manufacturers introduced computer-based measurement techniques. Computer-based data acquisition replaced paper records by digital data acquisition and storage. Moreover, computers allowed easy analysis and advanced display of the measurements. Figure 6.1 shows the main components of a computer-based monitoring system:

(a) data acquisition and instrument control;
(b) data analysis and storage;
(c) display of the results.

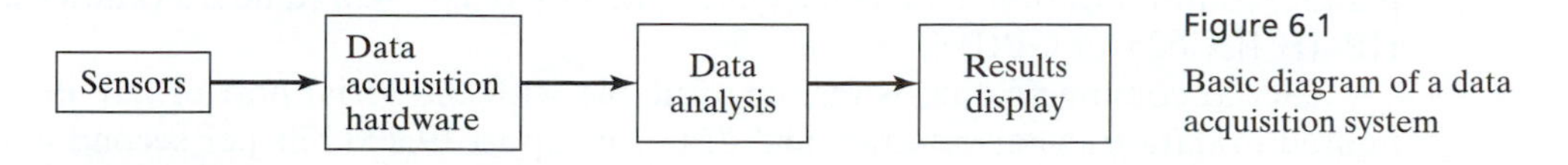

Figure 6.1

Basic diagram of a data acquisition system

These tasks are carried out by the following components:

(a) Acquisition of measurement data (signals) and control of instruments using
data acquisition board;
IEEE–488 standard for GPIB;
serial communication using RS-232 standard;
VXI standard, etc.

(b) Analysis of measurement data:
digital filtering;
digital signal processing;
statistical processing.

(c) Presentation of results:
graphical and numerical display and
hard copies, etc.

Programming for data acquisition can be achieved using textual programming languages or graphical programming languages. In this chapter, the data acquisition toolbox for MATLAB will exemplify textual programming and LabVIEW will exemplify graphical programming.

Measuring devices have a variety of forms, from simple transducers (that convert physical quantities into electric signals) to complex stand-alone instruments (multimeters, oscilloscopes, etc.) equipped with controls for adjusting the instrument to the particular measurement and displays for showing the results. Computer-based instrumentation provides specific solutions to various measuring devices. This presentation will focus on personal computer (PC) use for computerized instrumentation [99].

A simple solution for connecting transducers to a PC is the use of a plug-in data acquisition board, installed in an expansion slot in the PC, for transferring measurement data to the data bus and the memory of the PC. Most data acquisition boards are multifunctional (i.e., they accept both analog and digital signals). One of the advantages of this solution is that the electric signal of the transducer is directly transferred to the plug-in board. At the same time, this limits the distance of transmission, given that over long distances, electromagnetic interference can distort the measurement signal. Moreover, connecting a large number of transducers to the plug-in board results in an undesirable large number of wires. These plug-in data acquisition boards gained acceptance due to their low price and high flexibility provided by the associated software [82–87, 90]. This book focuses mainly on this solution for computer-based instrumentation.

A different category of solutions for computer-based measurement is the communication and control of instruments using digital communication between instruments and the computer—serial communication, based on RS–232 standard and parallel communication, based on a bus defined by IEEE–488 standard (known also as HP–IB, IEC625, or GPIB).

Serial communication is readily available with the serial port of any PC, but is limited in data transmission rate and distance (up to 19,200 bits per second and 15.2

meters for a PC using RS–232) and used to accommodate only one device connected to a PC. Universal serial bus (USB) permits serial communication of multiple instruments and a PC. The use of modems and telephone lines greatly extends the distances between RS–232 instruments and the PC.

Parallel communication, based on the IEEE–488 standard, can be achieved by a PC equipped with a GPIB board connected to the data bus and has a high transmission rate (nominal 1 MB per second) over a cable bus of maximum 20 meters connecting a maximum of 15 devices. This solution is not readily available because it is based on a standard different from the communication standard of the parallel port of the PC. The instruments designed for this solution are called GPIB instruments and are normally advanced instruments, which require a high transmission rate of data to the PC. This solution is more frequently used in laboratory applications and less frequently in industrial applications.

Measurement and control with computer-based instrumentation are based on software to obtain a complete virtual representation of the functionality of traditional instruments. Figure 6.1 shows a generic diagram of a computer-based instrumentation system for process monitoring.

The main characteristics of a computer-based instrumentation system are as follows:

integration of instruments based on various industrial standards (data acquisition DAQ boards, GPIB/IEEE–488.2, Serial port/RS–232, USB, etc);

various options for data acquisition and processing;

design of numerous new instruments using commercially available software and basic instrumentation components.

For process control, Fig. 6.2 shows a generic configuration for DAQ boards and DAQ pads, GPIB boards, and serial instruments interface. DAQ boards and DAQ pads share

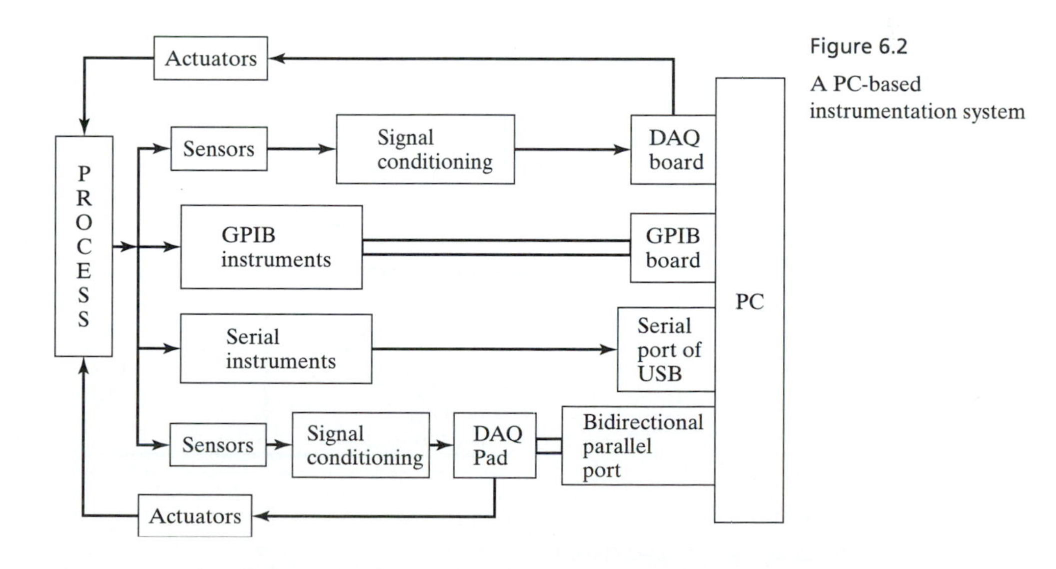

Figure 6.2
A PC-based instrumentation system

the structure shown in Fig. 6.3. The difference between them is that a DAQ Pad is physically located in a box, connected to a bi-directional parallel port of the PC, while the DAQ board is plugged into a PC slot [12, 99].

In the late eighties, a new VXI standard was issued, which allowed communications with transfer rates over 20 megabytes per second between VXI systems. VXI instruments are installed in a rack and are controlled by and communicate directly with a VXI computer. These VXI instruments do not have buttons or switches for direct local control and do not have the local display typical to traditional instruments. As a result, much more compact monitoring systems can be built [99].

Icon-based operating systems for personal computers facilitated present-day graphical programming for virtual instrumentation and enhanced the capabilities of monitoring and control systems.

A virtual instrument is a new type of computer-based system for acquisition, storage, processing, and presentation of measurement data and for controller implementation. A virtual instrument consists of hardware and software, which permit the transfer of signals from instruments to a computer and convert them in digital form for processing using graphical programming.

A virtual instrument can replace the front panel buttons and display of stand-alone traditional instruments by a virtual front panel on a PC monitor. Virtual instruments enable compact integration of the display, control, and centralization of complex measurement systems [36].

As a result of searching for a more natural communication between human operators and the machines, the first concepts of a programming language appeared at the beginning of the seventies based on the data-flow model. National Instruments was a pioneer in this domain and in 1986 launched on the market the first version of Lab-

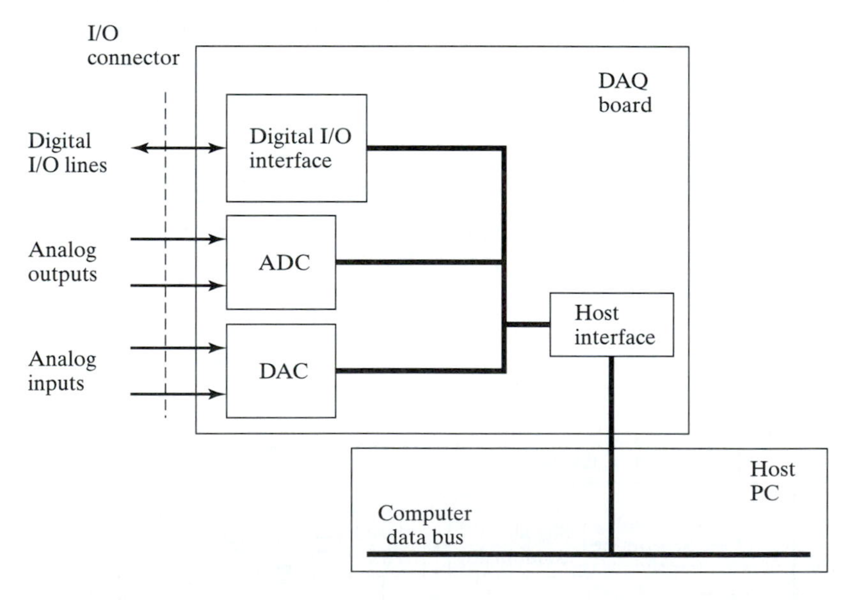

Figure 6.3

Block diagram of a DAQ board

VIEW, a first generation of graphical programming software for PC-based instrumentation. Initially developed for laboratory measurements, present-day LabVIEW is frequently adopted in industry for monitoring and control of production processes, for testing and for rapid prototyping. It contains tools for modern graphical programming and drivers for numerous instruments. Different from conventional textual programming languages, and based on sequential text program writing, graphical programming allows the development of programs using the graphical representation of the flow of data between various blocks of the actual system.

In the last ten years, PC-based monitoring and control systems using graphical programming became a frequently used approach. Graphical programming permits the utilization of graphical components for assembling systems and simplifies the development of monitoring and control programs such that scientists and engineers, as well as nonspecialists in computers, can easily develop PC-based monitoring or control systems. This is facilitated by a natural representation of the components of block diagrams and by avoiding complex syntax rules. The intuitive nature of the graphical programming languages reduces significantly the time required for learning, programming of the prototype, and realization of the final product.

Textual programming for computer-based instrumentation retains, however, specific advantages. In particular, the user familiarity with computer-aided engineering packages—for example, MATLAB—allows easy adoption of textual programming for the Mathworks data acquisition toolbox [100]. Moreover, in applications requiring the computation of measurement uncertainty, textual programming can be advantageous due to the explicit instructions used in data processing [91].

Examples of commercial packages for computer-based monitoring and control are the following:

LabVIEW, National Instruments [36];
data acquisition toolbox for use with MATLAB, MathWorks [100];
HPVEE, Hewlett Packard [15]; and
Labtech Notebook,Omega [14].

LabVIEW and data acquisition toolbox for use with MATLAB are presented in the next sections.

6.2 LABVIEW PROGRAMMING FOR VIRTUAL INSTRUMENTATION

6.2.1 LabVIEW Basics

LabVIEW (Laboratory Virtual Instrument Engineering Workbench) is a development environment for computer-based data acquisition and instrument control programs based on graphical programming. LabVIEW was released in 1986 for Macintosh and in 1992 for Windows [62] and, in the present form, can create compiled programs that are stand-alone executables. A LabVIEW program is a virtual instrument (VI) that replaces a large part of a traditional instrument by software. Moreover, LabVIEW contains libraries of ready-made components for data acquisition, analysis, storage, and display as well as for generating outputs for actuator control.

The basic components of a LabVIEW virtual instrument are the front panel and the block diagram. The front panel of a virtual instrument is a graphical user interface of the VI, which replaces the traditional instrument front panel by a computer monitor-based one. A block diagram of a virtual instrument is a graphical source code. This graphical code processes the flow of data inputs from front panel and transducers through various stages of data processing, up to the generation of outputs for the display and storage of the measurement results and for the control of actuators. The graphical links for data flow in a VI block diagram replace wiring for the electric signal transmission in traditional instruments.

Front panel and block diagram design and editing are achieved with the objects and tools available in a graphical form in the tools palette (Fig. 6.4(a)), controls palette (Fig. 6.4(b)), and functions palette (Fig. 6.4(c)) [73]. They are accessed from the menu bar, as follows:

Windows → Show Tools palette.

The tools palette contains tools used for designing both the front panel and the block diagram objects. The controls palette contains front panel controls (for providing inputs from the user) and indicators (for providing output results to the user). The functions palette contains a variety of objects for programming the VI, as for example numerical and logical operations, instrument input and output, and data acquisition functions.

Two examples of LabVIEW virtual instruments will be presented next—temperature conversion and liquid level height measurement.

(a)

(b)

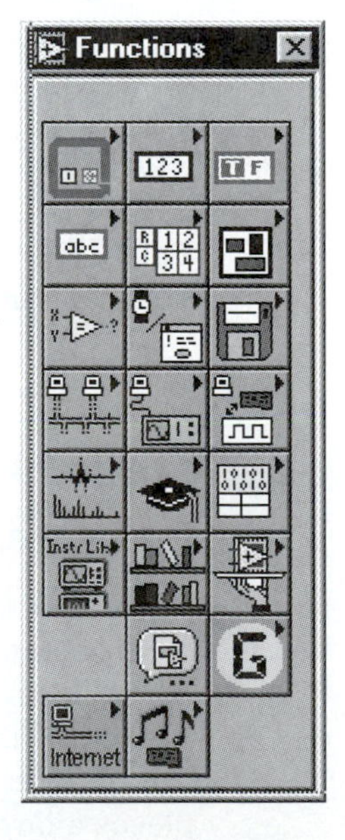

(c)

Figure 6.4

Tools palette (a), controls palette (b), and functions palette (c) (Source National Instruments [73])

Figure 6.5 shows the front panel of a virtual instrument for temperature conversion, Conversion Celsius to Fahrenheit.vi.

Figure 6.6 shows the block diagram of Conversion Celsius to Fahrenheit.vi.

In Fig. 6.5, the front panel control, a digital control labeled temperature in degrees Celsius, permits the user to provide the input value of the temperature in degrees

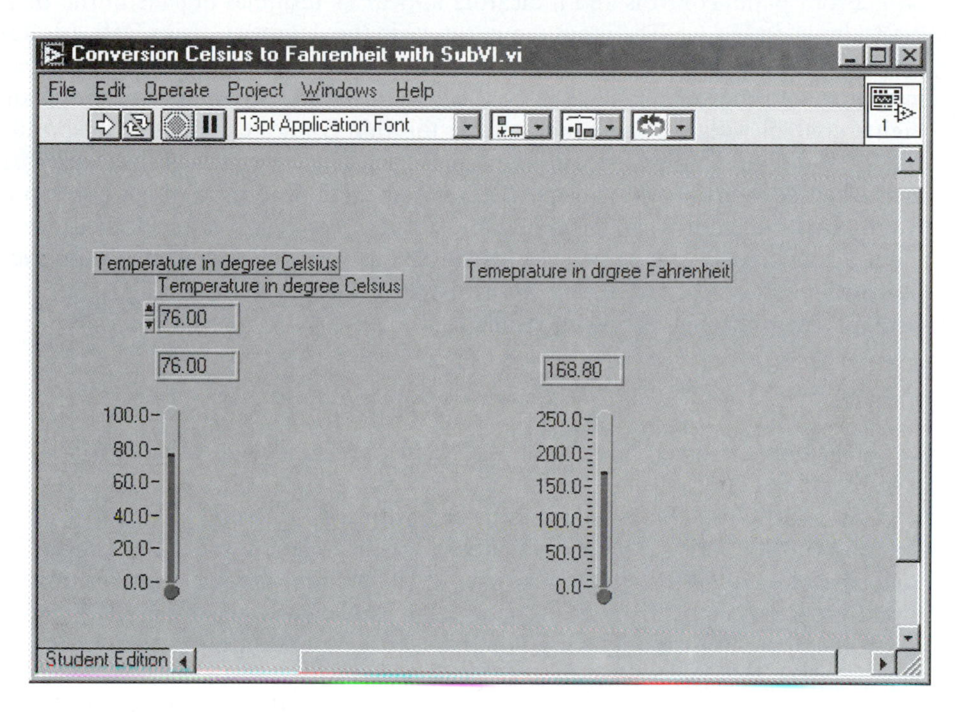

Figure 6.5

Front panel of the Conversion Celsius to Fahrenheit.vi

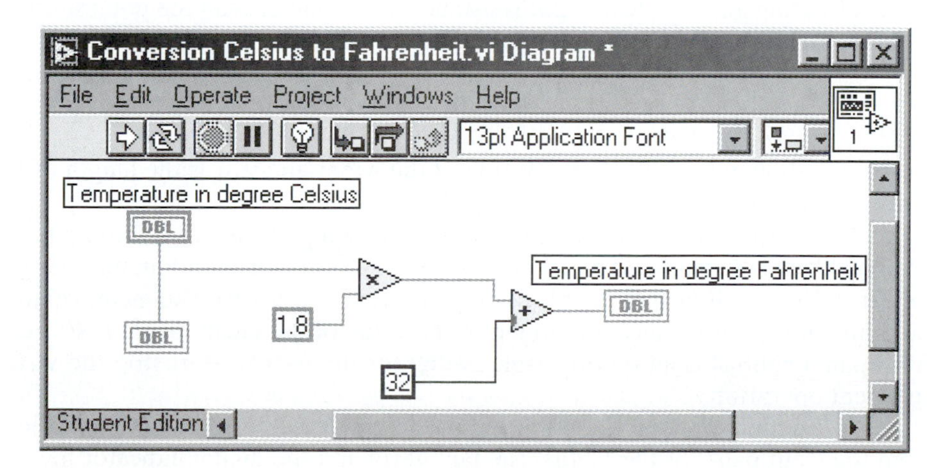

Figure 6.6

Block diagram of the Conversion Celsius to Fahrenheit.vi

Celsius. Front panel indicators are one thermometer—under the same label temperature in degrees Celsius—and a second thermometer labeled temperature in degrees Fahrenheit. The first thermometer displays the input value of the temperature in degrees Celsius, while the second thermometer displays the result of its conversion into degrees Fahrenheit.

Front panel controls and indicators appear as terminal objects in the block diagram shown in Fig. 6.6. The input temperature in the degrees Celsius DBL (double precision floating-point number) terminal corresponds to front panel control and front indicator with the same label. The bold frame terminal corresponds to the front panel digital control with the same label, while the thin frame terminals correspond to the front panel digital indicators with the same label. The conversion results are displayed in the temperature in the degrees Fahrenheit DBL terminal, which corresponds to front panel indicator with the same label. Other objects in the block diagram are the numeric constants (with the values 1.8 and 32, respectively), as well as numeric multiply and numeric add. These objects are linked by wires representing data flow in accordance with the conversion equation

$$(\text{temperature in degrees Fahrenheit}) = (\text{temperature in degrees Celsius}) \cdot 1.8 + 32$$

The conversion computation can be enclosed in a C→F subVI for a more compact block diagram.

In Fig. 6.7(a), the virtual instrument with a C→F subVI is shown. The block diagram of the C→F subVI is presented in Fig. 6.7(b). The creation of a subVI helps keep the block diagram simple and is used to facilitate the replacement of simulated inputs by actual inputs from transducers, a topic discussed in Section 6.2.3.

The virtual instrument for a liquid level height measurement system is shown in Fig. 6.8 [99]. A liquid surface height is first measured by the vertical position of a floater, connected by a rigid structure to the wiper of the potentiometer. The floater moves the wiper vertically, as a result of liquid level changes. The potentiometer is supplied with a DC voltage U_s. As shown in Section 2.5, if the potentiometer voltage U is measured using an ADC with input resistance much higher than the potentiometer resistance, then the following linear relationship is valid:

$$v_o = (v_i/y_{\max})y.$$

In this equation, y is the vertical position of the wiper and $y_{\max}$ is the length of the potentiometer, as shown in Fig. 2.22.

The measurement data of a randomly moving liquid surface y above a minimum level of 40 cm are simulated by constant value of 40 cm and a random number generator for 0–1 range, shown in the block diagram from Fig. 6.9(b). The mean value of the random numbers is 0.5 and consequently, the value of the mean height is 40.5 cm. The front panel control is an on/off toggle switch for the user to start/stop the virtual instrument operation.

An example of results is displayed in the two indicators, the tank and the graph from the front panel of Fig. 6.9(a). The tank level and the digital indicator height [cm]

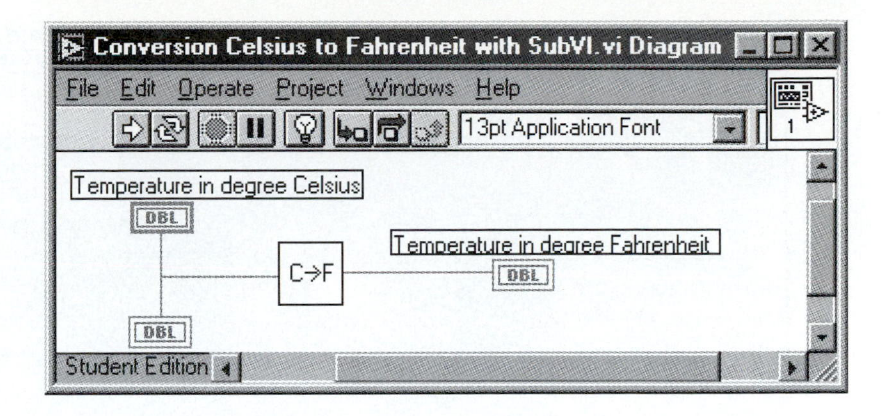

(a)

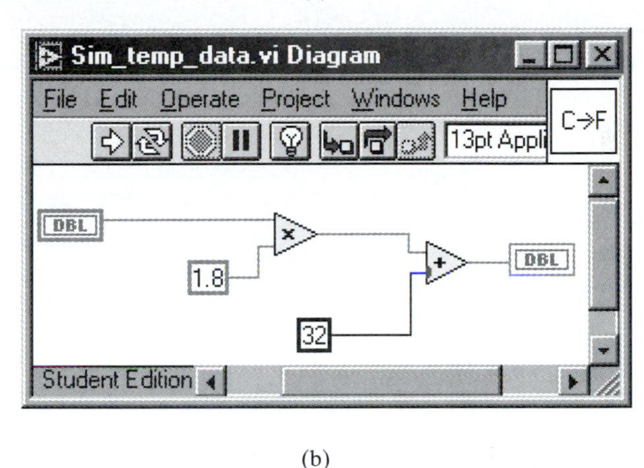

(b)

Figure 6.7

Block diagram (a) of Conversion Celsius to Fahrenheit.vi with C→F subVI (b)

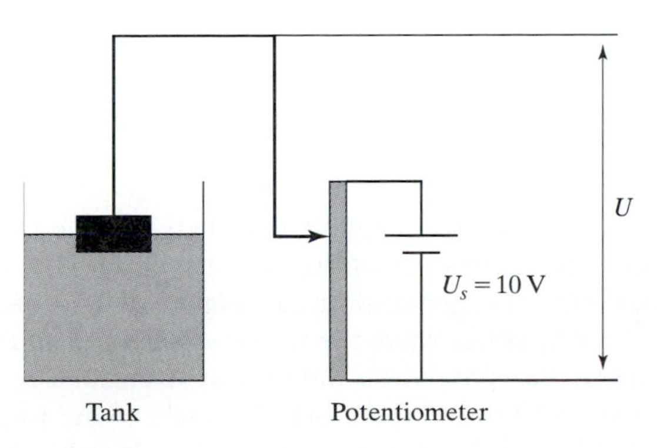

Figure 6.8

Liquid level height measurement with a potentiometer transducer

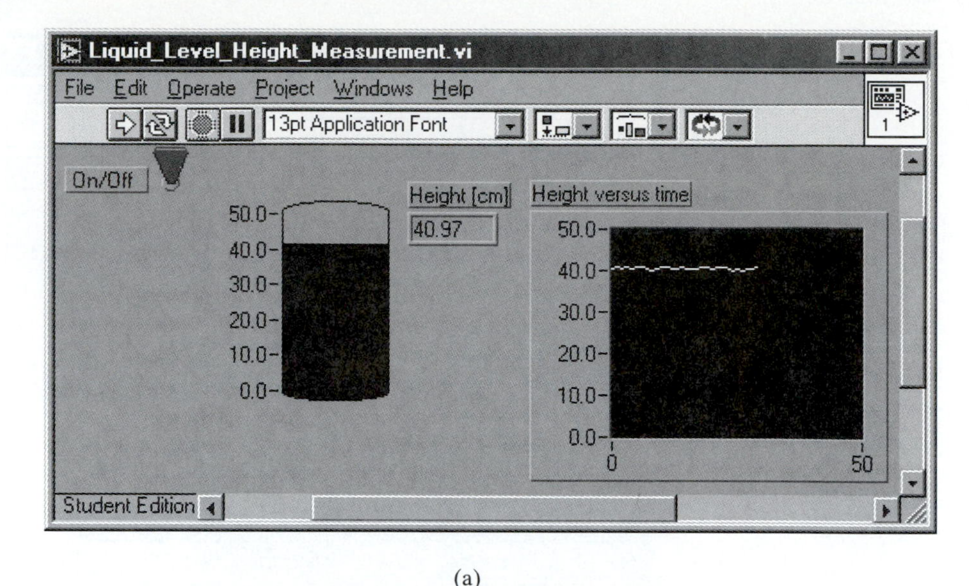

(a)

(b)

Figure 6.9

Front panel (a) and block diagram (b) of Liquid_Level_Height_Measurement.vi

show the current value, while the waveform graph, labeled height versus time, shows a window of the recent past up to present values of the height.

The repeated execution (for the generation of height versus time values) is achieved by the while loop, which operates when the toggle switch is on and stops at off.

The part corresponding to the simulation data input from measurements can be replaced by a Sub VI. Figure 6.10 shows the block diagrams of the Liquid_Level_Height_Measurement with_SubVI.vi (a) and of the Sub VI Sim_Height_Data.vi (b).

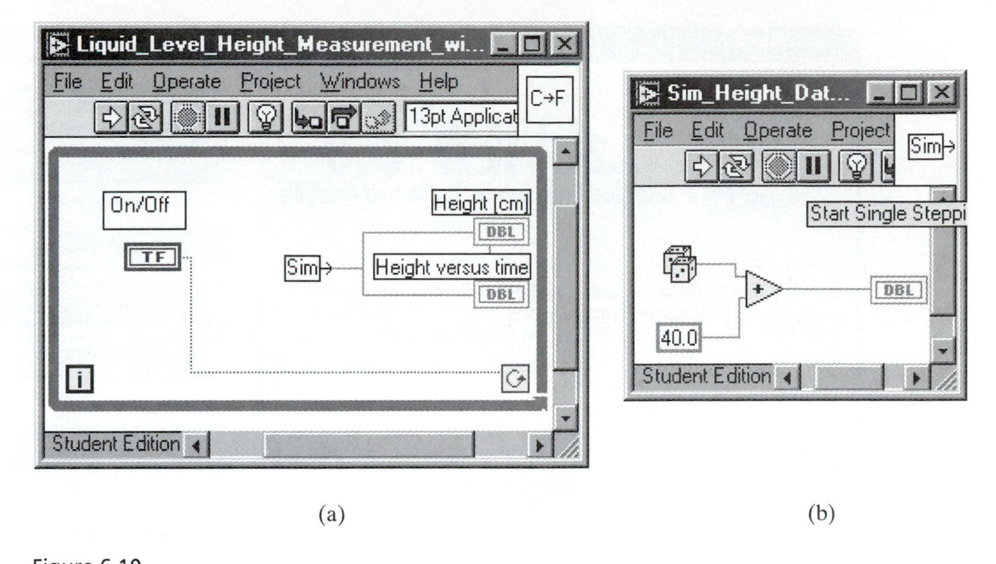

(a) (b)

Figure 6.10

Block diagrams of the Liquid_Level_Height_Measurement with_SubVI.vi (a) and of the Sub VI Sim_Height_Data.vi (b)

6.2.2 Example of Building a LabVIEW Virtual Instrument

Detailed presentation of LabVIEW programming is given in the National Instruments LabVIEW user manual [86] and various books about LabVIEW [40, 62, 83]. Next, an illustration will be presented of the programming of a virtual instrument for a liquid level height measurement system, shown in Fig. 6.8 [99].

After LabVIEW is launched, the LabVIEW dialog box shown in Fig. 6.11 appears. Clicking on New VI, a front panel appears. After selecting File →Save As from the tool bar, Liquid_Level_A file name can be chosen and placed in the desired folder.

The front panel (Fig. 6.9 (a)) of the Liquid_Level_Height_Measurement.vi is designed as follows:

The control is obtained from the controls palette

Controls → Boolean → Vertical Toggle Switch

and, after clicking on the icon posed in the front panel, it can be labeled by accessing

Show → Label

and writing on/off. (See Liquid_Level_A.vi Fig. 6.12.)

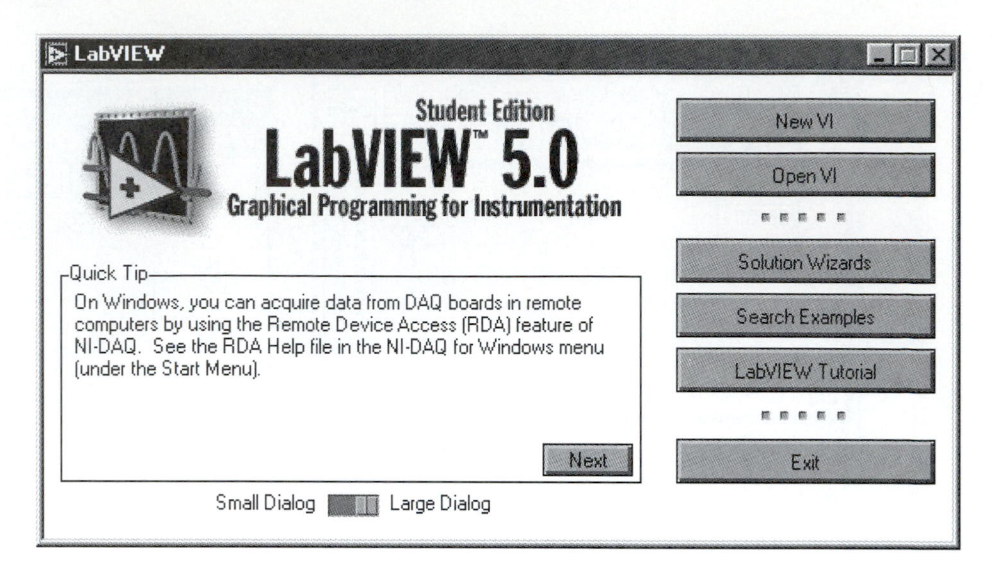

Figure 6.11

LabVIEW dialog box (Source National Instruments [73])

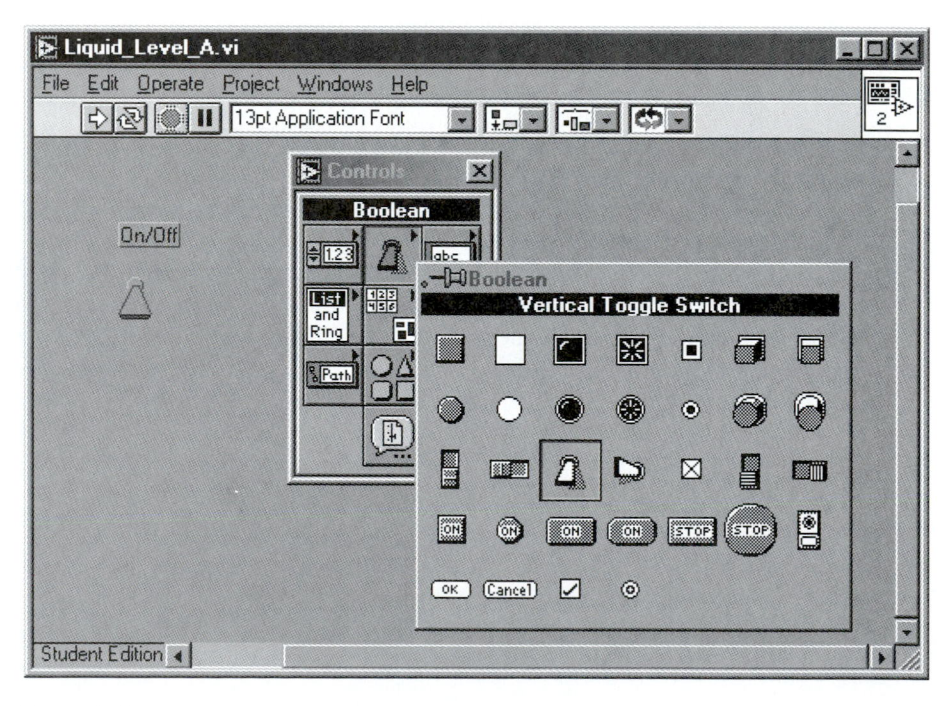

Figure 6.12

Liquid_Level_A.vi

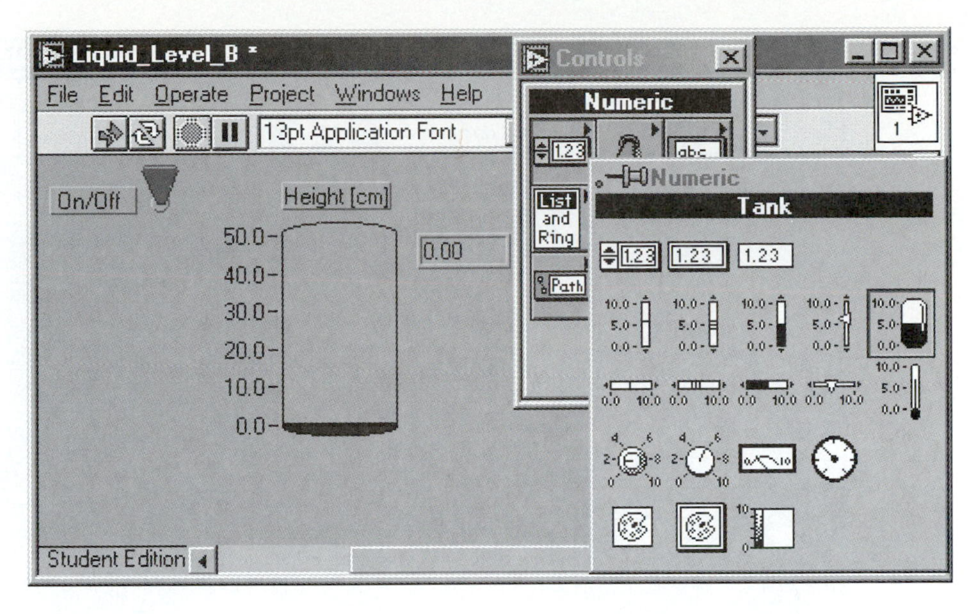

Figure 6.13

Liquid_Level_B.vi

The two indicators, tank and waveform graph, are obtained from the controls palette, as shown in Fig. 6.13 and Fig. 6.14:

Controls → Numeric → Tank

Controls → Graph → Waveform Graph

They can be labeled, following the same procedure, by writing height [cm] and height versus time.

In order to complete the block diagram of the Liquid_Level_Height_Measurement.vi, shown in Fig. 6.9(b), the objects for the simulated measurement data input have to be added. After the design of the front panel, the associated block diagram already contains the terminals of the control, on/off, and of the two indicators, height [cm] and height versus time. The block diagram is accessed from the top menu bar, as follows:

Windows → Show Diagram.

Figure 6.15 shows the selection of the random number generator for 0–1 for the Liquid_Level_D. Diagram:

Functions → Numeric → Random Number (0–1)

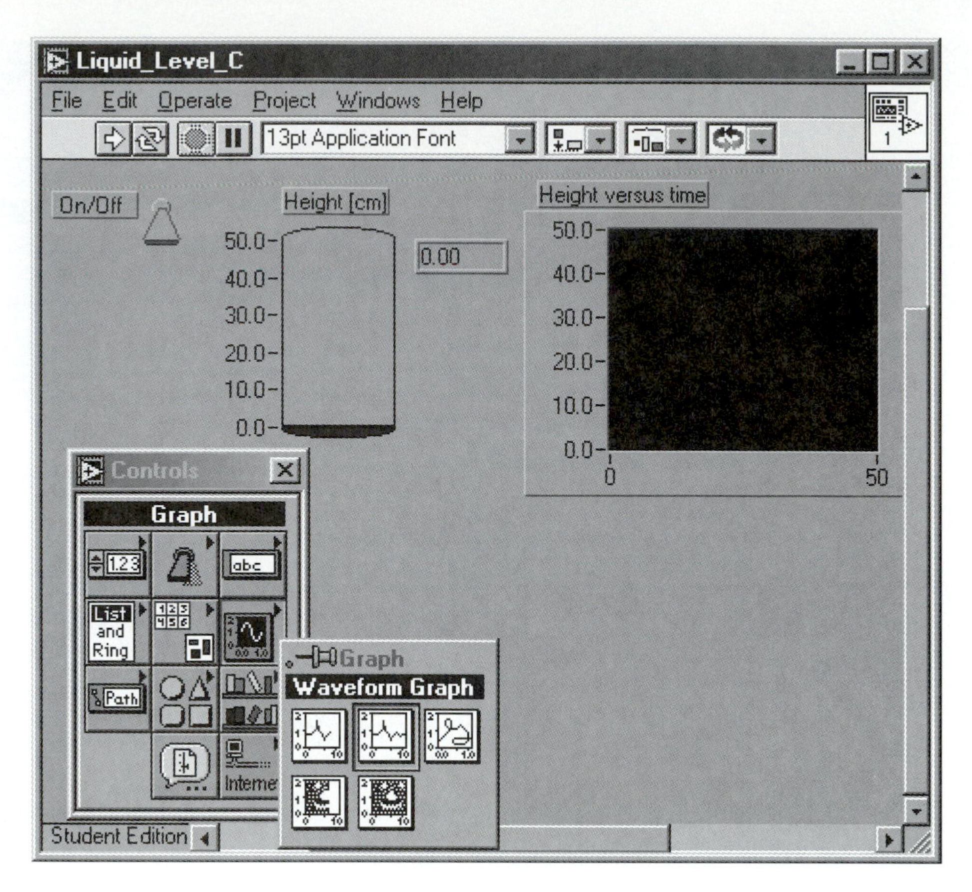

Figure 6.14

Liquid_Level_C.vi

The functions palette is accessed as follows:

Windows → Show Functions Palette.

The object add is selected as follows:

Functions → Numeric → Add.

The constant value of 40 cm is included as follows:

Functions → Numeric → Numeric Constant.

After placing the icon on the diagram, the constant value 40.0 can be included.

Figure 6.16 shows the resulting Liquid_Level_E. Diagram. Wiring is achieved with the connect wire tool from the tools palette:

Tools → Connect Wire.

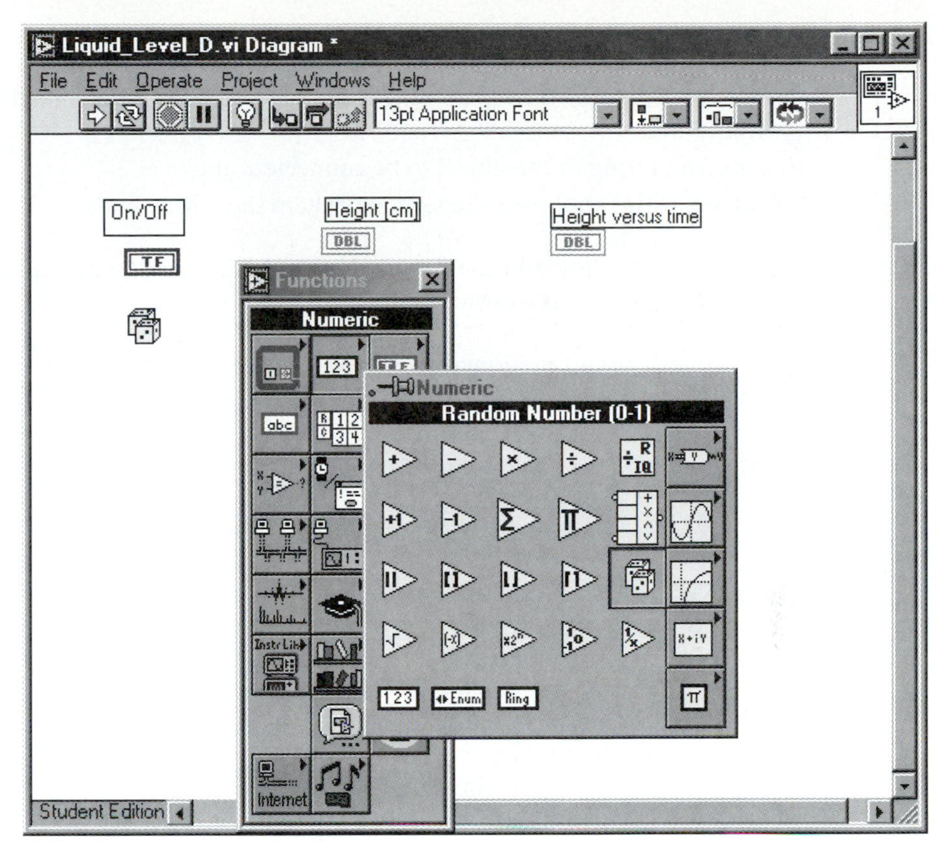

Figure 6.15

Liquid_Level_D. Diagram

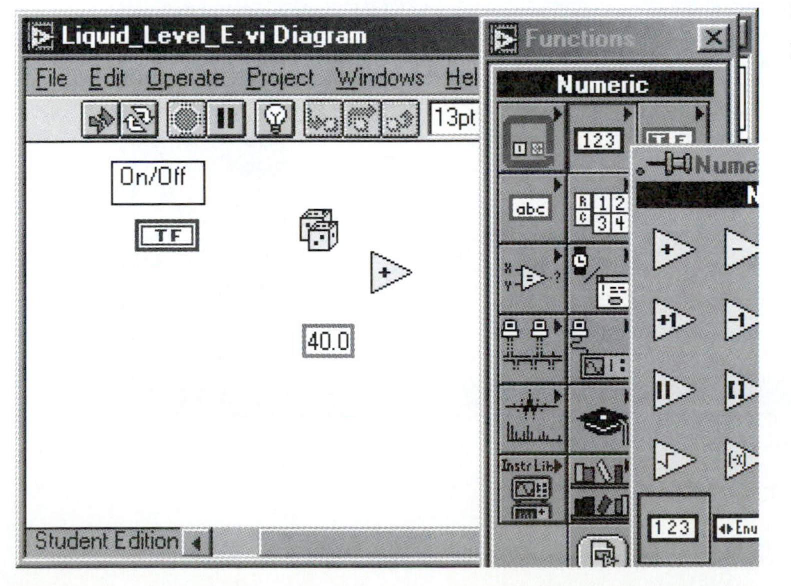

Figure 6.16

Liquid_Level_E. Diagram

Actual wiring of the objects of the block diagram is carried out by [101]

placing the wiring tool over an object to make visible the terminal where the wiring starts,
dragging the wiring to the object to be connected, and
placing the wiring tool over the terminal where the wiring ends.

The result of wiring is shown in the Liquid_Level_F. Diagram, from Fig. 6.17. In this figure, the following wiring is shown:

numeric constant and random number (0–1) to the add object, and
add object to the two terminals, height [cm] and height versus time.

A while loop is added as shown in Fig. 6.18:

Functions → Structures → While Loop.

By dragging a corner of the loop, all the previous objects can be enclosed in this while loop. The on/off terminal is wired to the conditional terminal of the while loop, as shown in Fig. 6.18. The while loop achieves repetitive generation of measurement data, as long as the vertical toggle switch is in the on position. The execution of this LabVIEW graphical program starts by clicking the vertical toggle switch to the on position and the run button on the tool bar.

The part corresponding to "simulate data from measurement" can be replaced by a Sub VI as shown in Fig. 6.19 [40].

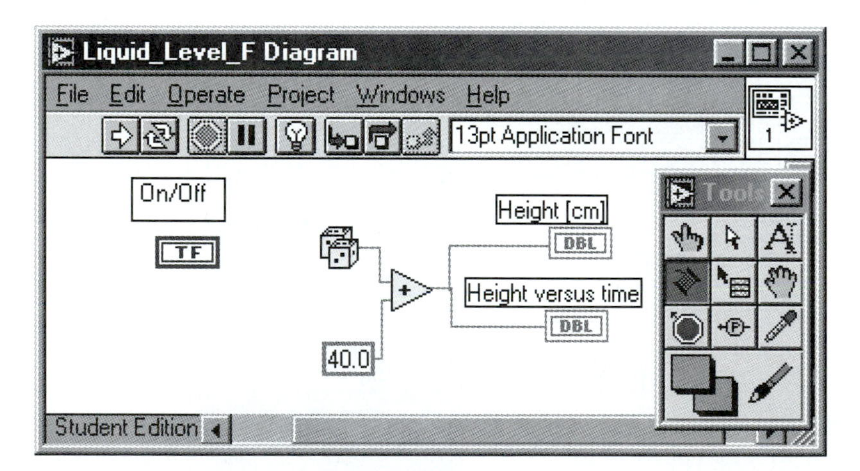

Figure 6.17
Liquid_Level_F.Diagram

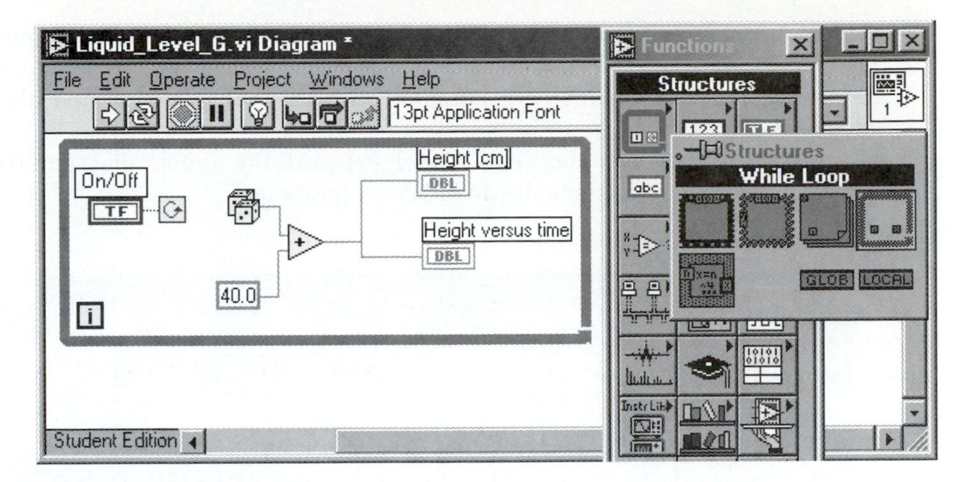

Figure 6.18

Liquid_Level_G. Diagram

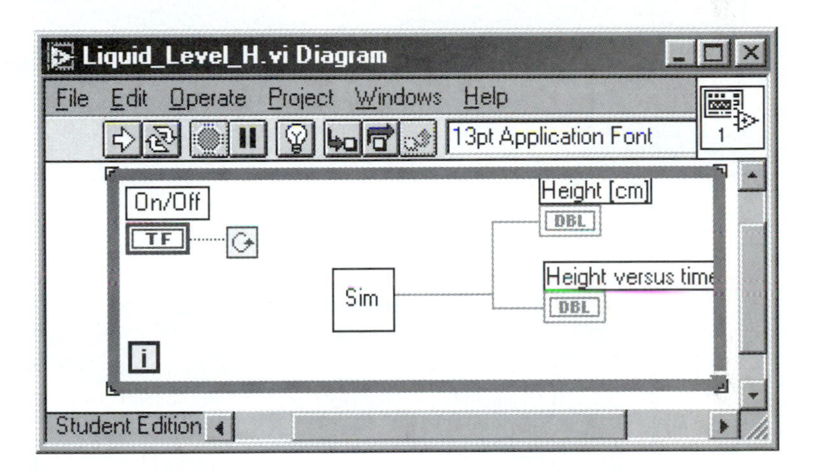

Figure 6.19

Liquid_Level_H. Diagram

This is obtained as follows:

Select the objects to be included in the Sub VI with the positioning tool from the tools palette;

Choose create Sub VI from edit pull-down menu;

Open the Sub VI to be created by double-clicking on the icon with the default label Untitled 1 and select Label from the pop-up menu Show;

Type Sim.

Further improvement of the virtual instrument for liquid level measurement can be obtained by the creation of an indicator danger, which will signal the occurrence of values of the height higher than a maximum acceptable value. In this case, the selected value is 40.99 cm.

Figure 6.20 shows the front panel (a) and the block diagram (b) of the Liquid_Level_I.vi, which includes an indicator for danger.

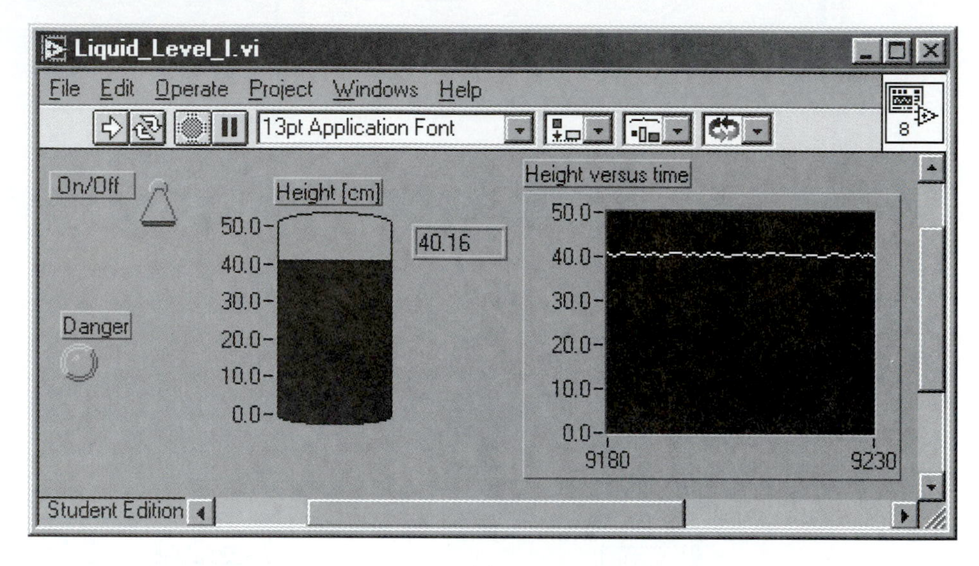

(a)

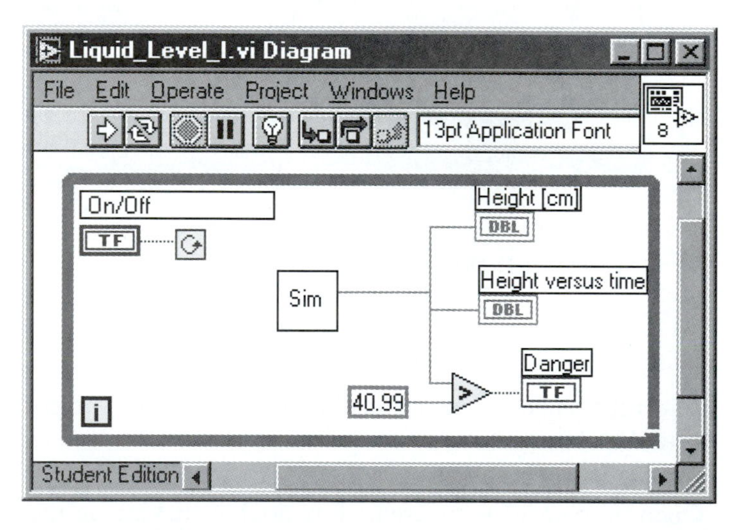

(b)

Figure 6.20

Inclusion of danger warning in Liquid_Level_I.vi

The changes in the front panel of this virtual instrument are obtained as follows:

(a)

Controls → Boolean → Round LED.

Label danger and then place it on the front panel. A TF (True/False) terminal will appear in the block diagram:

(b)

Functions → Comparison → Greater?

Place it in the block diagram:

(c)

Functions → Numeric → Numeric constant.

Place it in the block diagram and write in 40.99, the maximum acceptable value:

(d)

Tools → Connect Wire.

With the connect wire tool, the new objects, including the TF danger terminal, can be wired as shown in Fig. 6.20(b). The danger round LED will be activated whenever the height will take a value grater than 40.99 cm.

The random values of the height versus Time, shown on the waveform graph, can be also be used to calculate a mean value for time window of the simulation. This can be carried out as follows:

(a)

Functions → Analysis → Probability and Statistics → Mean vi.

Place Mean.vi in the block diagram, outside the while loop, as shown in Fig. 6.21(b):

(b)

Controls → Numeric → Digital Indicator.

Place the digital indicator on the front panel, as shown in Fig. 6.21(a) and label it mean height. In the block diagram, using the positioning tool from the tools palette, position the DBL terminal of the digital indicator, labeled mean height, outside the while loop:

(c)

Tools → Connect Wire.

Wire the new objects, as shown in Fig. 6.21(b).

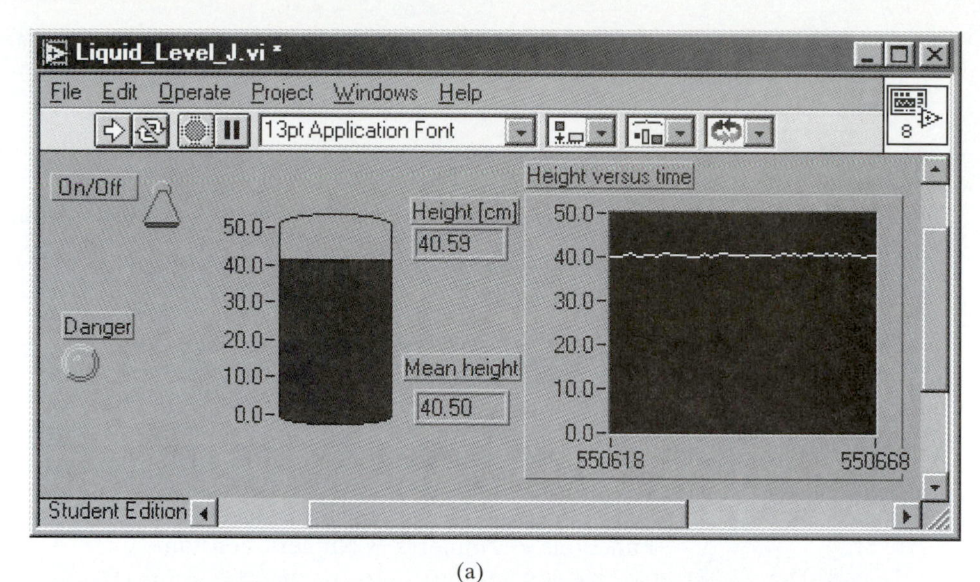

(a)

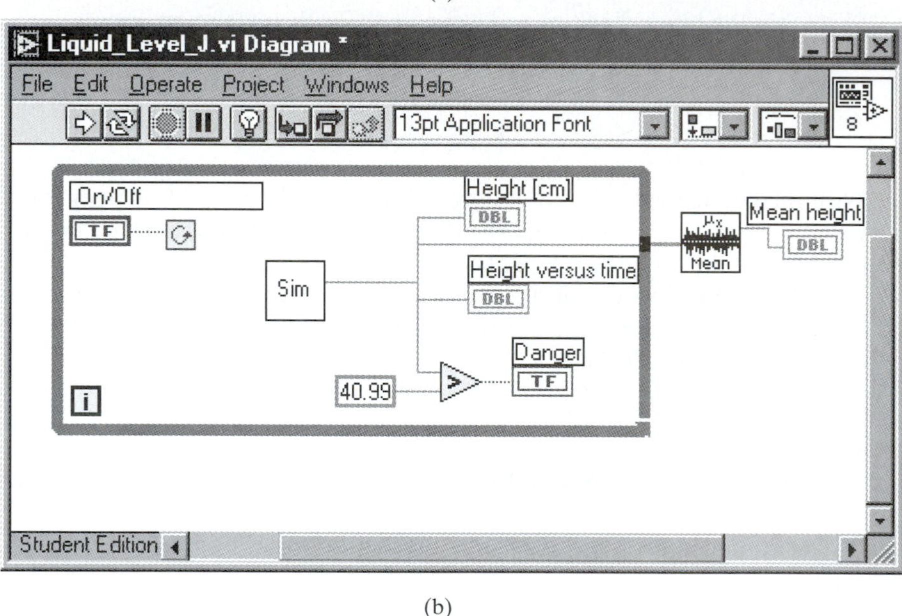

(b)

Figure 6.21

Inclusion of Mean.vi in Liquid_Level_J.vi

In order to permit the data collected to pass to the Mean.vi, after the while loop terminates, the dashed wire from the while loop to Mean.vi has to be changed in solid orange wire as follows:

right-click on the black tunnel of the while loop;
click on enable Indexing from the pop-up menu [101].

The execution of the virtual instrument Liquid_Level_J.vi is triggered by clicking the run button from the tool bar. The results are shown on the indicators from Fig. 6.21(a).

6.2.3 Adding Analog Input to the Virtual Instrument

Actual data acquisition with a DAQ board requires board installation and configuring, in accordance with the LabVIEW Data Acquisition Basics Manual [36] or LabVIEW Evaluation Package [101]. DAQ board installation and configuring is performed once for all later applications [99]. Configuring the board defines the connections between signal inputs and the channels of the board. The process is facilitated by the Solution Wizard. From the LabVIEW dialog box shown in Fig. 6.11, the Solution Wizard is activated by reaching

Solution Wizard → DAQ Solution Wizard → Launch Wizard

and then selecting

Welcome to the DAQ Solution Wizard → Go to DAQ Channel Wizard → New.

The information required by the DAQ Channel Wizard is first the type of channel (analog input, analog output, or digital I/O). For the illustrations presented in this book, analog input is selected.

The following further information is requested for the Analog Input Configuration window:

1. Name your channel.
2. Give it a description.
3. Choose the sensor type which best matches your measurement.

For the example presented in this section, the following information has to be typed:

1. Height
2. Potentiometer
3. Voltage Measurement

Next, the following items have to be completed in the Define the Physical Quantity window:

1. What is your unit?
2. What is the measurement range?

And the following has to be typed in the dialog box:

1. cm
2. min 0.00 cm
3. max 50.00 cm

Next, the following items have to be completed in the Define the Sensor's Scaling and Ranges window:

1. How does the sensor scale values from the physical range to the sensor output range?
2. What is the sensor's output range?

And the following has to be typed in the dialog box:

1. map ranges
2. min 0.00 V
3. max 10.00 V

Next, the following items have to be completed in the Select the Hardware window:

1. What DAQ hardware will read this signal?
2. Which channel on your DAQ hardware?
3. What analog input mode will be used?

And the following has to be typed in the dialog box:

1. Dev1: PCI-MIO-16XE-10
2. 0
3. Referenced single ended

This finishes the selection of data acquisition hardware settings and permits adding analog input to the virtual instrument built before with simulated data input from Sim Sub VI (Fig. 6.21).

For the liquid level height measurement, the AI Sample Channel.vi, shown in Fig. 6.22, is chosen as follows:

Functions → Data Acquisition → Analog Input → AI Sample Channel. vi

for acquiring one point (a single, untimed, nonbuffered measurement) from channel number 0.

Input limits do not have to be entered [36]. The replacement of Sim Sub VI by AI Sample Channel.vi, in the Liquid_Level_J.vi Diagram, from Fig, 6.21(b), is achieved as follows:

Click in Sim object and in the pop-up menu select

Replace → Functions → Data Acquisition → Analog Input → AI Sample Channel. vi.

Select

Tools → Connect Wire

and place the pointer on the AI Sample Channel.vi to make visible channel (0) terminal.

Right-click the terminal and, from the pop-up menu, select Create Constant and type Height as the assigned name for Channel (0).

Entering channel name—in this case, height—will return the value in physical units, cm. This is obtained by converting the voltage measurement analog input values into height using sensor scaling data.

Figure 6.23 shows the resulting Liquid_Level_K.vi Diagram.

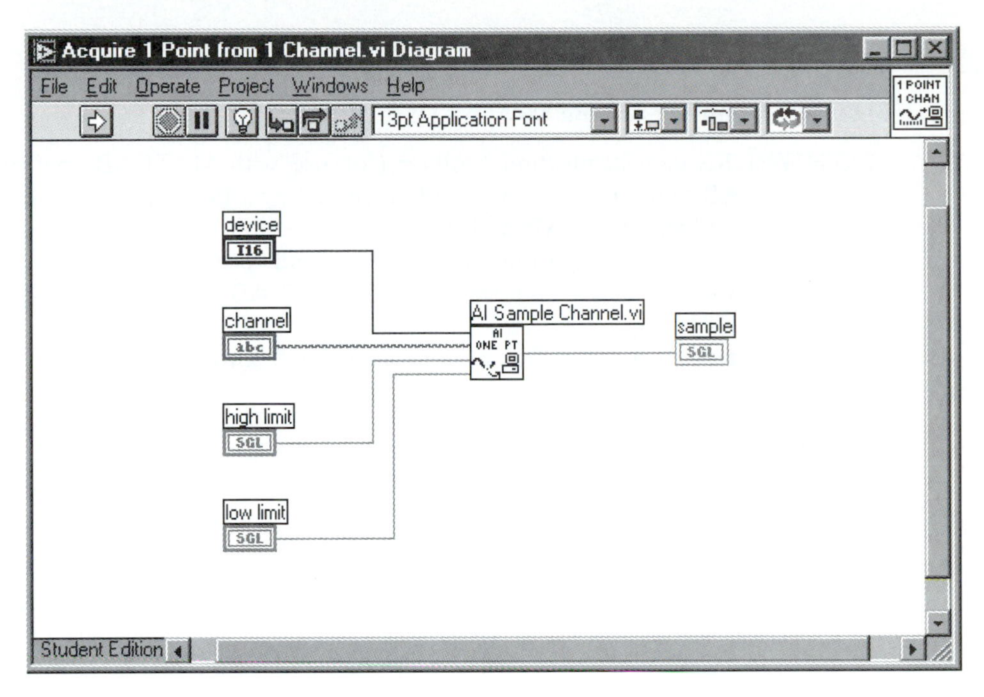

Figure 6.22

Acquire 1 Point from 1 Channel.vi

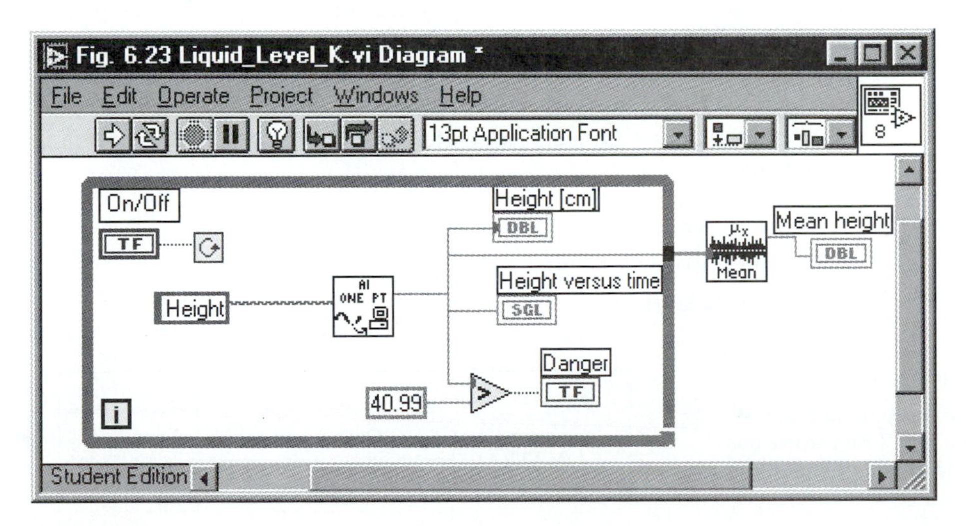

Figure 6.23

Liquid_Level_K.vi Diagram

Running this VI, by switching the vertical toggle switch at on position, leads to the execution of the while loop and the display of the measurement data on the height-versus-time waveform graph.

In Chapter 7, we present real-time applications with LabVIEW RT [96].

6.3 MATLAB DATA ACQUISITION TOOLBOX

6.3.1 Basic Characteristics of the MATLAB Data Acquisition Toolbox

MATHWORKS data acquisition toolbox (for use with MATLAB) permits access from MATLAB to the measurement data received by a PC equipped with a data acquisition board [90]. MATLAB and the data acquisition toolbox form an integrated environment for data acquisition, data analysis, and results visualization. Real-time data acquisition and control applications with MATLAB toolboxes will be presented in Chapter 7 [102–104].

The data acquisition toolbox has the following features:

- access to analog input, analog output, and digital I/O from a variety of commercially available data acquisition plug-in PC boards;
- single and multichannel data acquisition;
- various data acquisition rates;
- hardware and software triggering;
- integration of data acquisition and data processing within MATLAB and toolboxes (signal processing, system identification, control system, optimization, model predictive control, fuzzy logic, neural network, etc).

Figure 6.24 shows the data acquisition system based on MATLAB data acquisition toolbox. Shown in this figure is a PC, fitted with a plugged-in data acquisition board and an installed MATLAB data acquisition toolbox. The data acquisition board can receive analog and digital inputs from sensors and can send analog and digital outputs to actuators.

MATLAB and the data acquisition toolbox are based on object technology and textual programming. Objects associated with the data acquisition board can be used to select the

- sampling rate;
- channel properties; and
- triggering settings.

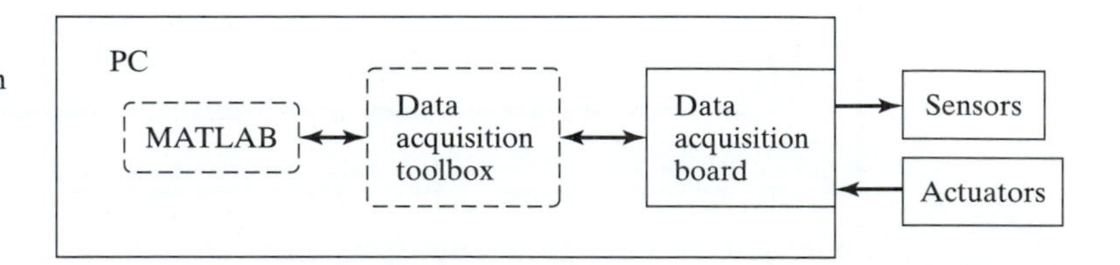

Figure 6.24

Data acquisition system based on MATLAB data acquisition toolbox

In the example for data acquisition with LabVIEW, presented in Section 6.2, one analog input channel object was created. In this section, a similar MATLAB analog input object is designed by setting the values for the parameters needed to execute an analog input task.

The data acquisition toolbox contains the following main commands for analog input [100]:

analoginput (to create an analog input object);
addchannel (to add channels to the analog input object);
getdata (to return the acquired data samples);
peekdata (to preview the most recent data).

For other applications, analog output and digital I/O objects are available.

For analog output, the data acquisition toolbox contains the following main commands [100]:

analogoutput (to create an analog output object);
addchannel (to add channels to the analog output object);
putdata (to provide data samples for the output object);
putsample (to provide a single sample to the output object immediately).

For digital input/output, the data acquisition toolbox contains the following main commands [100]:

digitalio (to create a digital I/O object);
addline (to add lines to the I/O object);
getvalue (to read line values);
putvalue (to write line values).

The data acquisition toolbox contains several commands common to analog input, analog output, and digital I/O. The following commands are illustrative:

daqfind (to find a specified data acquisition object);
daqread (to read data acquisition toolbox data files);
daqreset (to delete and unload all data acquisition objects);
obj2code (to convert a data acquisition object to MATLAB code).

Data acquisition tasks are initiated by various events, including

start and stop events;
triggers;
number of samples acquired.

Data acquired can be converted in specific engineering units. MATLAB permits both linear scaling and nonlinear calibration.

During acquisition, current status of the device object and channel as well as hardware information can be displayed.

6.3.2 Example of Data Acquisition with MATLAB Data Acquisition Toolbox

Figure 6.25 shows the experimental set-up for acquiring a musical sound with MATLAB data acquisition toolbox. In this experimental set-up, the data acquisition board is using an already existing Windows sound card. The sensor in this case is a microphone and the physical quantity measured is a high do produced by an electronic piano. After the electronic piano produces a high do, the acquisition is triggered manually, for a duration set by the textual program.

Figure 6.26 shows the listing of the Getsound.m program [100]. This textual program uses the commands available for MATLAB and the data acquisition toolbox. On the other end of the command lines are explanations of the commands, which start with % and, obviously, are not executable.

The results for the acquired data for a high do sound for an acquisition duration of ad = 1 [sec] are plotted in Fig. 6.27.

For a better view of the results, the first 1000 samples can be retained as follows:

select Edit from the toolbar;
reset the time scale at 1000 values.

The results for the first 1000 samples are shown in the plot from Fig. 6.28.

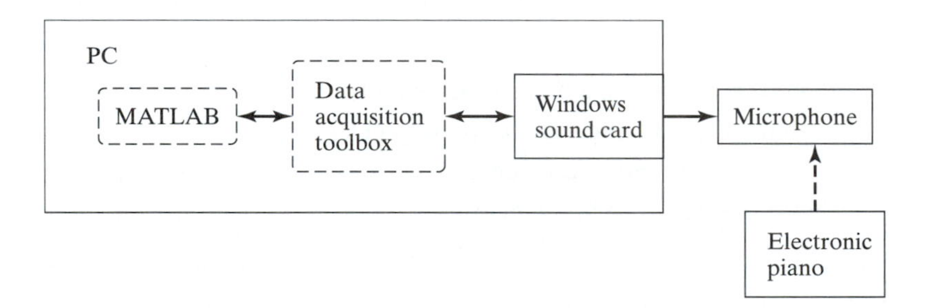

Figure 6.25

Experimental set-up for acquiring a musical sound

```
% Getsound.m
AI = analoginput ('winsound') ;              % creation of AI object
addchannel (AI, 1) ;                         % configuration of channel 1
sr = 10000 ;                                 % sampling rate
[samples/sec]
set (AI, 'SampleRate', sr)
ad = 1 ;                                     % acquisition duration [sec]
set (AI, 'SamplesPerTrigger', ad*sr) ;
start (AI)                                   % starts data acquisition
data = getdata (AI) ;                        % measurement data retrieval
delete (AI)                                  % deletion of AI object
data = data (1:end) ;
plot (data)                                  % plotting command for data (sampled amplitudes)
                                               versus time
                                             % in [sample]
```

Figure 6.26

Getsound.m program for data acquisition

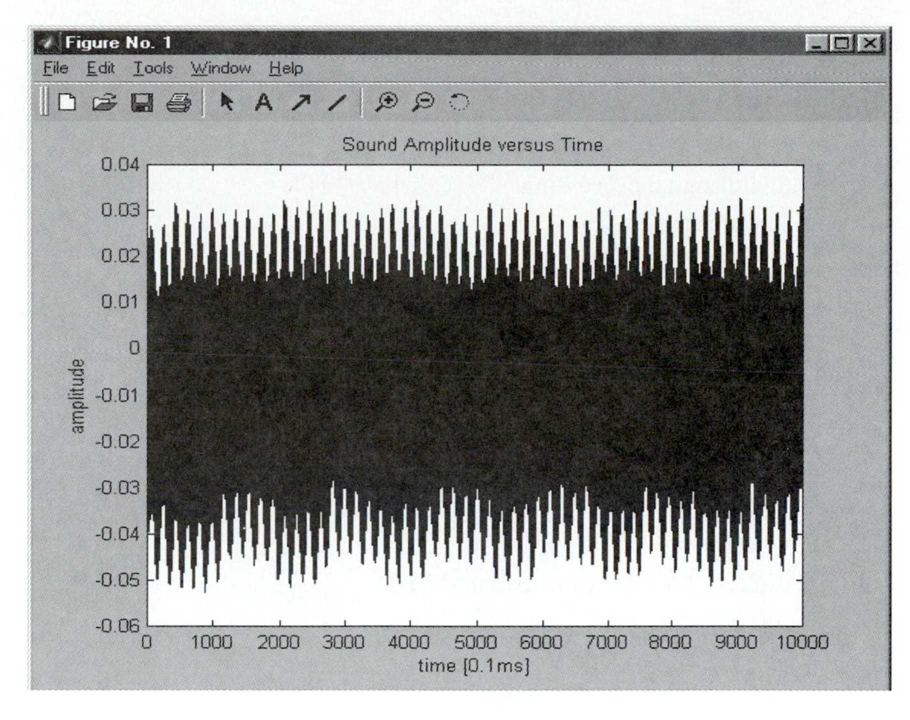

Figure 6.27

Plot of the results of Getsound.m program

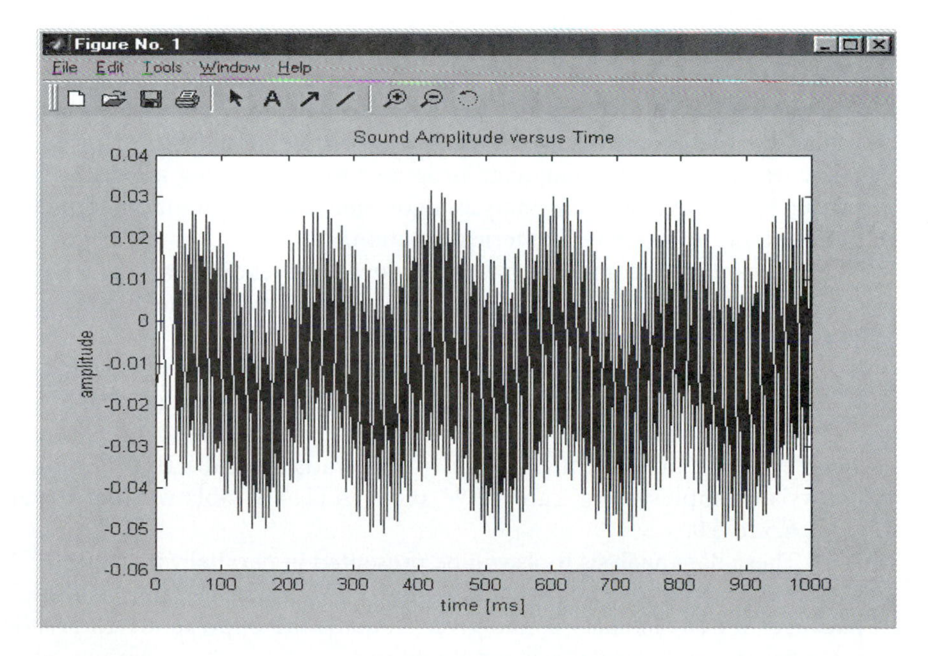

Figure 6.28

Plot of the results of Getsound.m program for the first 1000 samples

6.4 DATA ANALYSIS TOOLS

Data analysis tools in LabVIEW are [86] the following:

signal generation;
digital signal processing;
smoothing windows;
spectrum analysis and control;
filtering;
curve fitting;
linear algebra;
probability and statistics.

Signal-processing tools in MATLAB are [90] as follows:

waveform generation;
discrete Fourier transform and FFT;
filters;
linear system model;
statistical signal processing;
windows;
curve fitting and interpolation;
numerical analysis, etc.

Before the appearance of graphical programming languages, the development of signal analysis applications assumed specialized knowledge, normally available to electrical engineers, for postprocessing data acquired with a system monitoring package. In these cases, writing high-level language programs for monitoring applications required advanced knowledge of signal analysis algorithms, their assumptions, domains of applications and the appropriate numerical solution.

Advances in computer-based engineering, leading to the use of Object Oriented and graphical programming, greatly facilitated the use of signal analysis tools.

The preceding data analysis tools in LabVIEW and signal-processing tools in MATLAB are notable examples of present day easy-to-use signal analyses tools.

Some of these tools were already used in previous sections (e.g., Section 6.2, LabVIEW Mean.vi, from probability and statistics LabVIEW tools). Several data analysis examples using LabVIEW and MATLAB tools will be presented in Sections 6.5 to 6.9.

These data analysis tools will be presented in parallel with LabVIEW and MATLAB programming examples. This approach permits the use of these well-known packages for signal analysis and gives students the opportunity to practice programming using readily available student editions.

6.5 SIGNAL GENERATION

LabVIEW contains several virtual instruments for simulating function generators. These simulated periodic signals can be used to test algorithms before acquiring real signals with a DAQ board. The frequency of these periodic signals can be modified as needed for testing data acquisition codes with simulated signals. Real-time signal generation from a DAQ board is not guaranteed due to lack of control of the interval between successive samples [10, 36]. LabVIEW RT, which will be presented in chapter 7.3, contains a Timer.vi, which can be included in a while loop to synchronize the cycle time, [96]. This permits real-time control the of the interval between successive samples.

Nyquist theorem or sampling theorem imposes a condition to be satisfied in signal generation, which requires that the sampling interval dt [sec] be smaller than half of the analog signal cycle T. For signal generation, this means that the sampling interval dt should be small enough to provide at least two samples per analog signal cycle T of the periodic signal to be constructed, or $dt < T/2$. With two samples per cycle, a sinusoidal wave is generated as a square wave. A more accurate sinusoidal wave generation obviously requires more than two samples per cycle.

In terms of sampling rate r [samples/sec], where $r = 1/dt$, and the highest frequency f [cycles/sec] or [Hz], where $F = 1/T$ of the analog signal, the Nyquist theorem requires that $r > 2F$ or $F_N > F$, where $F_N = r/2$ is called the Nyquist frequency. This is another form of the Nyquist theorem requiring the analog signal maximum frequency F to be lower than the Nyquist frequency $F_N = r/2$.

Another way of verifying if the Nyquist theorem is not violated is based on normalized frequency f [cycles/sample], $f = F/r$.

Nyquist theorem requires that $f < 0.5$ [cycles/sample] or, using the reciprocal, requires that the analog periodic signal is sampled at least 2 [samples/cycle].

LabVIEW wave generation will be illustrated for sine, square, and sawtooth waves.

Sine Wave Generation.vi is shown in Fig. 6.29. The front panel contains digital controls for sine wave amplitude = 5 , frequency = 2.00 [cycles/sec], phase = 45 [degrees], sampling rate = 100.00 [samples/sec], and duration of trigger = 2 [sec]. Also on the front panel are three digital indicators for displaying the results of the block diagram computation. The results displayed are

$$\text{number of samples} = (\text{sampling rate}) \cdot (\text{duration of trigger}) = 200.00\ [\text{samples}],$$

$$\text{normalized frequency} = f = (\text{frequency } F)/(\text{sampling rate}) = 0.02\ [\text{cycles/sample}],$$

$$\text{number of cycles} = 4\ [\text{cycles}].$$

Normalized frequency of 0.02 [cycles/sample] is, obviously, much smaller than the required 0.5 [cycles/sample].

The front panel also contains a Waveform Graph for displaying sine wave over 200 samples. Time spacing is 1/(sampling rate) = 1/100 [sec/sample] such that the x-axis from 0 to 200 [samples] corresponds to a time scale from 0 to 200(1/100) = 2 [sec]. The

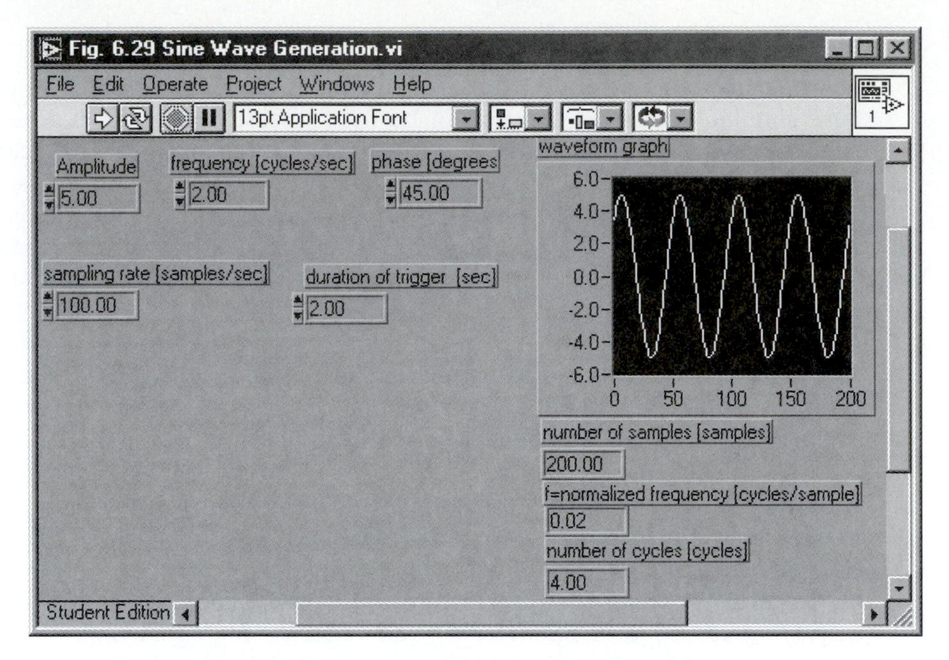

(a)

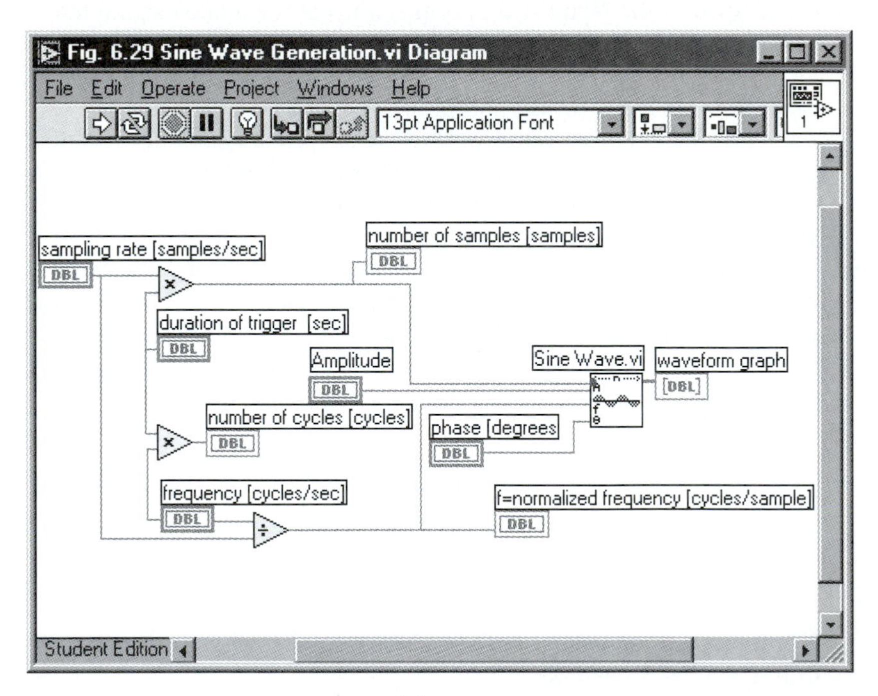

(b)

Figure 6.29

Sine Wave.vi front panel (a) and block diagram (b)

block diagram contains the Sine Wave.vi, which permits us to choose not only sine amplitude and frequency, but also the initial phase.

Square Wave Generation.vi is shown in Fig. 6.30. The front panel contains digital controls for producing a square wave:

$$\text{amplitude} = 5;$$
$$\text{frequency} = 2.00\ [\text{cycles/sec}];$$
$$\text{phase} = 45\ [\text{degrees}];$$
$$\text{sampling rate} = 100.00\ [\text{samples/sec}];$$
$$\text{duration of trigger} = 2\ [\text{sec}].$$

and two digital indicators for displaying the number of samples = 200.00 [samples], normalized frequency = 0.02 [cycles/sample], and the number of cycles = 4.00 [cycles] The front panel also contains also a Waveform Graph for displaying the square wave over 200 samples equivalent to a time scale from 0 to 2 [sec].

Sawtooth Wave Generation.vi is shown in Fig. 6.31. The front panel contains digital controls for producing a square wave—

$$\text{amplitude} = 5;$$
$$\text{frequency} = 2.00\ [\text{cycles/sec}];$$
$$\text{phase} = 0\ [\text{degrees}];$$
$$\text{sampling rate} = 100.00\ [\text{samples/sec}];$$
$$\text{duration of trigger} = 2\ [\text{sec}].$$

and two digital indicators for displaying the number of samples = 200.00 [samples], normalized frequency = 0.02 [cycles/sample], and the number of cycles 4.00 [cycles]. The front panel also contains a Waveform Graph for displaying the sawtooth wave over 200 samples equivalent to a time scale from 0 to 2 [sec].

MATLAB waveform generation will be illustrated for the same waves:

sine wave;
square wave;
sawtooth wave.

Figure 6.32 shows the MATLAB script M-file of the sinus.m program for sine wave generation. The comments following the % symbol explain the MATLAB commands. The plot of the resulting sine wave is shown in Fig. 6.33. The figure shows a Waveform Graph displaying the sine wave over 200 samples. The time spacing is 1/(sampling rate) = 1/100 [sec/sample] such that the time axis from 0 to 200 [samples] corresponds to a time scale from 0 to 2 [sec].

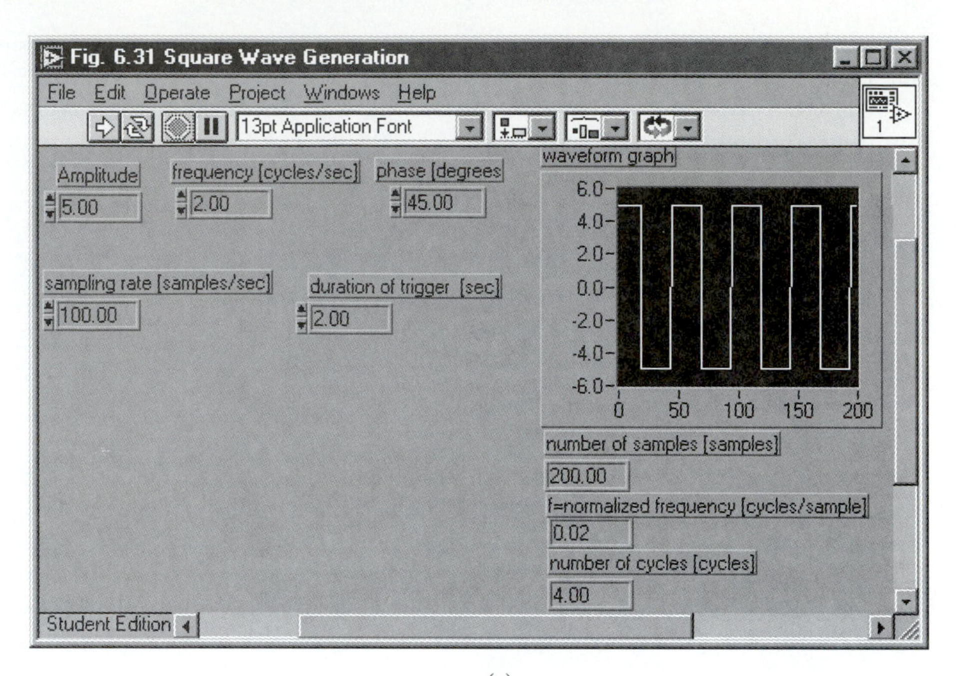

(a)

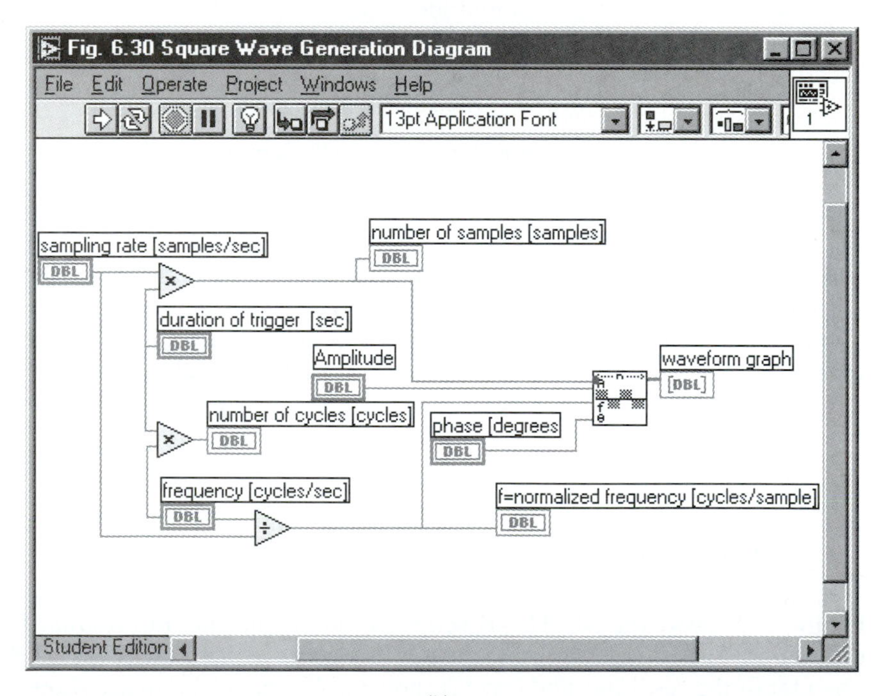

(b)

Figure 6.30

Square Wave.vi front panel (a) and block diagram (b)

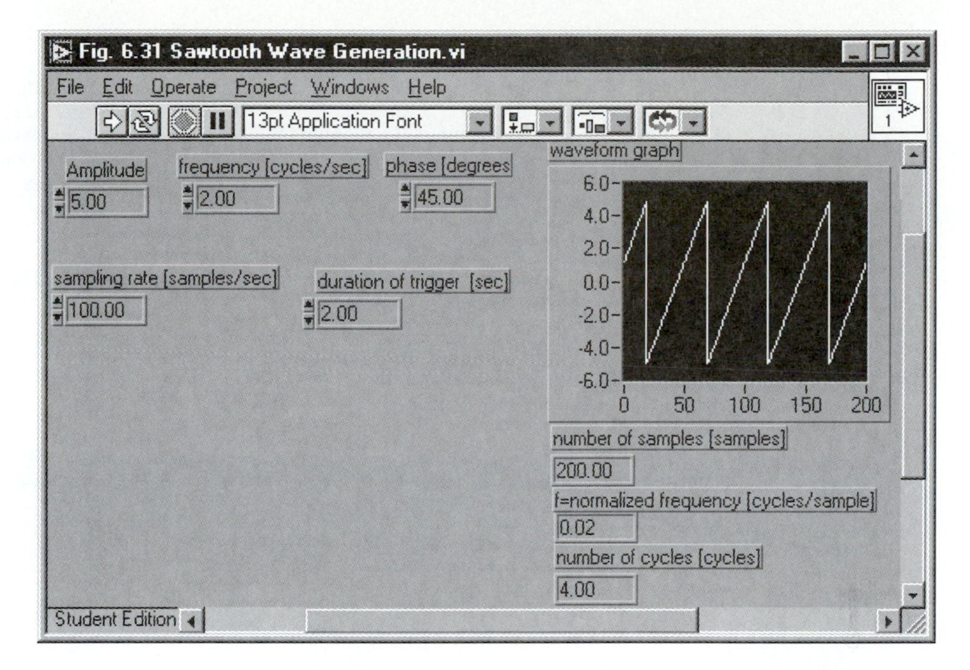

(a)

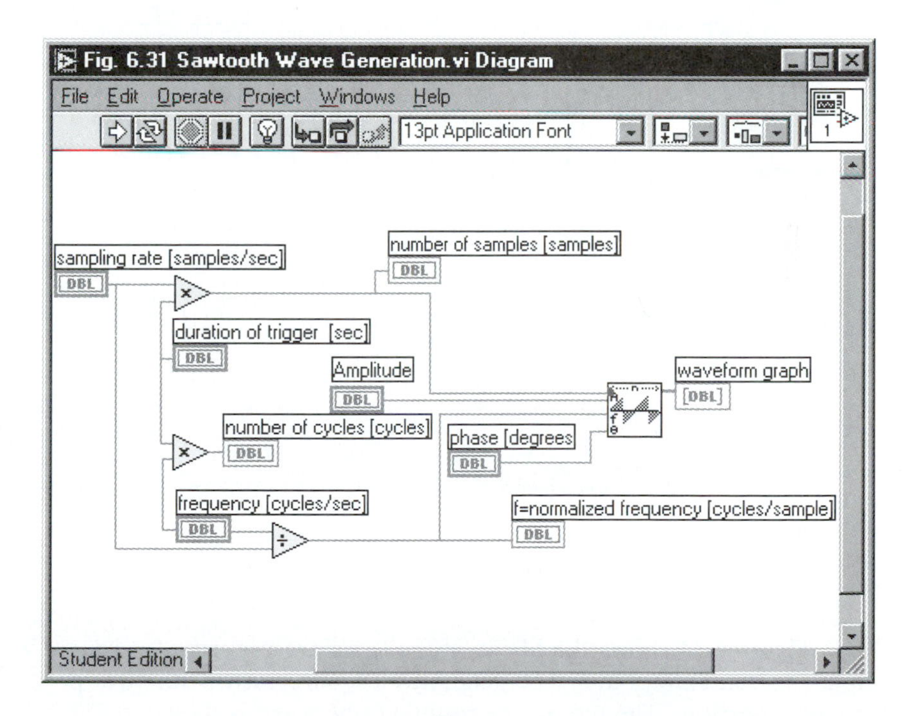

(b)

Figure 6.31

Sawtooth Wave.vi front panel (a) and block diagram (b)

```
% sinus.m (Sine Wave)
phase=45;                    % phase [degrees]
F=2;                         % frequency [cycles/sec]
d=2;                         % duration of trigger [sec]
r=100 ;                      % sampling rate [samples/sec]
n=r*d;                       % number of samples [samples]
t=0:1:n;                     % current time [samples]
phi=pi*phase/180;            % phase [rad]
A=5.0;                       % amplitude
k=4;                         % number of cycles [cycles]
omega = 2*pi*k/n;            % angular velocity [rad/sec]
y=A*sin (omega*t+phi);
plot (y, 'b'); grid;
title ('Sinus Wave');
ylabel ('Amplitude') ; xlabel ('Time');
```

Figure 6.32

MATLAB sinus.m program listing

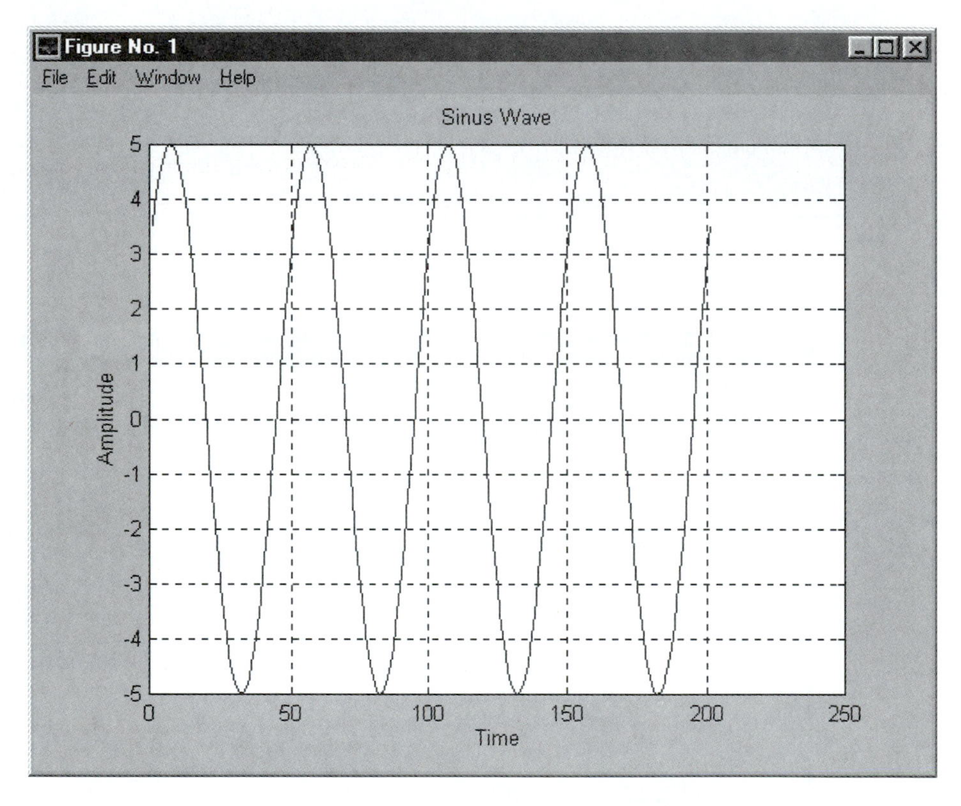

Figure 6.33

Sine wave generated by sinus.m program

Figure 6.34 shows the M-file of the square.m program for square wave generation. The command duty = 50 generates a square wave with 50% positive and 50% negative portions. The plot of the resulting sine wave is shown in Fig. 6.35. The figure shows a Waveform Graph displaying the sine wave over 200 samples, which corresponds to a time scale from 0 to 2 [sec].

```
% square.m
cd c:\matlab\toolbox\signal ;
d=2;                    % duration of trigger [sec]
r=100;                  % sampling rate [samples/sec]
n=r*d;                  % number of samples [samples]
t=0:1:n;                % current time
k=4;                    % number of cycles [cycles]
duty=50;
A=5. 0;                 % amplitude
omega = 2*pi*k/n;       % angular velocity [rad/sec]
y=A*square (omega*t, duty);
plot (t, y, 'r'); grid;
title ('Square Pattern');
xlabel ('Time');
ylabel ('Amplitude')
```

Figure 6.34

MATLAB sinus.m program listing

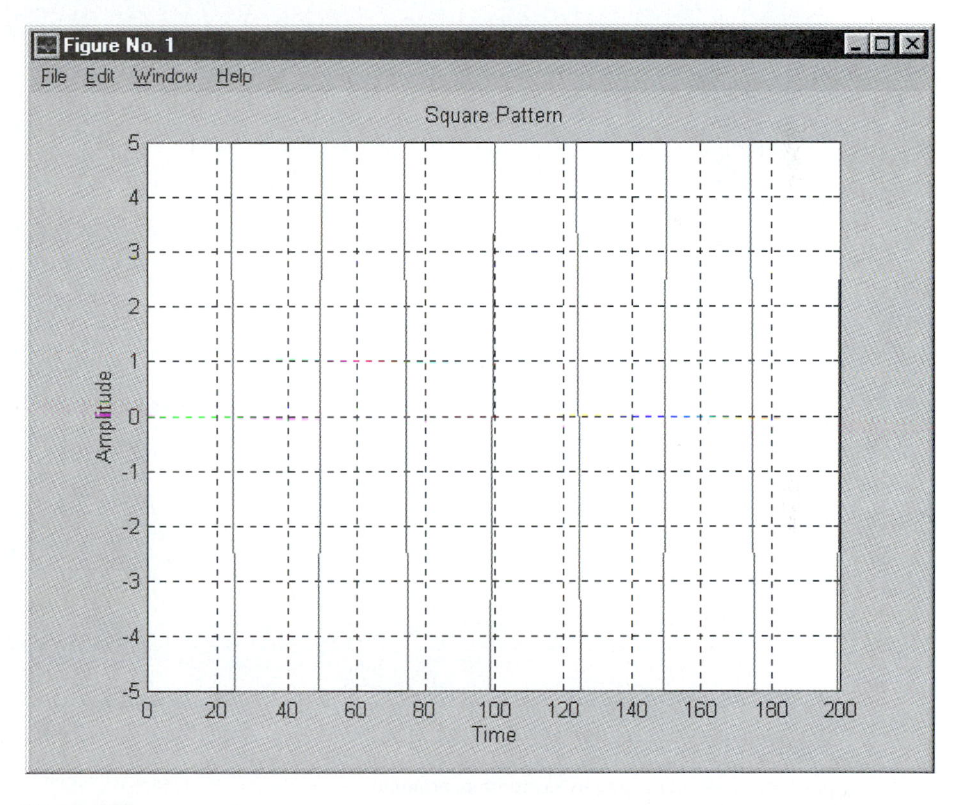

Figure 6.35

Square wave generated by square.m program

Figure 6.36 shows the M-file of the sawtooth.m program for sawtooth wave generation. The plot of the resulting sine wave is shown in Fig. 6.37. The graph shows a waveform graph displaying the sine wave over 200 samples or a time scale from 0 to 2 [sec].

```
%sawtooth.m
cd c:\matlab\toolbox\signal;
d=2;
r=100;
n=r*d;
t=0:1:n;
k=4;
A=5. 0;
omega = 2*pi*k/n;
y=A*sawtooth (omega*t);
plot (t, y, 'r'); grid;
title ('Sawtooth Pattern');
xlabel ('Time');
ylabel ('Amplitude');
```

Figure 6.36

MATLAB sawtooth.m program listing

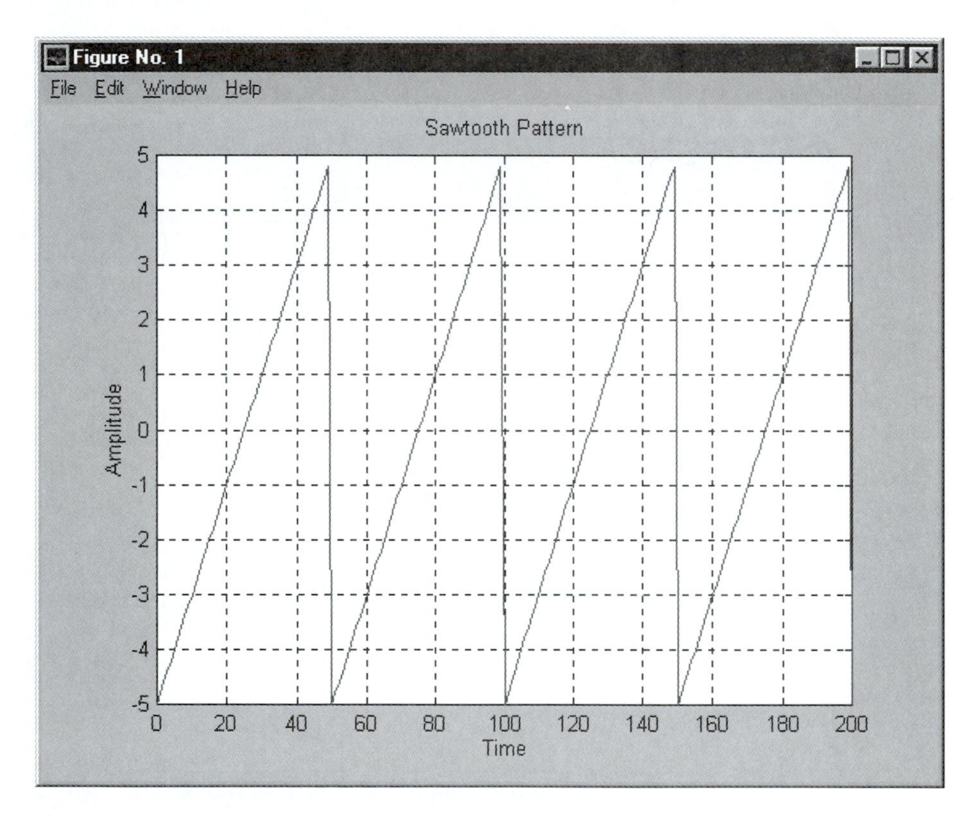

Figure 6.37

Sawtooth wave generated by sawtooth.m program

6.6 DIGITAL SIGNAL PROCESSING FOR THE FOURIER TRANSFORM

The frequency content of a signal $y(t)$ can be calculated from the n samples of the sampled signal x_i (for $i = 0,1,\ldots,n - 1$, where n is the number of samples), using the digital signal processing algorithm for discrete Fourier transform (DFT) [86, 90]. In the time domain of the sampled signal x_i, the time interval between samples, the sampling interval dt [sec/sample], is the inverse of the sampling rate r [samples/sec] or [Hz]:

$$dt = 1/r.$$

The discrete Fourier transform (DFT) of the sampled signal x_i (for $i = 0,1,\ldots,n-1$) is given by

$$X_k = \sum_{i=0}^{n-1} x_i \left[\cos\left(\frac{2\pi k}{n} i\right) - j\sin\left(\frac{2\pi k}{n} i\right)\right],$$

where j is the symbol for the imaginary part of complex numbers and $k = 0,1,2,\ldots,n-1$ (i.e., the frequency domain representation X_k (for $k = 0,1,\ldots,n-1$) has the same number of samples n as the time domain sampled signal x_i (for $i = 0,1,\ldots,n-1$). Similar to the sampling interval dt for the n values of the sampled signal, the n values of the frequency domain representation have a frequency spacing df between any X_k and X_{k+1} values given, in this case, by

$$df = 1/(ndt) = r/n.$$

For a real valued input signal x_i (for $i = 0,1,\ldots,n-1$), this is called a real DFT. The result of the DFT transform is always a complex number. The transformation gives the complex numbers in (real part) + $j \cdot$ (imaginary), which can be converted into polar form as (amplitude) and (phase).

A property of the frequency domain representation X_k (for $k = 0,1,\ldots,n-1$) is that the second half of the values are a mirror image of the first half of the values with regard to a middle point on the frequency scale $k = 0,1,\ldots,n-1$. The middle point on the frequency scale is $(n/2) \cdot df = (n/2) \cdot (r/n) = r/2 = F_N$, the Nyquist frequency. If n is an even number, for example $n = 10$, the middle point of the frequency scale is is $n/2 = 5$ [sample] and the elements X_k for $k = 6$ to 10 are mirror images of the elements X_k for $k = 1$ to 4. When n is an odd number, for example, $n = 9$, the middle point of the frequency scale is is $n/2 = 4.5$ [sample] and the elements X_k for $k = 5$ to 9 are mirror images of the elements X_k for $k = 1$ to 4. In this case, the Nyquist frequency F_N is not an element X_k.

Frequency components beyond the Nyquist frequency F_N are called negative frequencies and, being mirror images, do not contain any new frequency domain information and can be ignored. Frequency components below the Nyquist frequency F_N are called positive frequencies and contain all the frequency domain information.

The DFT requires approximately n^2 operations. In case the number of samples n is a power of 2, in other words,

$$n = 2^m \text{ for } m = 1, 2, 3, 4, \ \ldots.$$

The DFT can be computed by a fast algorithm, called fast Fourier transform (FFT), which requires fewer operations, approximately $n\log_2(n)$ operations [86].

Several examples will illustrate the real FFT of sinus and square waves using LabVIEW realFFT.vi and MATLAB *fft*(y) function.

Shown in Fig. 6.38 is the FFT-Sine.vi for FFT calculation of a sine wave (generated by Sine Wave Generation.vi shown in Fig. 6.29). The block diagram from Fig. 6.29 (b) is completed with RealFFT.vi found in

Functions → Analysis → Digital Signal Processing → RealFFT.vi.

The complex number output of the RealFFT.vi is converted in polar form using

Functions → Numeric → Complex → Complex To Polar. vi.

The front panel from Fig. 6.29 (a) is completed in Fig. 6.38 (a), with a waveform graph labeled realFFT-amplitude that plots the amplitude versus frequency of the RealFFT.vi output. Also, the front panel is completed with three more indicators for displaying the results of the computations, shown in the block diagram from Fig. 6.38 (b):

the number of samples n [samples] given by the product of the sampling rate [samples/sec] and the duration of trigger [sec];

f = normalized frequency [cycle/sample] given by the division of the frequency [cycles/sec] and sampling rate [samples/sec];

the number of cycles [cycles] given by the product of the duration of trigger [sec] and the frequency [cycles/sec].

The front panel shown in Fig. 6.38 (a) displays the same values for the time domain of the sine wave as in Fig. 6.29 (a): frequency $F = 2$ [cycles/sec], sampled at a sampling rate $r = 100$ [samples/sec], and the duration of trigger = 2 [sec].

The frequency axis has $n = 200$ [samples} of X_k (for $k = 0,1,\ldots,199$) with a frequency spacing of $df = r/n = 100/200 = 0.5$ [Hz] (i.e., for a frequency scale from 0 to $n \cdot df = 200 \cdot 0.5 = 100$ [Hz]). The amplitude X_0 corresponds to the frequency 0.5 [Hz] and the amplitude X_{199} corresponds to the frequency 100 [Hz].

The Nyquist frequency is F_N is $n/2 = 200/2 = 100$ or $r/2 = 100/2 = 50$ [Hz]. Given that the first value is $k = 0$, Nyquist frequency corresponds to $k = 99$ [samples], the midpoint on the frequency axis of the realFFT-amplitude graph. In the positive frequencies range of the of the real FFT amplitude graph, the frequency $F = 2$ [Hz] or 3 [samples], can be seen.

The front panel shown in Fig. 6.39 corresponds to the same data used for Fig . 6.38, but for a duration of trigger of 2.1 [sec]. As a result, the number of cycles in the trigger is 4.2, a noninteger number, which explains smearing in the real FFT-amplitude diagram. In the negative frequencies range, the frequency of $199 - 3 = 196$ [sample] or $50 - 2 = 48$ [Hz] can be seen.

The normalized frequency value of 0.02 [cycles/sec] is much lower than the Nyquist normalized frequency of 0.5 [cycles/sec]. The amplitude of X_3 in the realFFT-amplitude graph is 500 and requires rescaling by dividing by the number of points in the input signal array [87].

In Fig. 6.38, the sine wave had an integer number of 4 cycles. In case the number of cycles of the sine wave undergoing FFT is not an integer number, spectral leakage occurs [86]. This results in FFT amplitudes that are spread on a wide range about the frequencies of the input analog signal (i.e., in erroneous or fictitious frequencies).

In Fig. 6.39 is shown such a case of an input sine wave with all values identical to Fig. 6.38, but for a duration of trigger of 2.1 [sec]. This value of the duration of trigger

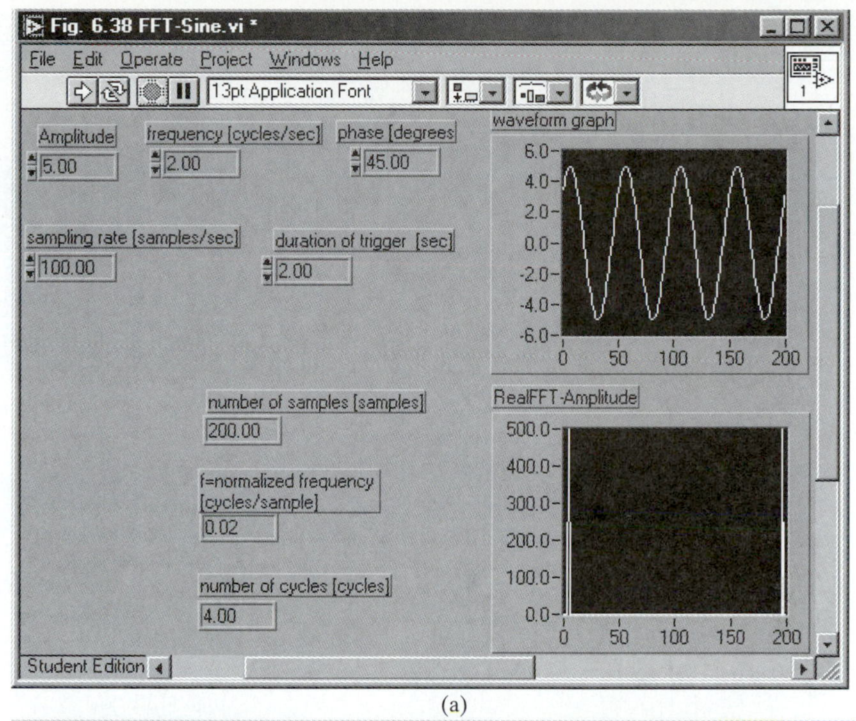

(a)

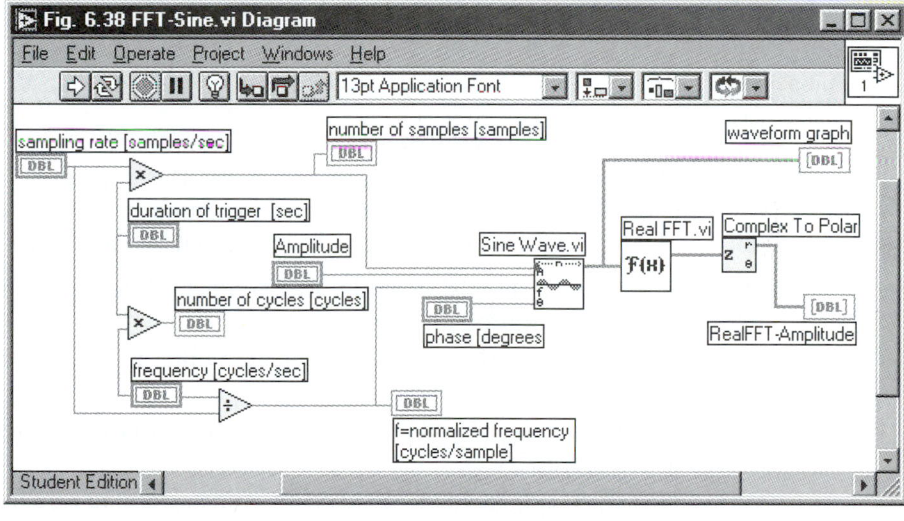

(b)

Figure 6.38

Front panel (a) and block diagram (b) of FFT -Sine.vi for FFT for 2 [sec]

gives a number of cycles of 4.2 [cycles] (i.e., a noninteger number of cycles). It is obvious from the realFFT-amplitude results that spectral leakage occurred, even if the sine wave frequency did not change. The remedy for avoiding spectral leakage is signal windowing, an approach presented in Section 6.4.4.

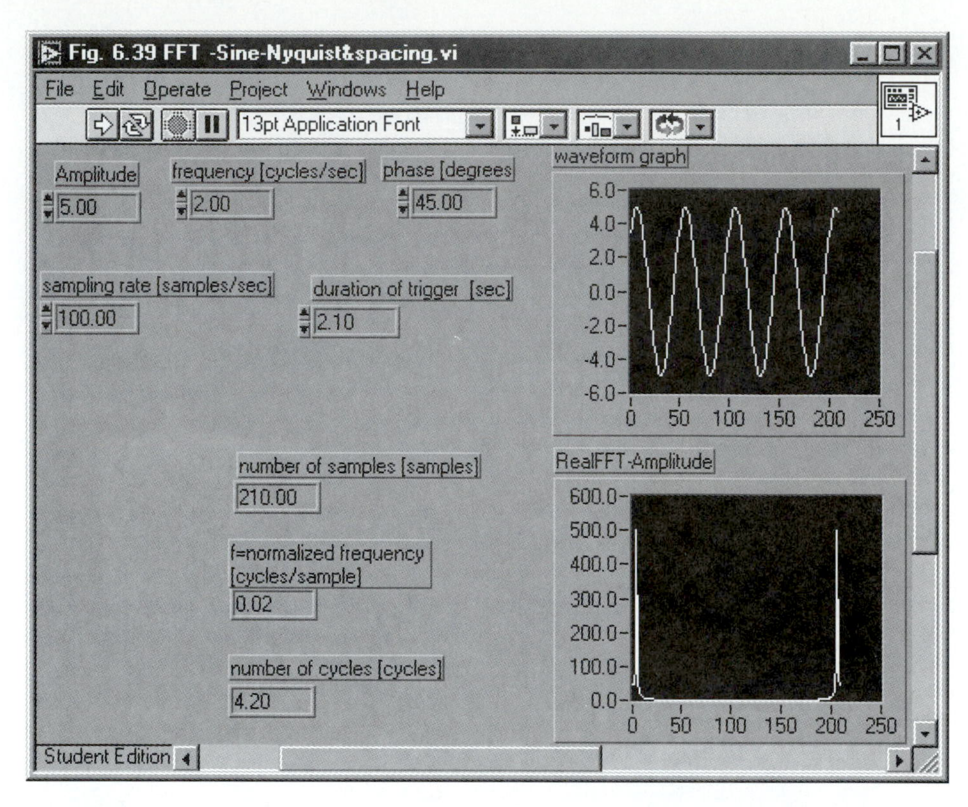

Figure 6.39

Front panel of FFT-Sine.vi for FFT for 2.1 [sec]

For a square wave, The front panel from Fig. 6.30 (a) is completed with a waveform graph labeled RealFFT-Amplitude and with five more indicators for displaying the results of the computations in the Block Diagram from Fig, 6.40 (b):

number of samples [samples] given by the product of the sampling rate [samples/sec] and the duration of trigger [sec];

f = normalized frequency [cycle/sample] given by the division of the frequency [cycles/sec] and sampling rate [samples/sec];

number of cycles [cycles] given by the product of the duration of trigger [sec] and the frequency [cycles/sec];

Nyquist frequency [Hz] given by the division of the sampling rate by 2; and

frequency spacing [Hz] given by the division of the sampling rate [samples/sec] and number of samples [samples].

The front panel shown in Fig. 6.40 (a) shows the same values for the time domain of the sine wave as in Fig. 6.30 (a), frequency $F = 2$ [cycles/sec], sampled at a sampling rate $r = 100$ [samples/sec] for a duration of trigger = 2 [sec]).

The Frequency axis has $n = 200$ [samples} of X_k (for $k = 0,1,\ldots,199$) with a frequency spacing of $df = 0.5$ [Hz] (i.e., for a frequency scale from 0 to 100 [Hz]).

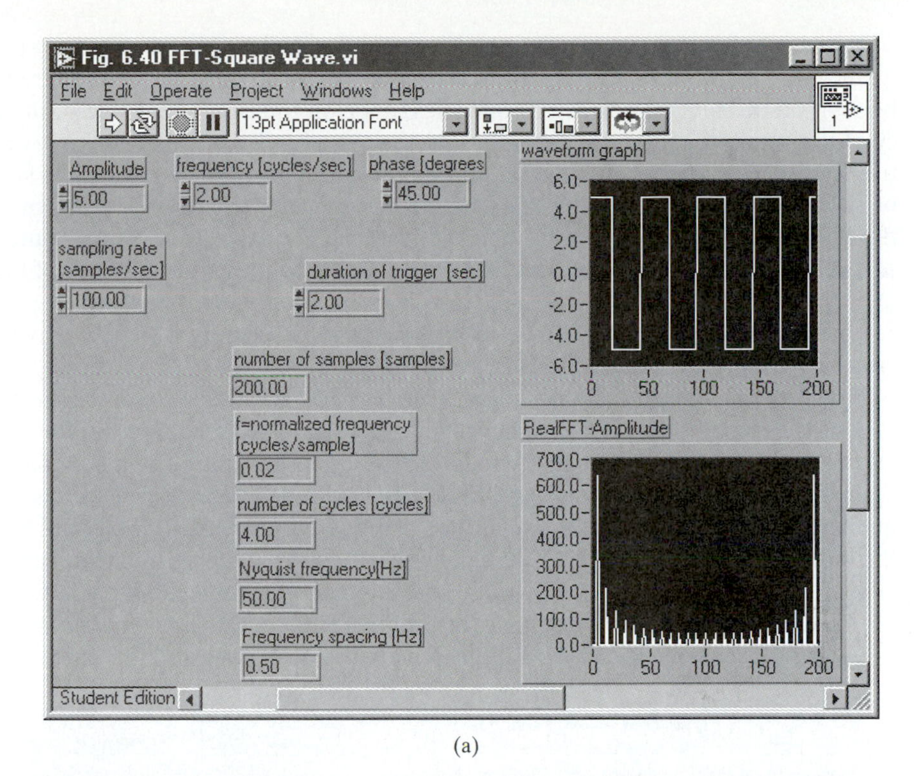

(a)

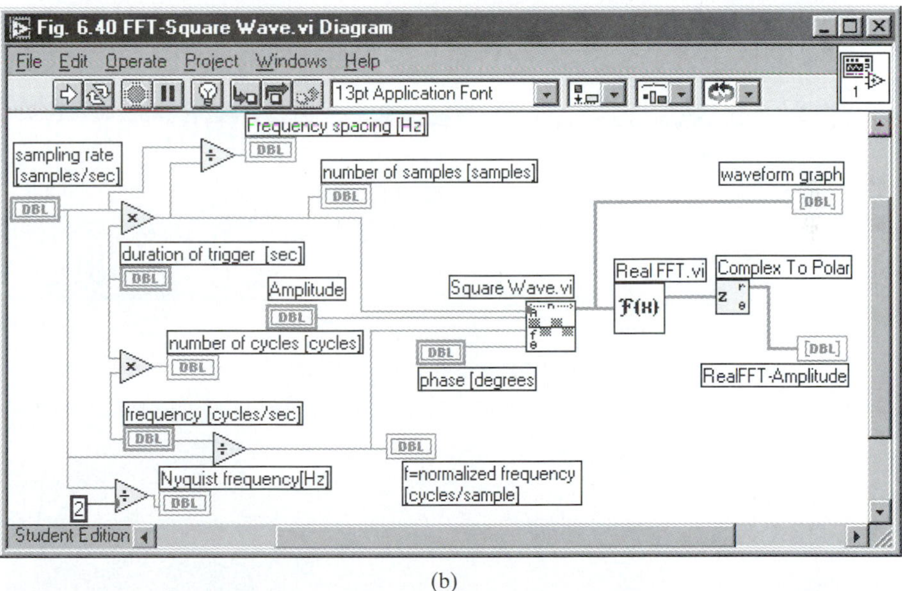

(b)

Figure 6.40

Front panel (a) and block diagram (b) of FFT-Square Wave.vi

Nyquist frequency is F_N is 50[Hz] or 100 [samples] and is the midpoint on the frequency axis of the RealFFT-Amplitude graph. The frequencies of the square wave can be seen on the several positive frequencies range on the right hand side of the frequency of the wave of 2 [Hz] or 3 [samples]. In the negative frequencies range several negative frequencies on the left hand of the $199 - 3 = 196$ [sample] or $100 - 2 = 98$ [Hz] can be seen. It can be observed that they are mirror images of the positive frequencies about the Nyquist frequency of 100 [samples] or 50[Hz]. The larger frequency content of the square wave, compared to the sine wave case, is due to the larger number of higher frequencies required to recover square wave from frequency components.

Figure 6.41 illustrates a case of digital signal processing of a sine wave of a frequency of $F = 110$ [Hz] sampled at a sampling rate of 100 [Hz]. This is a case of violation of Nyquist theorem, given that sampling rate $r = 100$ is lower than the highest frequency F of the signal, rather than higher than 2 F. Normalized frequency in this case is $f =$ 1.1 [cycles/sample], higher than the required 0.5 [cycles/sample]. The result is an alias frequency of $110 - 100 = 10$ [Hz] or 20 [samples], with its mirror image as a negative frequency at $199 - 20 = 179$ [samples]. The positive frequency of 10 [Hz] does not exist in the input sine wave of 100 [Hz] (i.e., an alias frequency). Alias frequencies can be avoided by respecting the requirement of Nyquist theorem. An analog low-pass filter with cutoff frequency equal to Nyquist frequency of 50 [Hz] would filter out the input analog signal with frequency of 110 [Hz] and eliminate the occurrence of alias frequencies.

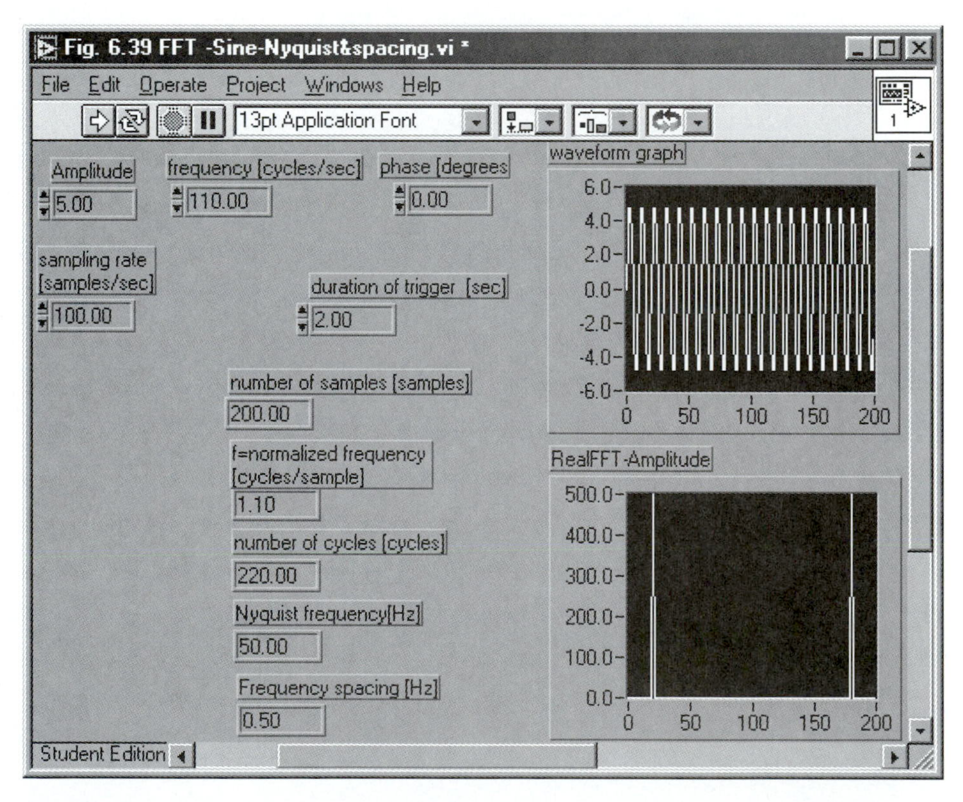

Figure 6.41

Front panel of FFT-sine wave.vi for $F = 110$ Hz and $r = 100$ Hz

The following section presents MATLAB signal processing programs for FFT using *fft(y)* function. Figure 6.42 shows the listing of the FFT-sinus.m program for fast Fourier transform of a sine wave. This program contains a sine wave generation part already described in Fig. 6.32. The input *y* to the *fft(y)* function, consists of real numbers from sampled sinus function. The number of samples in time domain is $n = r \cdot d = 100 \cdot 2 = 200$ [samples]. FFT output *yy* contains complex numbers X_k for $k = 0$ to $n - 1$ or 0 to 199 (i.e., 200 complex numbers representing sampled FFT). The amplitude *m* of the complex number FFT output *yy* is calculated by the function *abs(yy)*. Figure 6.43 shows the plot sine wave versus Time in [samples], with sampling interval $dt = 1/r = 0.01$ [sec/sample] and the plot RealFFT-sinus versus Frequency in [samples], with fre-

```
%FFT_sinus.m (Sine Wave)
phase=45;                 % phase [degrees]
d=2;                      % duration of trigger [sec]
r=100;                    % sampling rate [samples/sec]
n=r*d;                    % number of samples [samples]
t=0:1:n;                  % current time [samples]
phi=pi*phase/180;         % phase [rad]
A=5.0;                    % amplitude
k=4;                      % number of cycles [cycles]
F=k/d;                    % frequency [cycles/sec]
omega = 2*pi*k/n;         % angular velocity [rad/sec]
y=A*sin (omega*t+phi);
subplot (211) ; plot (y, 'b'); grid; hold on;
title ('Sinus Wave');
xlabel ('Time');
ylabel ('Amplitude');
yy=fft (y);
m=abs (yy);
f=(0:length (yy)-1);
subplot (212); plot (f, m); grid;
title ('Real FTT-Sinus');
ylabel ('Amplitude'); xlabel ('frequency')
```

Figure 6.42

FFT-sinus.m MATLAB program

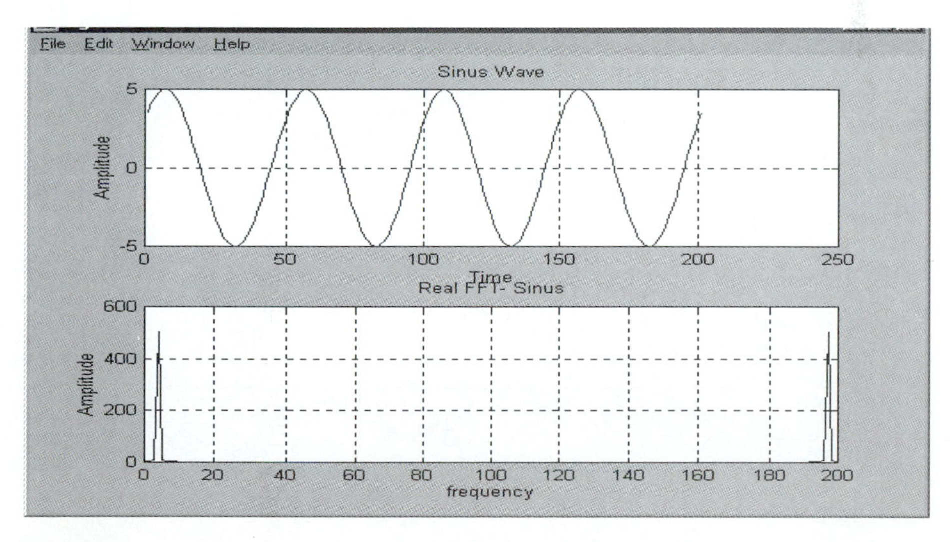

Figure 6.43

Plots of FFT-sinus.m MATLAB program

```
%FFT_square.m
cd c:\matlab\toolbox\signal;
d=2;                          % duration of trigger [sec]
r=100;                        % sampling rate [samples/sec]
n=r*d;                        % number of samples [samples]
t=0:1:n;                      % current time
k=4;                          % number of cycles [cycles]
duty=50;
A=5.0;                        % amplitude
omega = 2*pi*k/n;             % angular velocity [rad/sec]
y=A*square (omega*t, duty);
subplot (211); plot (t, y, 'b'); grid; hold on;
title ('Square Pattern');
xlabel ('Time');
ylabel ('Amplitude');
yy=fft (y);
m=abs (yy);
f=(0:length (yy)-1);
subplot (212) ; plot (f, m); grid;
title ('Real FTT-Square Pattern');
ylabel ('Amplitude'); xlabel ('frequency');
```

Figure 6.44

FFT_square.m program

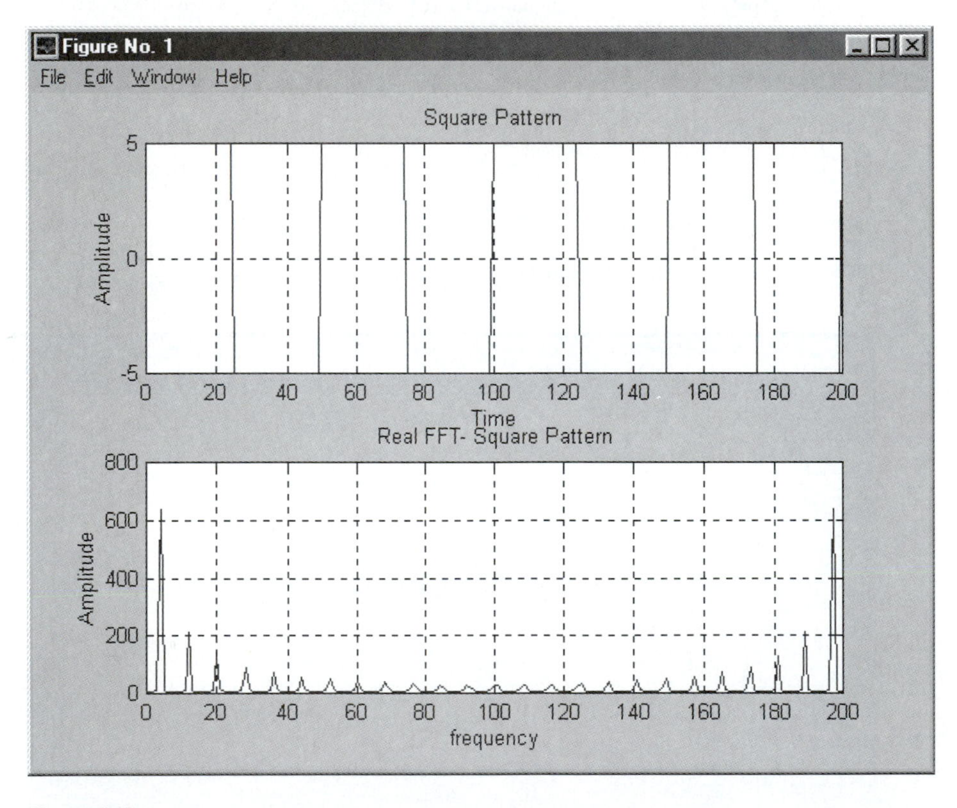

Figure 6.45

Square pattern amplitude vs. time and RealFFT-sinus amplitude vs. frequency

quency spacing $dF = r/n = 0.5$ [Hz/sample]. The frequency axis f is defined from 0 to (length(yy) − 1) = 199 [samples](i.e., more than 200 samples).

As expected, the FFT results in Fig. 6.43 contain the same positive and negative frequencies as in Fig. 6.38 (a).

Figure 6.44 shows the listing of the FFT_square.m program for Fast Fourier transform of a square pattern. This program contains a square pattern generation part already described in Fig. 6.34. The input y to the *fft*(y) function, consists of 200 [samples] of the square pattern. FFT output yy contains of 200 complex numbers representing sampled FFT with the amplitude m calculated by the function *abs*(yy). Figure 6.45 shows the plot square pattern versus time in [samples] and the plot RealFFT-sinus versus Frequency over 200 samples with 0.5 [Hz/sample], identical to the results shown in Fig. 6.40(a).

6.7 SIGNAL SPECTRUM

The results of FFT from LabVIEW RealFFT.vi and MATLAB *fft*(y) are obtained in amplitude units, which have to be rescaled by dividing the amplitude values by the number of points in the signal array and in frequency units that have to be rescaled by multiplying with frequency spacing [86, 87, 90]. Moreover, frequency components beyond Nyquist frequency $F_N = r/2$ do not contain any new information regarding FFT and can be discarded. Spectrum analysis VIs achieve this scaling and discarding automatically, as shown in this section [86].

Figure 6.46 shows the front panel (a) and block diagram (b) of the Sine-Ampl&PhaseSpectrum.vi. This VI is obtained from the FFT-Sine.vi (Fig. 6.38) as follows:

replace RealFFT.vi by Amplitude and Phase Spectrum.vi by

clicking on RealFFT.vi → select Replace from the pop-up menu →
Analysis → Measurement → Amplitude and Phase. vi;

delete Complex to Polar.vi and RealFFT-Amplitude waveform graph;

calculate dt = sampling cycle [sec/sample] as reciprocal of sampling rate [samples/sec],

Functions → Numeric → Reciprocal;

and wire to sampling rate and bottom input to Amplitude and Phase.vi;

create two Bundle.vi from

Functions → Cluster → Bundle,

and click with right-hand side button and select add input from the pop-up menu;

create a 0 constant,

Functions → Numeric Constant;

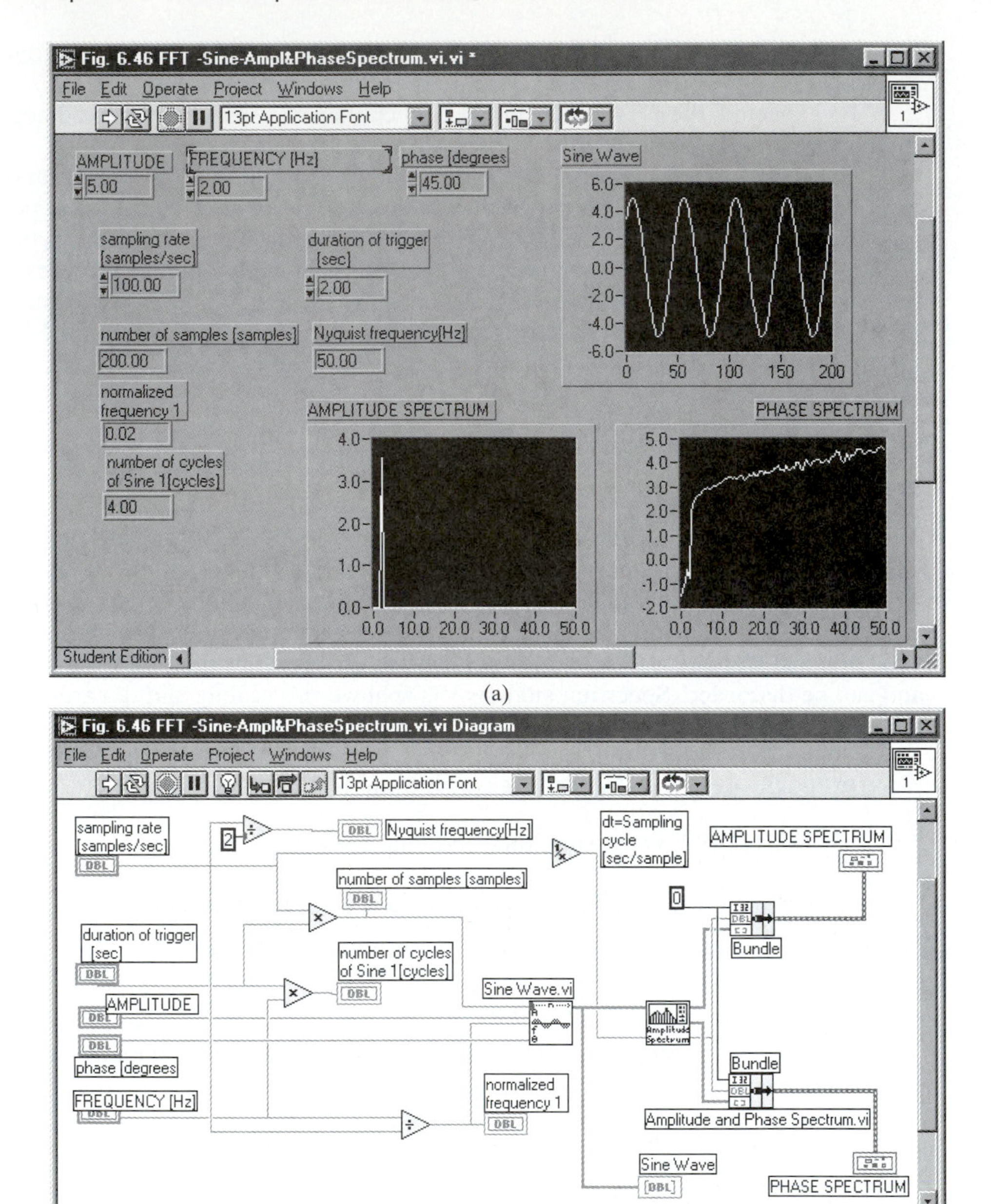

(a)

(b)

Figure 6.46

Front panel (a) and block diagram (b) of the Sine-Ampl&PhaseSpectrum.vi

and wire to the top inputs of Bundle.vi;

wire top output of amplitude and Phase Spectrum.vi to the bottom input of the top Bundle.vi;

wire middle output of Amplitude and Phase Spectrum.vi to the bottom input of the bottom Bundle.vi;

wire bottom output of Amplitude and Phase Spectrum.vi to the middle inputs of the two Bundle.vi;

delete realFFT-amplitude waveform chart;

create two waveform graphs, label them AMPLITUDE SPECTRUM and PHASE SPECTRUM, and wire them to the appropriate Bundle.vi.

For the same values as in Fig. 6.38, the frequency spectrum results are the same, but are represented in proper units for amplitude and for frequency [Hz]. The graph limited the frequency axis from 0 to the Nyqyist frequency $F_N = 50$ [Hz]. In Fig. 6.46, phase spectrum of the sine wave is also shown.

A sinusoidal signal of frequency = 90 [Hz], sampled at a sampling rate = 100 [samples/sec], has the amplitude and phase spectrum shown in Fig. 6.47. Given that the normalized frequency $f = 0.9$ [cycle/sample] is higher than 0.5, aliasing is expected.

The alias frequency is given by absolute (closest integer multiple of sampling rate−input signal frequency) and in this case is 100 − 90 = 10 [Hz]. Since signal fre-

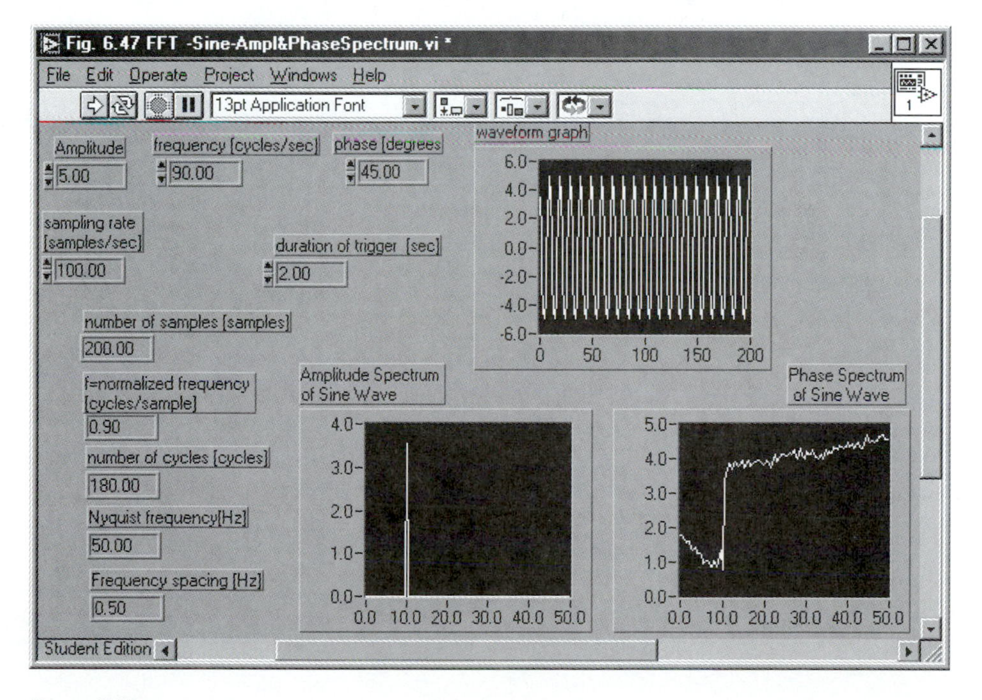

Figure 6.47

Sine-Ampl&PhaseSpectrum.vi for an aliasing case

quency of 90 [Hz] is higher than the Nyquist frequency of 50 [Hz], on the amplitude spectrum graph only the alias frequency of 10 Hz is displayed [86].

The frequency spectrum of the sum of two sinusoidal signals is shown in Fig. 6.48. The frequencies of the two summed-up sinusoidal signals, 2 [Hz] and 40 [Hz] are visible in the amplitude spectrum graph.

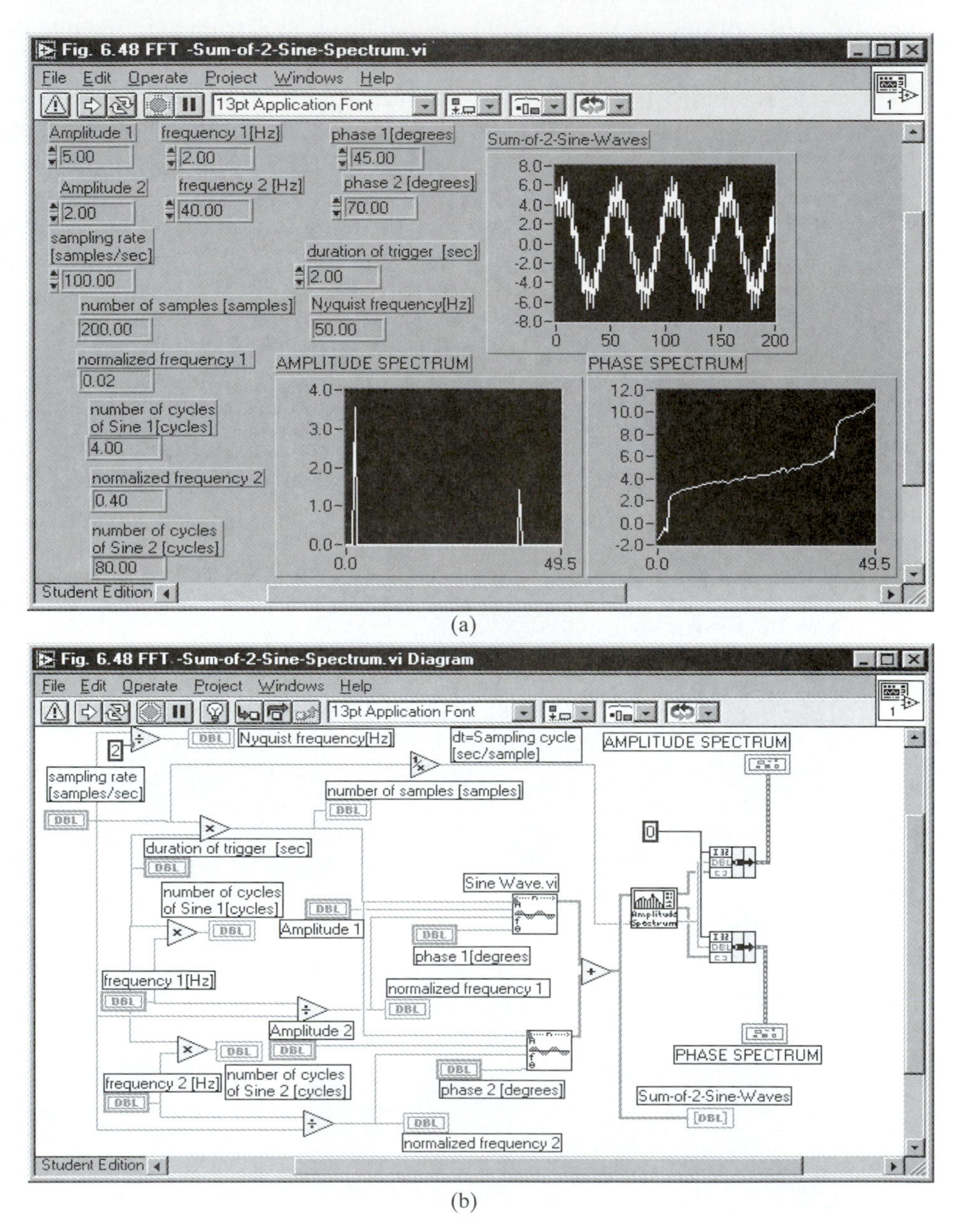

Figure 6.48

Front panel (a) and block diagram (b) of Sum-of-2-Sine-Spectrum.vi

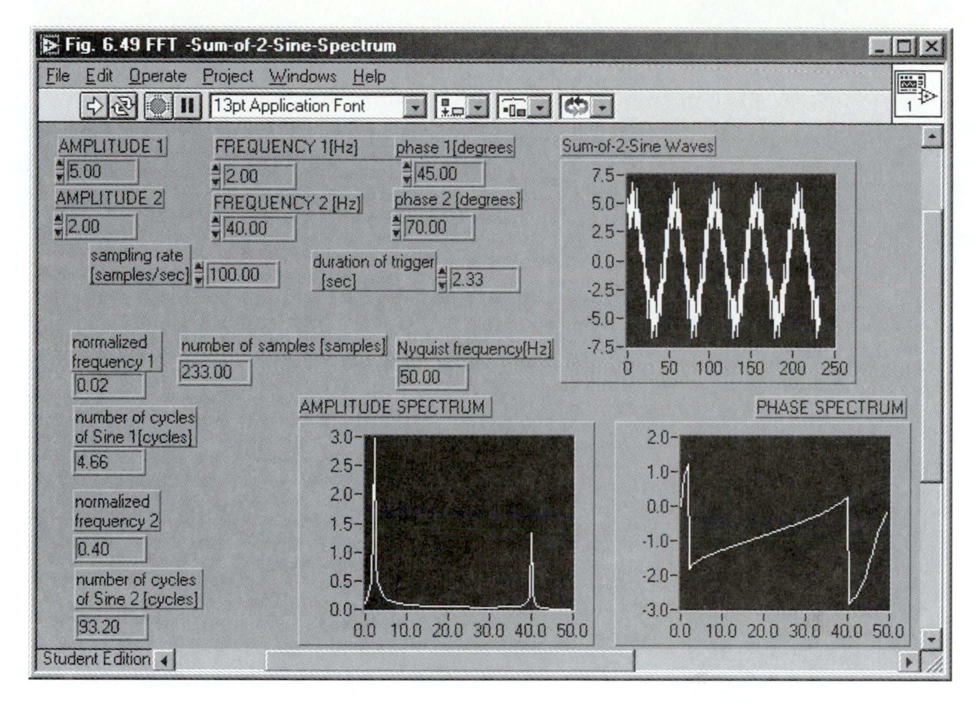

Figure 6.49

Sum-of-2-Sine-Spectrum.vi for a duration of trigger of 2.33 [sec].

Shown in Fig. 6.49 is the same case of the two summed-up sinusoidal signals, 2 [Hz] and 40 [Hz], but for a duration of trigger of 2.33 [sec]. This value of the duration of trigger gives numbers of cycles of 4.66 [cycles] of Sine 1 and 93.2 [cycles] for Sine 2 (i.e., a noninteger numbers of cycles). The amplitude spectrum graph shows spectral leakage about both input frequencies, even if the sine wave frequency did not change from the case shown in Fig. 6.48. The remedy for avoiding spectral leakage is signal windowing, an approach presented in the next section.

6.8 SMOOTHING WINDOWS

Spectral leakage, illustrated in Figs. 6.39 and 6.49, is caused by a noninteger number of cycles of periodic signals included in the duration of trigger. Spectral analysis of such signals results in a wider spectrum when compared to the signals containing an integer number of cycles in the duration of trigger. Spectral leakage distorts the frequency spectrum of the signal and has to be removed. In practical applications, it is unavoidable to have some frequency components represented by a noninteger number of cycles during the trigger. The example of Fig. 6.48 shows that for a duration of trigger to 2.00 sec. Both sine waves have an integer number of cycles retained in the trigger, but changing frequency 2 to, for example, 41.73 Hz, would not permit us to find a duration of trigger for which both frequencies would be retained in an integer number of samples.

Smoothing windows achieve the removal of spectral leakage by reducing the participation of the signal towards the tails of the duration of the trigger and thus reducing the effect of noninteger number of cycles in the spectral analysis. Smoothing windows can be seen as the signal value weighting in time with lower weights towards the tails of the window. For example, when using a generic window (e.g., a Hanning window), the sampled signal x_i for $i = 0,1,2,\ldots,n - 1$ is multiplied by a function $1 - \cos(2\pi i/n)$:

$$y_i = 0.5x_i\,[1 - \cos(2\pi i/n)] \qquad \text{for} \qquad i = 0, 1, 2, \ldots, n - 1.$$

As $1 - \cos(2\pi i/n)$ tends toward 0, for $i \to 0$ and $i \to n - 1$, y_i also will tend toward 0 as i takes values toward the two ends of the trigger. The result is a correction of the frequency content obtained by spectral analysis on the expense of modifying the amplitudes in the amplitude spectrum.

Figure 6.50 shows comparatively the results of amplitude spectrum of the original signal and of the windowed signal.

The FFT-Sine Spectrum Windowed.vi shown in Fig. 6.50 is obtained from the FFT-Sine Spectrum.vi from Fig. 6.46, as follows:

delete from the block diagram the Bundle.vi leading to phase spectrum of sine wave terminal;

delete phase spectrum of sine wave waveform graph from the front panel;

create three new waveform graphs in the front panel

Controls → Graph → Waveform Graph

and label them Windowed Sine Wave, Windowed Sine Wave, and Amplitude Spectrum of Windowed Signal;

create a Hanning window

Functions → Analysis → Measurement → Hanning Window.vi;

create an amplitude and phase data analysis virtual instrument

Analysis → Measurement → Amplitude and Phase vi;

create a Bundle.vi from

Functions → Cluster → Bundle;

click with the right-hand button, and select add input from the pop-up menu;

wire the new objects of the block diagram as shown in Fig. 6.50(b).

The results are shown in the front panel of Fig. 6.50(a) for the same input data as in Fig. 6.46, except for a duration of trigger of 2.33 sec, that results in a noninteger number of cycles = 4.66 (i.e., in the spectral leakage shown in the amplitude spectrum of original

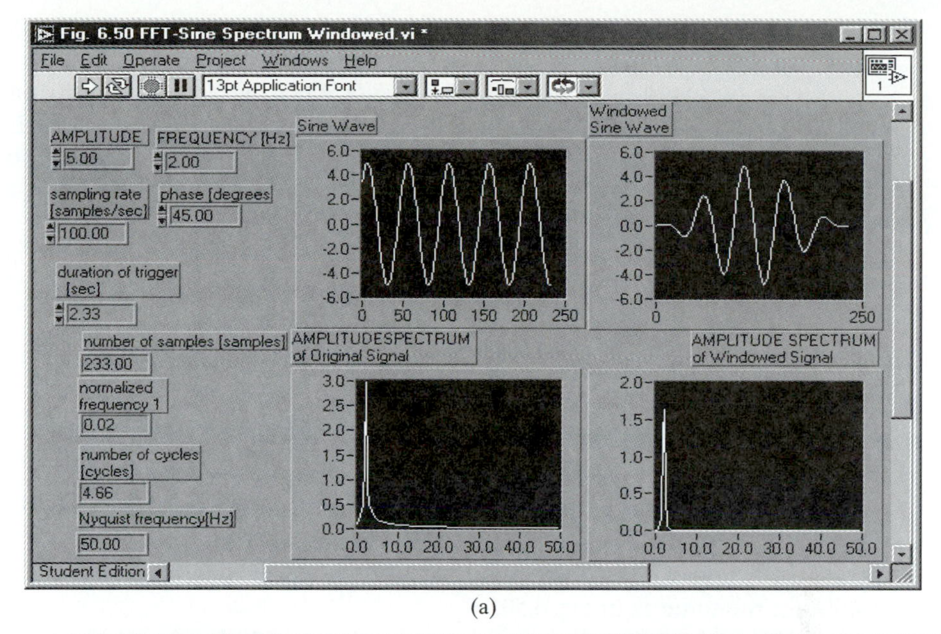

(a)

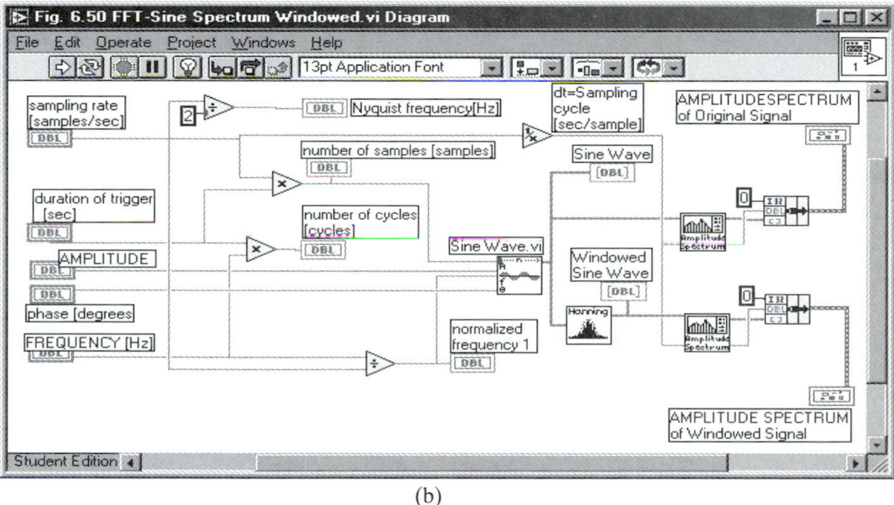

(b)

Figure 6.50

Amplitude spectrum of the original signal and of the windowed signal

signal). In time domain, Hanning window modifies the sine wave into a windowed sine wave with vanishing tails. The amplitude spectrum of the windowed signal has the spectral leakage removed and, as a result, the amplitude spectrum of the windowed signal has a frequency content similar to the frequency content shown in Fig. 6.46(b), which had an integer number of cycles in the trigger.

Spectral leakage presence and removal can also be illustrated in MATLAB applications.

Figure 6.51 shows the listing of the MATLAB program sinusSpectrum.m, which contains the same sinus signal generation from Fig. 6.42 for $k = 4$ [cycles]. In order to obtain graphs similar to the rescaled graphs produced by Amplitude and Phase.vi shown in Fig. 6.46, frequency spacing f, frequency unit rescaled $f1$, and FFT amplitude rescaled $m1$ are calculated [87, 90].

Nyquist frequency is equal to $r/2 = 50$ Hz and the results shown in Fig. 6.52 beyond this frequency correspond to negative frequencies and can be ignored. As expected, these results resemble those shown in Fig. 6.46 (a).

Sampling spacing $dt = 1/r$ in [sec/sample]is used for rescaling the x axis of the sine wave graph.

The results from Fig. 6.52 are obtained for an integer number of 4 cycles of the sinus wave during the trigger. In order to illustrate spectral leakage, program sinusSpectrum.m is modified into sinusSpectrum1.m by changing the duration of trigger to 2.33 [sec] , which results in a noninteger number of cycles 4.66 [cycles] during a trigger. The results of the program sinusSpectrum1.m, presented in Fig. 6.53, are plotted in Fig. 6.54. It can be observed that in this case, spectral leakage occurs and that the results are the same as in Fig. 6.50.

A Hanning window can be generated with the MATLAB program HanningWindow.m listed in Fig. 6.55 and shown in Fig. 6.56. The plot shows that in the middle of the window the amplitude is close to 1, while the ends of the tails reduce the input values to 0.

Figure 6.51

MATLAB program sinusSpectrum.m

```
%sinusSpectrum.m
phase=45;                     % phase [degrees]
d=2;                          % duration of trigger [sec]
r=100;                        % sampling rate [samples/sec]
n=r*d;                        % number of samples [samples]
dt=1/r;                       % sampling spacing in [sec/sample]
t=0:dt:d;                     % current time [samples]
phi=pi*phase/180;             % phase [rad]
A=5. 0;                       % amplitude
k=4;                          % number of cycles [cycles]
F=k/d;                        % frequency [cycles/sec]
omega = 2*pi*F;               % angular velocity [rad/sec]
y=A*sin (omega*t+phi);
subplot (211); plot (t, y); grid; hold on;
title ('Sinus Wave');
ylabel ('Amplitude');         xlabel ('Time [sec]') ;
yy=fft (y);                   % FFT of sinus wave
m=abs (yy);                   % Amplitude of FFT
f=(0:length (yy)-1);
df=r/n;                       % frequency spacing [Hz]
f1=f*df;                      % rescaling of frequency unit from sample to Hz
m1=m/n;                       % rescaling of FFT amplitude output unit
subplot (212) ; plot (f1, m1) ; grid;
title ('Real FFT-Sinus');
ylabel ('Amplitude'); xlabel ('frequency [Hz]');
```

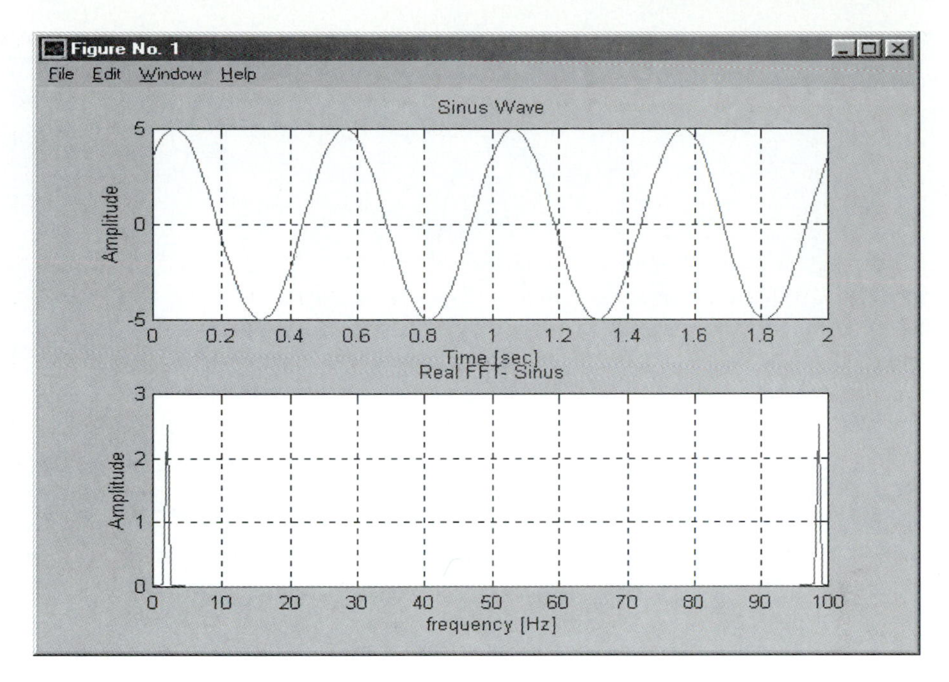

Figure 6.52

Results of MATLAB program sinusSpectrum.m

Figure 6.53

MATLAB program sinusSpectrum1.m

```
%sinusSpectruml.m
phase=45;                % phase [degrees]
d=2.33;                  % duration of trigger [sec]
r=100;                   % sampling rate [samples / sec]
n= r*d;                  % number of samples [samples)
dt=1/r;                  % sampling spacing in [sec / sample]
t=0 : dt: d;             % current time [sec]
phi=pi*phase/180;        % phase [rad]
A=5.0;                   % amplitude
F=2;                     % frequency [cycles / sec]
k=F*d;                   % number of cycles [cycles]
omega = 2*pi*F;          % angular velocity [rad / sec]
y=A*sin (omega*t+phi);
subplot (2,1,1); plot (t,y);  grid; hold on;
title ('Sinus Wave');
ylabel ('Amplitude'); xlabel ('Time [sec]');
yy=fft(y);               % FFT of sinus wave
m=abs (yy);              % amplitude of FFT
f=(0 : length(yy)-1);
df=r/n;                  % frequency spacing [Hz]
f1=f*df;                 % rescaling of frequency unit from sample to Hz
m1=m/n;                  % rescaling of FFT amplitude output unit
subplot (2,1,2); plot (f1, m1); grid;
title ('Real FFT- Sinus');
ylabel ('Amplitude'); xlabel ('frequency [Hz]');
```

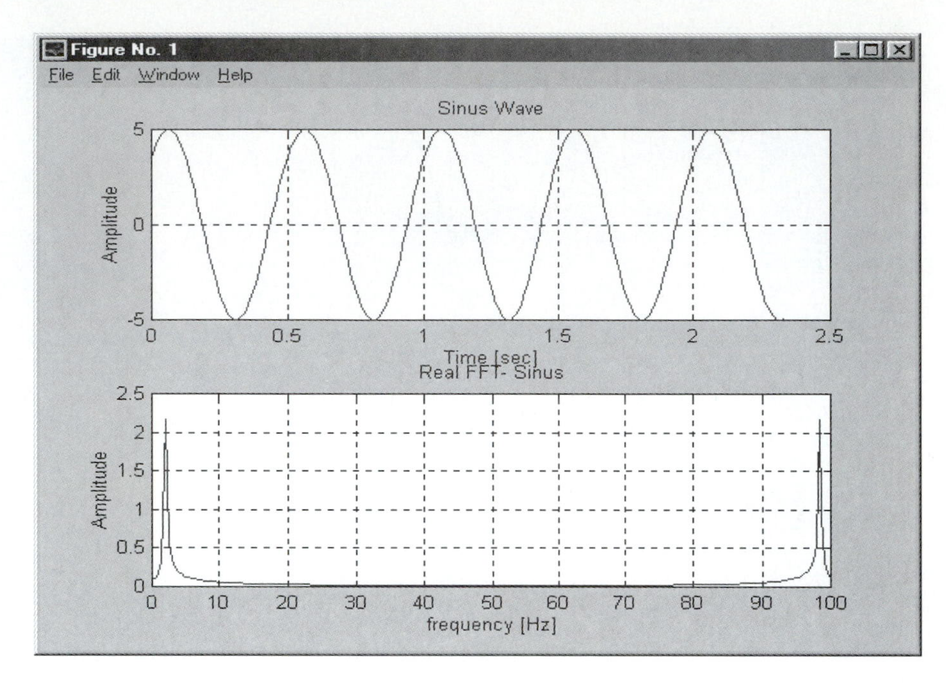

Figure 6.54

Results of MATLAB program sinusSpectrum1.m

Figure 6.55

HanningWindow.m program

```
%HanningWindow.m
phase=45;                 % phase [degrees]
d=2.33;                   % duration of trigger [sec]
r=100;                    % sampling rate [samples / sec]
n=r*d;                    % number of samples [samples]
dt=1/r;                   % sampling spacing in [sec / sample]
t=0:dt:d;                 % current time [sec]
Th=d;                     % period of Hanning window=duration of trigger [sec]
Fh=1/Th;
omegah=2*pi*Fh;
yh=0.5* (1-cos (omegah*t)); % Hanning window
plot (t, yh);
title ('Hanning Window');
ylabel ('Amplitude'); xlabel ('Time [sec]');
```

Windowing the input signal will reduce spectral leakage observed in the results shown in Fig. 6.54. In Fig. 6.57 the MATLAB program WindowedSinusSpectrum.m is listed, which completes the sinusSpectrum1.m program from Fig. 6.53 with a Hanning window.

The Hanning window equation $yh = 0.5 \cdot (1 - \cos(\text{omega}h \cdot t))$ defines yh used in $y1 = y \cdot yh$ to produce a windowed output. Figure 6.58 shows the results of the WindowedSinusSpectrum.m program.

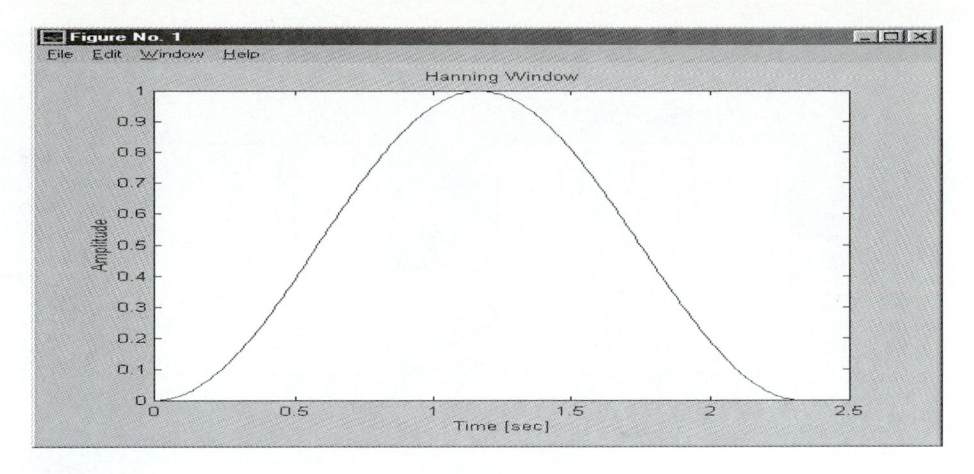

Figure 6.56

Results of HanningWindow.m program

Figure 6.57

MATLAB program WindowedSinusSpectrum.m

```
%WindowedSinusSpectrum.m
phase=45;                % phase [degrees]
d=2.33;                  % duration of trigger [sec]
r=100;                   % sampling rate [samples / sec]
n=r*d;                   % number of samples [samples]
dt=1/r;                  % sampling spacing in [sec / sample]
t=0:dt:d;                % current time [sec]
phi=pi*phase/180;        % phase [rad]
A=5.0;                   % amplitude
F=2;                     % frequency [cycles /sec]
k=F*d;                   % number of cycles [cycles]
omega=2*pi*F;            % angular velocity [rad /sec)
y=A*sin(omega*t+phi);
subplot (2, 2, 1); plot (t, y); grid; hold on;
title ('Sine Wave');
ylabel ('Amplitude'); xlabel ('Time [sec]');
yy=fft (y);              % FFT of sinus wave
m=abs(yy);               % amplitude of FFT
N=length (yy)-1;
f=(0:N);
df=r/n;                  %frequency spacing [Hz]
fHz=f*df;                % rescaling of frequency unit from sample to Hz
mm=m/n;                  % rescaling of FFT amplitude output unit
subplot (2, 2, 3); plot (fHz, mm); gri;
title ('AMPL SPECTRUM-Original Signal');
ylabel ('Amplitude'); xlabel ('frequency [Hz]');
Th=d;                    % period of Hanning window=duration of trigger [sec]
Fh=1/Th;
omegah=2*pi*Fh;
yh=0.5* (1-cos(omegah*t));  % Hanning window
y1=y.*yh;                   % windowed sinus wave
subplot (2, 2, 2); plot (t, y1);
title ('Windowed Sine Wave');
ylabel ('Amplitude'); xlabel ('Time [sec]');
yy1=fft (y1);               % FFT of windowed sinus wave
m1=abs (yy1);               % amplitude of FFT
mm1=m1/n;                   % rescaling of FFT  amplitude output unit
subplot (2, 2, 4) ; plot (fHz, mml); grid;
title ('AMPL SPECTRUM-Windowed Signal')
ylabel ('Amplitude') ; xlabel ('frequency  [Hz]') ;
```

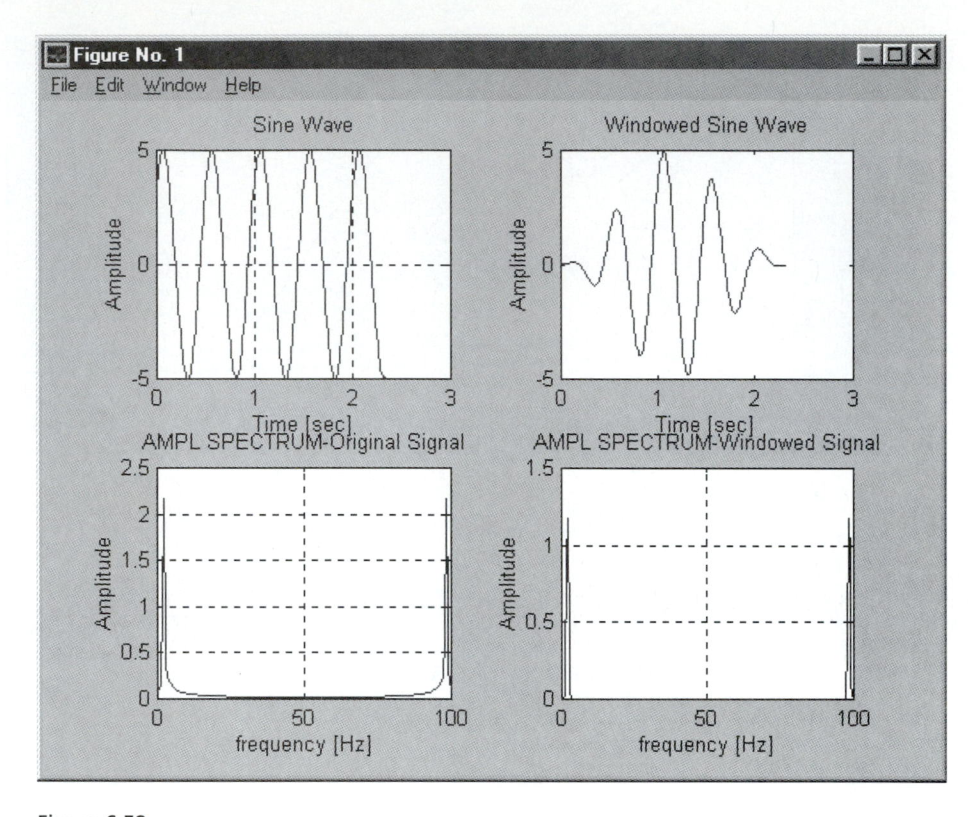

Figure 6.58

Results of the WindowedSinusSpectrum.m program

These results are the same as those shown in Fig. 6. 50(a) and confirm the elimination of spectral leakage by the Hanning window.

Besides the Hanning window, several other windows are available in LabVIEW and MATLAB. Figure 6.59 shows various windows from LabVIEW [86].

While the Hanning window is a general purpose window, the other windows are helpful in other special applications. For example, the rectangular window is suitable for transients, which are shorter than the duration of the window [86]. Kaiser–Bessel window has the advantage of adjustable shape by changing its parameter beta.

6.9 DIGITAL FILTERS

Practical applications require the use of a variety of filters for removing the noise from measurement results, for separating higher and lower frequency components of a signal, etc. This section will focus on some frequently used digital filters:

(a) median filter [87, 90];
(b) first order filter [89];
(c) Butterworth IIR (infinite impulse response) filter [87, 90].

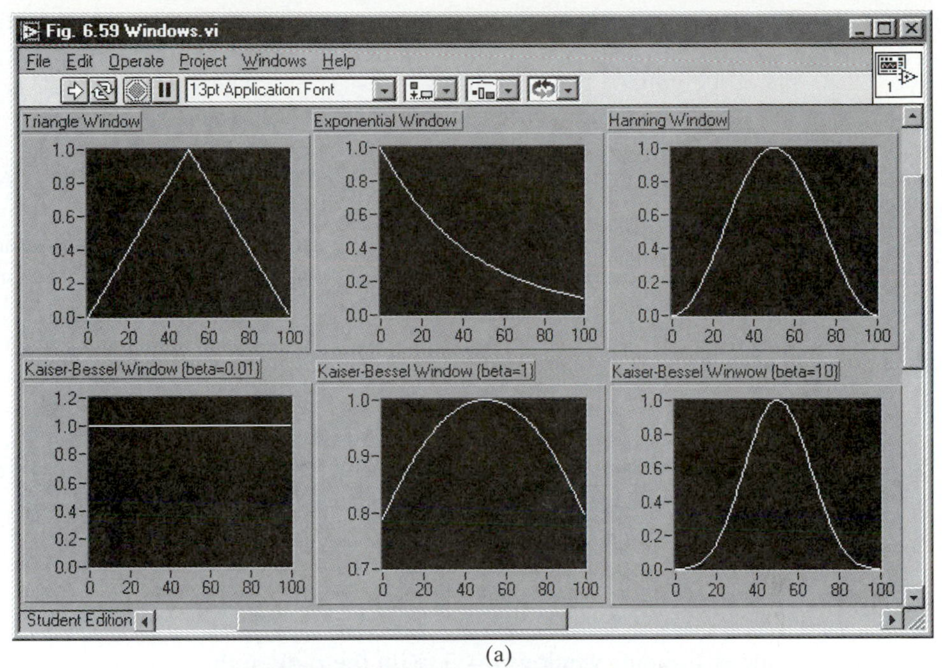

(a)

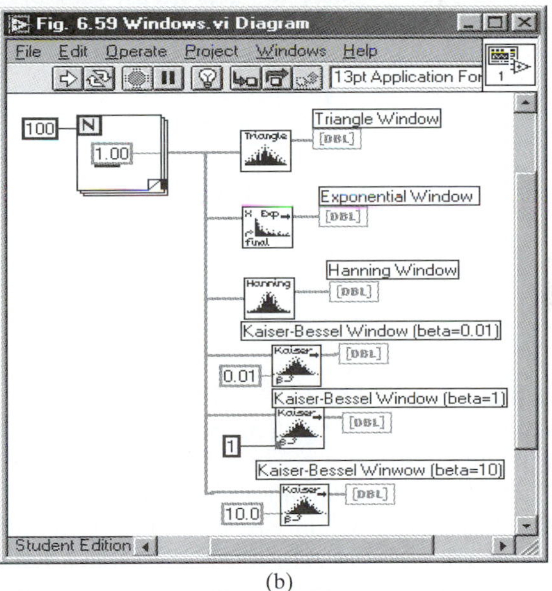

(b)

Figure 6.59

Various LabVIEW windows

A median filter is simple to apply given that it requires only sorting the input values in increasing value order and does not require any other mathematical operations. This filter, for two-dimensional window median calculation, has applications in digital image processing.

The median filter is applied to a N [elements] input sequence X. A sliding window of n [elements] can have n be an odd number or an even number. As expected, $n < N$.

An odd number n can be written as $n = 2r + 1$, where $r = (n - 1)/2$ is the rank of the filter. In this case, the sliding window content is denoted as

$$J_i = \{j_{i-r}, j_{i-r+1}, \ldots, j_{i-1}, j_i, j_{i+1}, \ldots j_{i+r-1}, j_{i+r}\} \quad \text{for} \quad i = 1, \ldots, N.$$

This sliding window is centered about the ith element j_i of the input sequence.

The values of the sliding window can be sorted in increasing value order can be denoted as

$$S_i = \{s_{i-r}, s_{i-r+1}, \ldots, s_{i-1}, s_i, s_{i+1}, \ldots, s_{i+r-1}, s_{i+r}\} \quad \text{for} \quad i = 1, \ldots, N,$$

centered about the ith element s_i. For n an odd number, s_i is the median.

For n an even number, the median is given by the average of the two central values of the sorted window in increasing value order.

For example, for $N = 7$, input values

$$X = \{7 \quad 2 \quad 6 \quad 2 \quad 9 \quad 1 \quad 3\},$$

and an odd size sliding window $n = 3$ (with the rank of the filter $r = (3 - 1)/2 = 1$), the following results are obtained:

i	J_i	S_i	median
1	0 7 2	0 2 7	2
2	7 2 6	2 6 7	6
3	2 6 2	2 2 6	2
4	6 2 9	2 6 9	6
5	2 9 1	1 2 9	2
6	9 1 3	1 3 9	3
7	1 3 0	0 1 3	1

The values for $i = 1$ and $i = 7$ are based on sliding windows with arbitrarily chosen elements 0 for the missing elements and can be ignored. The range of values in the values in the input sequence is 1 to 9. The effect of median filter is the reduction of the range of valued in the median filter output from 2 to 6.

Figure 6.60 shows the LabVIEW Median_Filter_for_Noisy_Square_Wave.vi. The block diagram from Fig. 6.60 (b) shows the summation of Square Wave.vi output with Gaussian White Noise.vi output to generate a noisy square wave. The result of the summation is shown in the waveform graph labeled noisy square wave and is provided as input to the Median filter.vi. The output of the Median filter.vi is shown in the waveform graph labeled median filter for noisy square wave from Fig. 6.60(a). A median filter of rank $r = 20$ outputs a wave with the Gaussian white noise removed, but the square wave results with rounded corners due to the filtering of some of higher frequency components of the square wave.

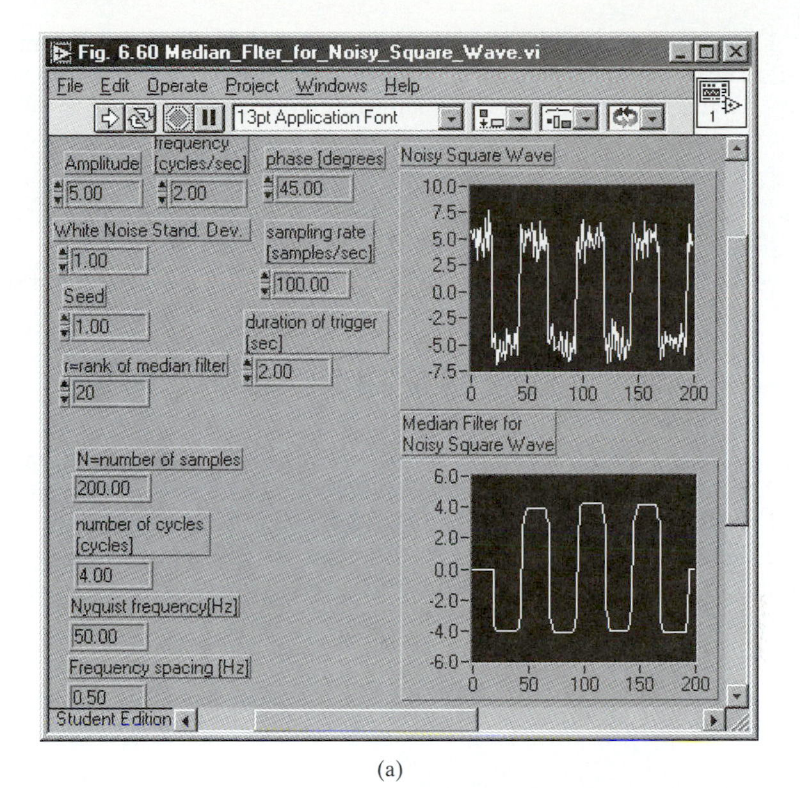

(a)

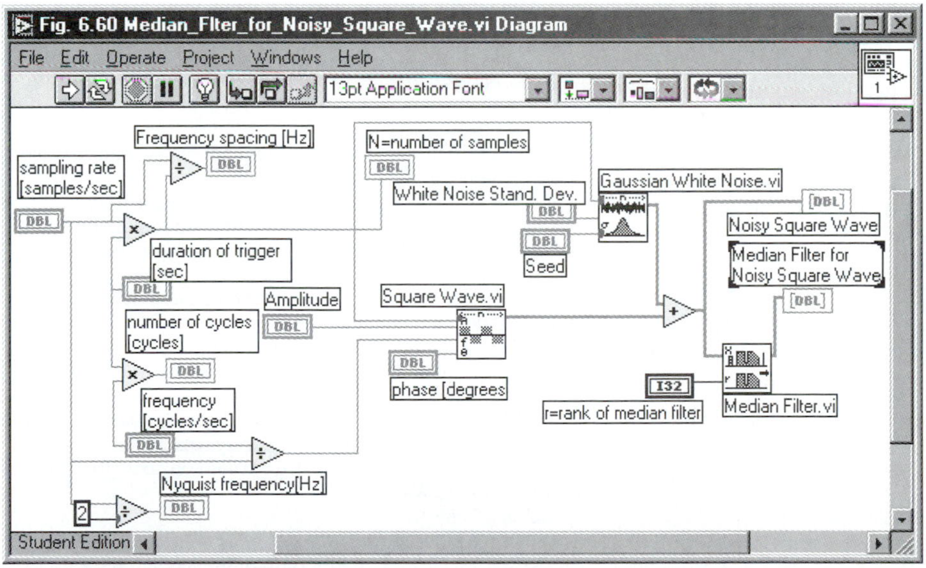

(b)

Figure 6.60

Median_Filter_for_Noisy_Square_Wave.vi

Figure 6.61 shows the listing of the MATLAB Median_Filter.m program. The program uses the function square() to generate a square wave: $y = A \cdot \text{square}(\text{omega} \cdot t, \text{duty})$. To this square wave, a Gaussian white noise is added:

$$yn = y + An * \text{randn}(\text{size}(t))$$

to generate a noisy square wave.

A median filter of rank $r = 10$ and a number of elements in the sliding window $n = 2 \cdot r + 1$ is obtained with the function

$$yy = \text{medfilt1}(yn,n).$$

Figure 6.62 shows the results of the MATLAB Median_Filter.m program.

The MATLAB Median_Filter.m program results closely resemble the results from Fig. 6.60 of the LabVIEW Median_Filter_for_Noisy_Square_Wave.vi. A typical analog first order filter [89] is the RC filter shown in Fig. 6.63. This is a low-pass filter with the cut-off frequency of $f_c = 1/(2\pi RC)$ [Hz], [89]. The model of this filter with no load is given by the current law of Kirchhoff— that is,

$$(X - Y)/R - C\,dY/dt = 0$$

or

$$X = RC\,dY/dt + Y,$$

Figure 6.61

Listing of the MATLAB Median_Filter.m program.

```
%Median_Filter.m  (Median Filter for Noisy Square wave)
cd c:\matlab\toolbox\signal;
d=2;                      % duration of trigger [sec]
r=100;                    % sampling  rate [samples/sec]
n=r*d;                    % number of samples [samples]
t=0:1:n;                  % current time
k=4;                      % number  of cycles [cycles]
duty=50;
A=5.0;                    % amplitude
omega=2*pi*k/n;           % angular velocity [rad/sec]
y=A*square(omega*t, duty);
An=1.0;                   % amplitude of Gaussan white noise
yn=y+An* randn (size (t)); % noisy square wave
subplot (211); plot (t, yn,'b'); grid; hold on; axis ([0  200 -7.5  7.5])
title ('Square Wave');
xlabel ('Time');
ylabel ('Amplitude');
r=10;                     % rank of the median filter
n=2*r+1;                  % number  of elements in the sliding window
yy=medfilt 1(yn, n);
subplot (212); plot (t, yy); grid; axis ([0  200 -6  6])
title ('Real FTT-Square  Pattern');
ylabel ('Amplitude');  xlabel ('frequency');
```

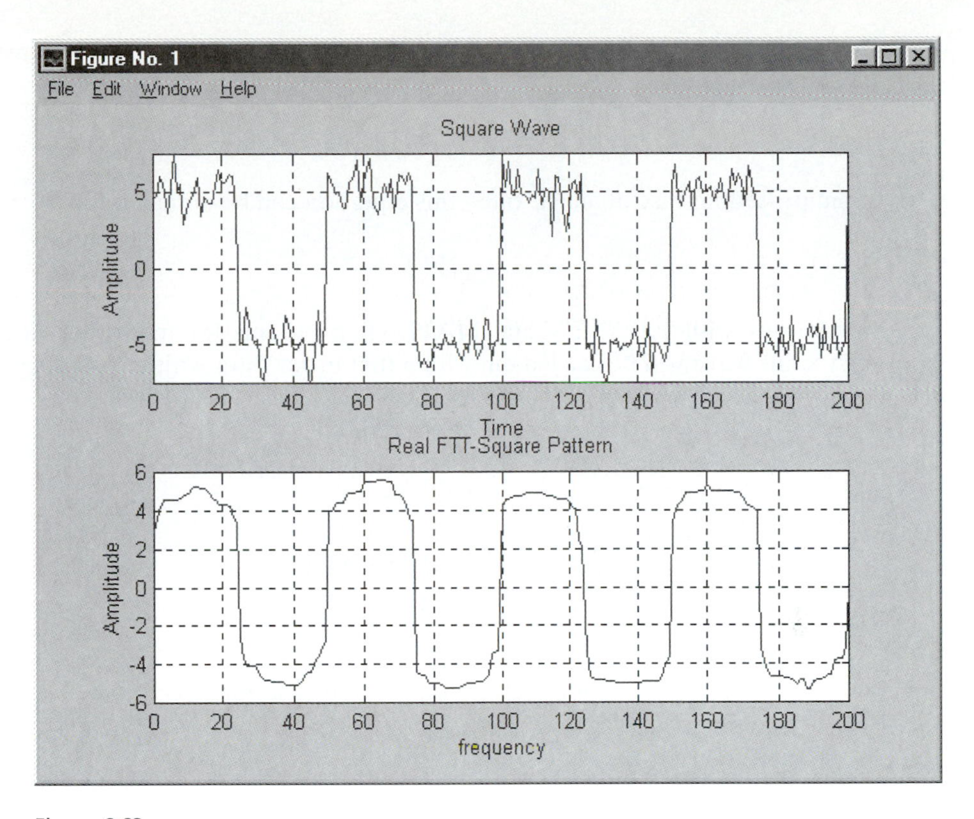

Figure 6.62

Results of the MATLAB Median_Filter.m program

where X is the input voltage and Y is the output voltage. The time constant of the filter is $T = RC = 1/(2\pi f_c)$.

For the backward finite difference approximation for a sampling interval T_s,

$$dY/dt = (y_n - y_{n-1})/T_s \qquad \text{for } n = 1, 2, \ldots,$$

the model of the filter can be written as the finite difference equation

$$x_n = T\,(y_n - y_{n-1})/T_s + y_n$$

or

$$y_n = (T_s/(T + T_s))\;x_n + (T/(T + T_s))\;y_{n-1}.$$

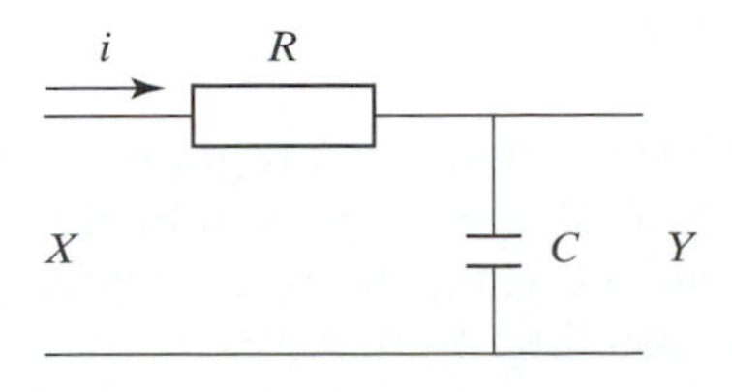

Figure 6.63

Low-pass analog RC filter

The general form of the digital first-order filter is given by [87, 89] the recursive finite difference equation

$$y_n = (b_0/a_0)\ x_n - (a_1/a_0)\ y_{n-1} \qquad \text{for } n = 1, 2, \ldots.$$

In the preceding case of an *RC* filter, this equation can be obtained for

$$T_s/(T + T_s).$$

Simulations with LaBVIEW and MATLAB programs for first-order filtering of a noisy sinus wave will be carried out with a first order filter with the parameters

$$f_c = 25\ [\text{Hz}]$$

and

$$T_s = 0.001\ [\text{sec}],$$

which give

$$\text{sampling rate} = 1/T_s = 1000\ [\text{samples/sec}],$$
$$\text{T} = 1/(2\pi f_c) = 0.064\ [\text{sec}],$$

and

$$y_n = 0.135x_n + 0.865y_{n-1} \qquad \text{for } n = 1,2,\ldots.$$

Figure 6.64 shows the LabVIEW program IIR_Filter_for_Noisy_Sinus_Wave.vi.

The block diagram from Fig. 6.64 (b) shows the summation of Sine Wave.vi output with Gaussian White Noise.vi output to generate a noisy sine wave [87]. The result of the summation is shown in the waveform graph, labeled noisy sine wave, and is provided as input to the IIR Filter.vi with $N = 1$ forward coefficient, b_0, and $M = 2$ reverse coefficient $a_0 = 1$ and a_1. The output of the IIR Filter.vi is shown in the waveform graph labeled IIR Filter for Noisy Sine Wave in Fig. 6.64 (a). A first-order IIR filter outputs a wave with the Gaussian white noise partly reduced.

Figure 6.65 shows the listing of the MATLAB First_Order _Filter.m program for filtering a noisy sine wave with a first-order filter. The first order filter is implemented with the recursive equation

$$yy(j + 1) = 0.865{*}yy(j) + 0.135{*}yn(j + 1), \ \text{for } j = 1{:}N.$$

The results of the MATLAB First_Order _Filter.m program are plotted in Fig. 6.66. Again, the noisy sinus wave has part of the noise removed by the first order filter. For the particular example of a noisy sine wave, this first-order filter has a lower performance in noise filtering than the median filter from Fig. 6.60.

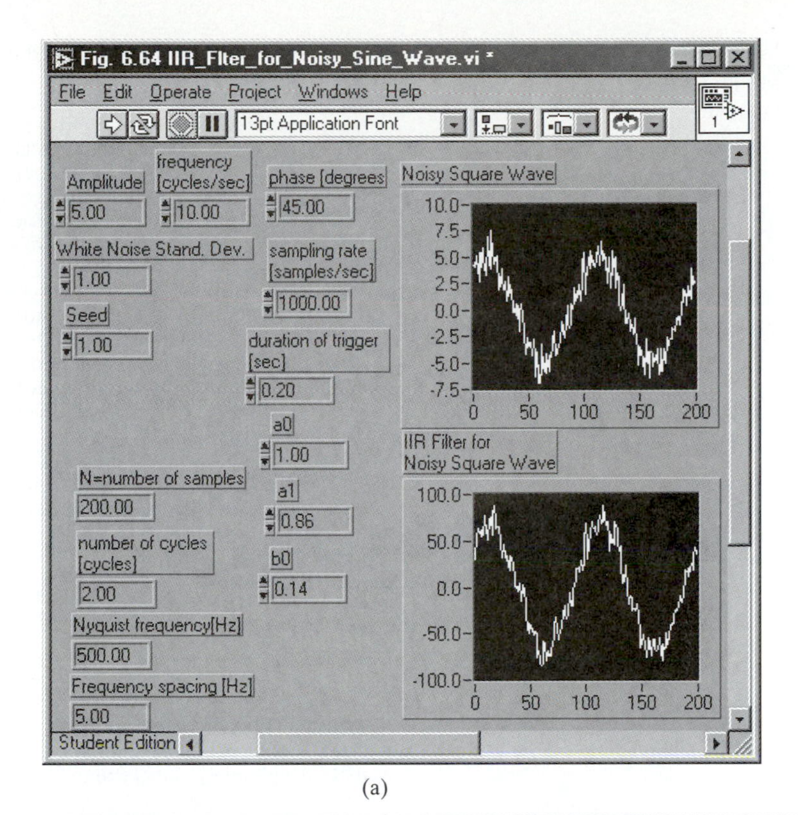

(a)

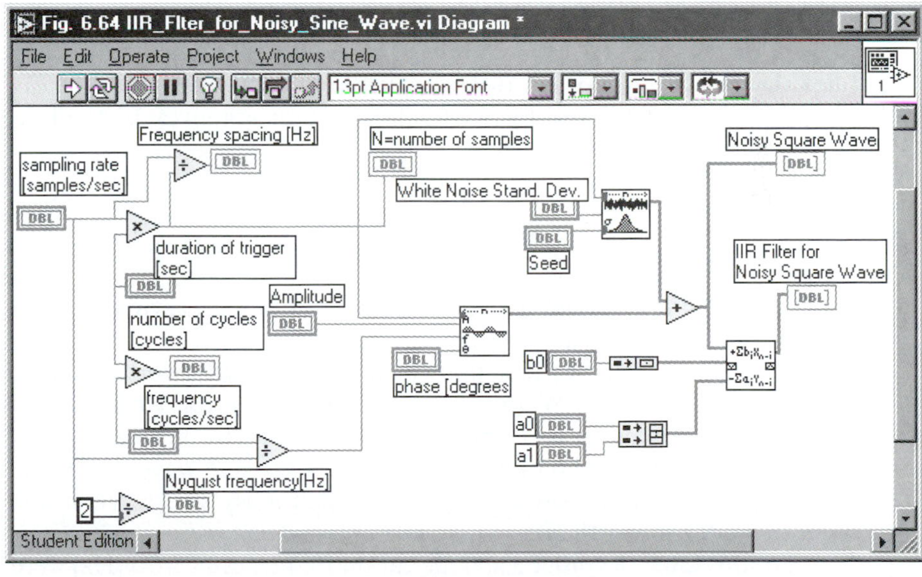

(b)

Figure 6.64

LabVIEW program IIR_Filter_for_Noisy_Sinus_Wave.vi.

Figure 6.65

MATLAB First_Order _Filter.m program.

```
%First_Order_Filter.m (First Order Filter for Noisy Sinus Wave)
cd c:\matlab\toolbox\signal;
t=linspace (0, 0.2, 201);
x=linspace (0, 0.2, 199);
phase=45;                  % phase [degrees)
d=0.2;                     % duration of trigger [sec]
r=1000;                    % sampling rate [samples/sec]
w=r/2;                     % Nyquist frequency
n=r*d;                     % number of samples [samples]
dt=1/r;                    % sampling spacing in [sec/sample]
i=0:n;                     % current time [samples]
phi=pi*phase/180;          % phase [rad]
A=5.0;                     % amplitude
k=2;                       % number of cycles [cycles]
F=k/d;                     % frequency [cycles/sec]
omega=2*pi*k/n;            % angular velocity [rad/sec]
y=A*sin (omega* i+phi);
An=1.0;                    % amplitude of Gaussian white noise
yn=y+An* randn (size (i)); % noisy sinus wave
subplot (211); plot (t, yn, 'b'); grid; hold on;
title ('Noisy Sinus Wave');
xlabe1('Time [sec]');
ylabe1('Amplitude');
N=n-2;
yy(1)=yn (1 ;
  for j=1:N;
    yy(j+1)=0.865*yy(j)+0.135*yn (j+1);
  end
  subplot (212); plot (x, yy); grid;
  title ('First Order Filter-Sinus Wave');
ylabel ('Amplitude'); xlabel ('time [sec]');
```

A better filtering result can be obtained with a higher order Butterworth IIR filter. There are several IIR filters available in LabVIEW and MATLAB (Butterworth, Chebyshev, Elliptic, Bessel, etc.). IIR filters are given by the following difference equation for $i = 1,2, \ldots$ [87]:

$$y_i = \frac{1}{a_0} \sum_{j=0}^{N-1} b_j x_{i-j} - \sum_{k=1}^{M-1} a_k y_{i-k}.$$

For $N = 1$ and $M = 2$, the first-order filter used in the previous section is obtained. Usually, the coefficient $a_0 = 1$.

A filter with good generic performance is the Butterworth filter [87, 90]. Figure 6.67 shows the LabVIEW program Butt_IIR_Filter_for_Noisy_Sine_Wave.vi. The block diagram from Fig. 6.67 (b) shows the summation of Sine Wave.vi output with Gaussian White Noise.vi output to generate a noisy sine wave. The result of the summation is shown in the waveform graph labeled noisy sine wave and is provided as input to the Butterworth Filter.vi.

For this filter, the order 5 and the cut-off frequency $f_c = 25$ [Hz] were chosen. This cut-off frequency lets pass the sine input frequency of 10 Hz. The output of the Butterworth Filter.vi is shown in the waveform graph labeled Butterworth Filter for Noisy Sine Wave in Fig. 6.67(a). A fifth order Butterworth filter outputs a wave with the Gaussian white noise removed, but the first part of the sine wave is partly distorted.

Figure 6.68 shows the listing of the MATLAB Butterworth_Filter.m program for filtering a noisy sine wave with a fifth-order filter. The coefficients a and b of the low-pass Butterworth filter of order 5 with cut-off frequency 25 [Hz] are obtained with the function

$$[b,a] = \text{butter}(5,\ 25/w).$$

The Butterworth filter is given by the function

$$yy = \text{filter}(b,a,yn),$$

where

$$yn = y + An * \text{randn}(\text{size}(i))$$

is the noisy sinus wave.

The results of the MATLAB Butterworth_Filter.m program are plotted in Fig. 6.69. The noisy sinus wave has the noise removed, similarly to the case shown in Fig. 6.67(a).

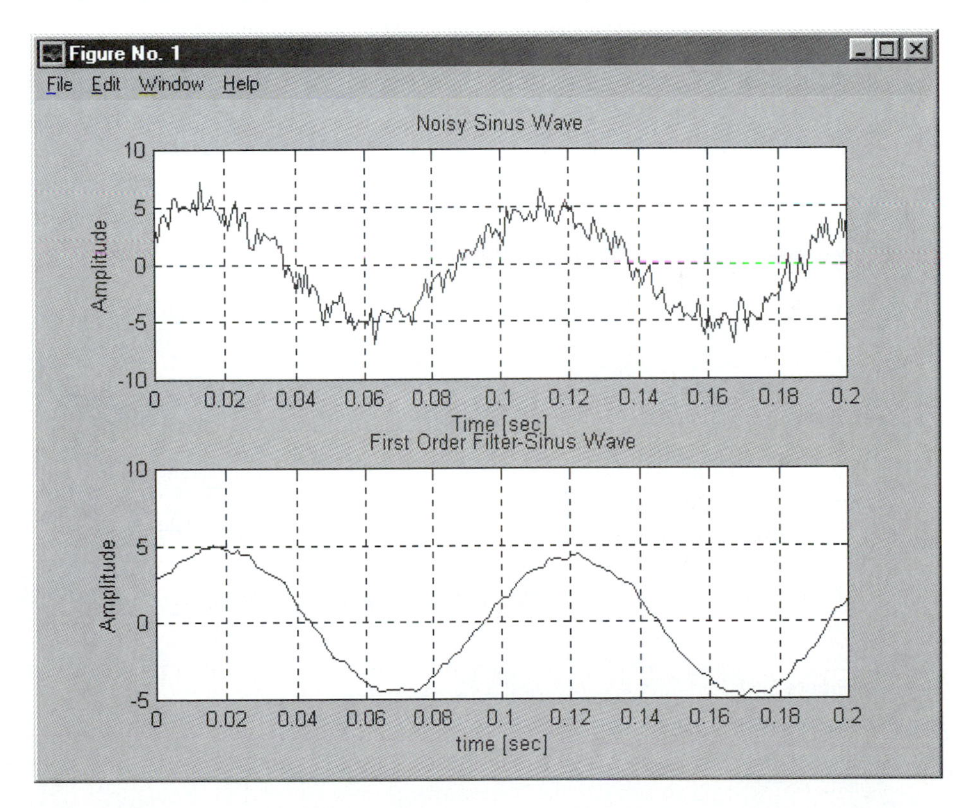

Figure 6.66

Results of the MATLAB First_Order _Filter.m program

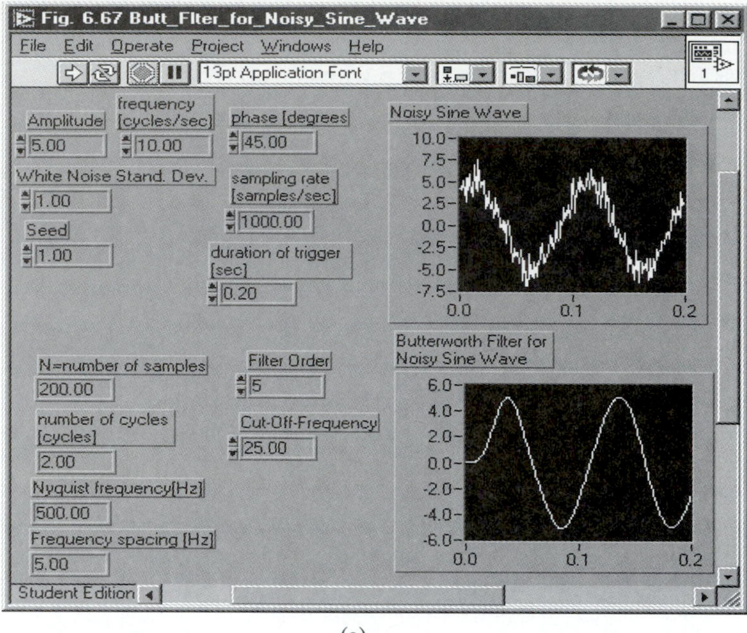

(a)

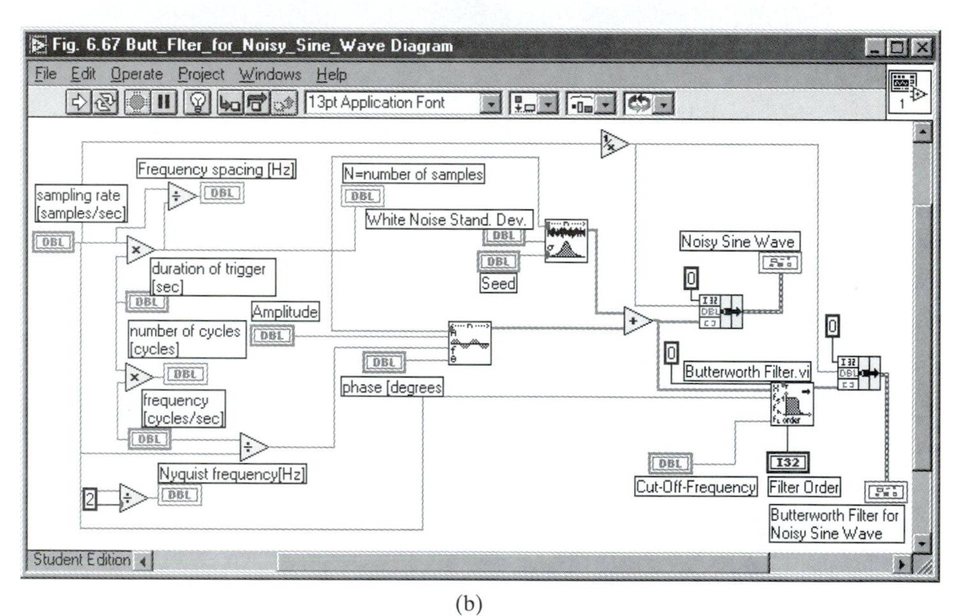

(b)

Figure 6.67

LabVIEW program Butt_IIR_Filter_for_Noisy_Sine_Wave.vi.

```
%Butterworth_Filter.m (Butterworth Filter for Noisy Sinus Wave)
cd c:\matlab\toolbox\signal;
t=linspace (0, 0.2, 201);
x=linspace (0, 0.2, 199);
phase=45;                    % phase [degrees]
d=0.2;                       % duration of trigger [sec]
r=1000;                      % sampling rate [samples /sec]
w=r/2;                       % Nyquist frequency
n=r*d;                       % number of samples [samples]
dt=1/r;                      % sampling  spacing in [sec /sample]
i=0:n;                       % current time [samples]
phi=pi*phase/180;            % phase [rad]
A=5.0;                       % amplitude
k=2;                         % number of cycles [cycles]
F=k/d;                       % frequency [cycles /sec]
omega=2*pi*k/n;              % angular velocity [rad /sec]
y=A*sin (omega* i+phi);
An=1.0;                      % amplitude of Gaussian white noise
yn=y+An* randn (size (i)); % noisy sinus wave
subplot (211); plot (t, yn, 'b'); grid; hold on;
title ('Noisy Sinus Wave');
xlabel('Time [sec]');
ylabel('Amplitude');
[b, a]=butter (5, 25/w);     %coefficients for 25-Hz lowpass Butterworth filte
                               order 5,
yy=filter (b, a, yn);        %Butterworth filtering of noisy sinus wave
subplot (212); plot (t, yy); grid;
   title ('Butterworth Filter-Sinus Wave');
      ylabel ('Amplitude'); xlabel ('time [sec]');
```

Figure 6.68

MATLAB Butterworth_Filter.m program for filtering a noisy sine wave

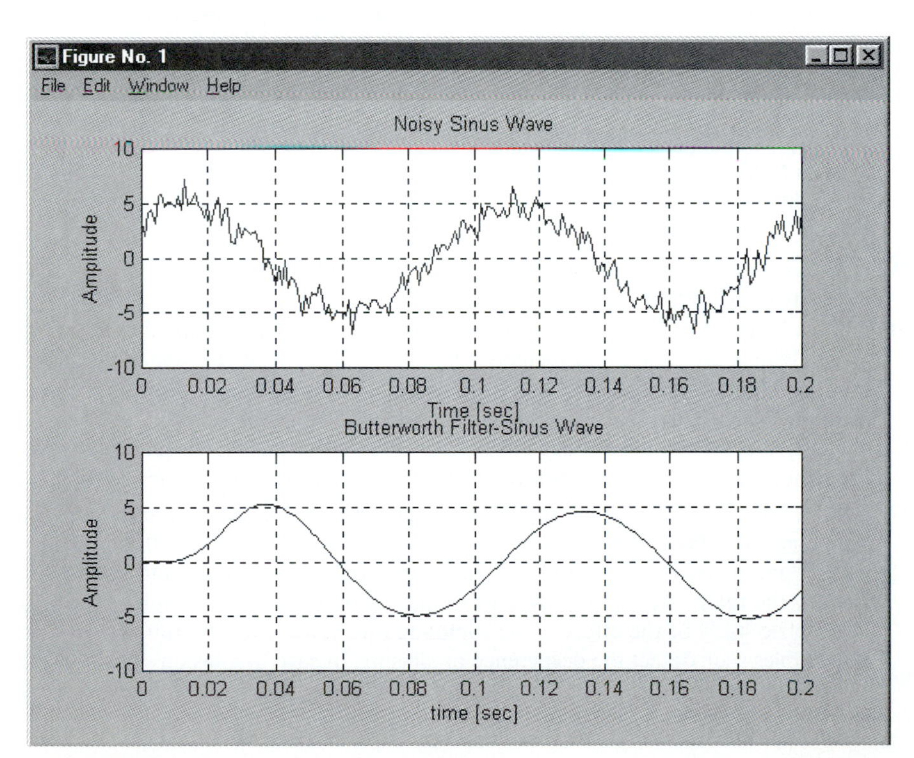

Figure 6.69

Results of the MATLAB Butterworth_Filter.m program

PROBLEMS

6.1. Build a front panel and block diagram of a LabVIEW virtual instrument for velocity conversion from miles per hour to meters per second.

6.2. Build a front panel and block diagram of a LabVIEW virtual instrument for temperature measurement with a thermometer containing numerical and waveform graph display. Using simulated data source with a randomness of $\pm 4°$ Celsius about a constant 25° Celsius, replace the simulated data source by a Sub VI.

6.3. Add a Mean.vi to filter out the random effect in the temperature measurement virtual instrument built in problem 6.2 and display both unfiltered and filtered results.

6.4. Write a MATLAB program for acquiring a signal with frequency of less than 300 Hz. Justify the choice of the sampling rate.

6.5. Build the front panel and the block diagram of a LabVIEW virtual instrument for generating a sine wave of amplitude 0.1 frequency of 10 Hz and phase 90°, added to a square wave of amplitude 0.5 frequency 1 Hz and phase 0°. Design the front panel with controls for the amplitudes, frequencies, phase, sampling rate, and duration of trigger as well as indicators for the number of samples, the number of cycles for each wave, and the waveform graph of the signal generated.

6.6. Write a MATLAB program for generating a sine of amplitude 0.1 frequency of 10 Hz and phase 90°, added to a square wave signal of amplitude 0.5 frequency 1 Hz and phase 0°. Compare the graphical display of the signal generated with the results from problem 6.5.

6.7. Build a front panel and a block diagram of a LabVIEW virtual instrument for generating a sine wave of amplitude 0.2 frequency of 20 Hz and phase 120°, added to a triangle wave of amplitude 0.75 frequency 2 Hz and phase 0°.

6.8. Upgrade the virtual instrument built for problem 6.7 with a Fourier transform block, for the signal generated, using Real FFT.vi, and display the realFFT amplitude versus Frequency. Calculate the frequency spacing.

6.9. Write a MATLAB program for generating a sine wave of amplitude 0.2 frequency of 20 Hz and phase 120°, added to a sawtooth wave signal of amplitude 0.75 frequency 2 Hz and phase 0°.

6.10. Upgrade the MATLAB program, developed for problem 6.9, with a *fft*(y) function and display the amplitude versus frequency. Calculate the frequency spacing.

6.11. Write a MATLAB program for generating a sine wave of amplitude 0.1 frequency of 10 Hz and phase 90° and include a Fourier transform of the signal generated using *fft*() function and display the amplitude versus frequency. Perform a parametric study of the effect of the sampling rate value over the range from 5 to 30 [samples/sec], Detect the occurrence of aliasing. In case of aliasing, calculate the alias frequency and compare the displayed results using computed frequency spacing.

6.12. Build a LabVIEW virtual instrument for generating a sine wave of amplitude 0.1 frequency of 10 Hz and phase 90° added to a sine wave of amplitude 0.2 frequency of 7 Hz and phase 0° and include Fourier transform of the signal using realFFT.vi. Display the realFFT amplitude versus frequency in a waveform graph. Perform a parametric study of the effect of the sampling rate value over the range from 5 to 30 [samples/sec], detect the occurrence of aliasing. In case of aliasing, calculate the alias frequency and compare the displayed results using computed frequency spacing. Compare the results with those from problem 6.9.

6.13. Use the LabVIEW program from problem 6.12 and replace the realFFT.vi by amplitude and Phase.vi. Use Bundle.vi for processing amplitude and phase versus frequency results to be displayed. Compare the displayed results of frequency analysis with those from problem 6.12 using the computed frequency spacing.

6.14. Perform a parametric study of the effect of the duration of trigger value, which can result in sampling a noninteger number of cycles of the sine wave and detect the occurrence of spectral leakage for the virtual instrument built in problem 6.13. Add a Hanning window for windowing the signal and an Amplitude and Phase.vi for the windowed signal. Repeat the parametric study of the effect of the duration of trigger value, which can result in sampling a noninteger number of cycles of the windowed sine waves and compare with the results for nonwindowed sine wave.

6.15. Perform a parametric study of the effect of the duration of trigger value, which can result in sampling a noninteger number of cycles of a sine wave and detect the occurrence of spectral leakage for the MATLAB program from problem 6.11. Add a Hanning window for windowing the sum of sine waves. Repeat the parametric study of the effect of the duration of trigger value, which can result in sampling a noninteger number of cycles of the windowed signal, and compare with the results for nonwindowed signal.

6.16. For the 15 values {5 9 6 1 3 4 9 5 3 6 3 6 4 2 7} of a sampled signal, calculate the output values of a median filter of rank 2.

6.17. Use the LabVIEW program from problem 6.12 and add to the signal a white noise of amplitude 0.01 using Gaussian White Noise.vi. Include an IIR Filter.vi of order 1 and display the noisy and the filtered signals.

6.18. Use the MATLAB program from problem 6.11 and add to the signal a white noise of amplitude 0.01 using randn () function. Include an first order filter of order 1 and display the noisy and the filtered signals.

6.19. Use the LabVIEW program from problem 6.17 and replace IIR Filter.vi by a Butterworth Filter.vi and display the noisy and the filtered signals. Perform a parametric study of the effect of the filter order on cut-off frequency on the filtered signal and choose suitable values.

6.20. Use the MATLAB program from problem 6.18 and replace first-order filter by a Butterworth filter using butter() function and display the noisy and the filtered signals. Perform a parametric study of the effect of the filter order on cut-off frequency on the filtered signal and compare the results with those obtained in problem 6.19.

CHAPTER 7

Real-Time Monitoring and Control: PC-Based and Embedded Microcontrollers

7.1 SOLUTIONS FOR REAL-TIME APPLICATIONS

Various solutions for computer-based monitoring and control, outlined in Section 1.2.1, can be distinguished in terms of feasibility for real-time (RT) applications. Interfacing a PC with the system requires the use of the PC microprocessors for regular PC tasks as well as for data acquisition and control tasks. Examples of PC interfacing with the system were presented in Section 4.1 using DAQ boards. The DAQ boards are plugged into the extension slots of the PC and provide an interface between PC data bus on one side and digital inputs and outputs as well as analog inputs and outputs on the other side. Such systems are called nondeterministic. A system is called deterministic if for given inputs, the next step outputs are determined and the response time is known [93]. Windows is characterized as nondeterministic, given that the response time to an event cannot be guaranteed due to device drivers and interrupt priorities [96]. As a result, Windows-based PCs can have response time with delays of close to a second [96]. Many RT applications require shorter response times and can be implemented only on computer-based systems specifically designed for real-time applications. An example is shown in Fig. 7.1. These RT applications use a host PC for developing, debugging, and downloading the RT code for the RT target computer included in the RT DAQ board. The host PC is composed of a dedicated CPU, RAM, and ROM. The RT target computer executes the RT code and has no direct user interface for viewing the results during the execution or for changing the parameters of the RT code. A host interface on the RT DAQ board serves the communication between the host and the target computers during the development, debugging, and downloading of the RT code. This host interface can also be used offline for viewing the results on the

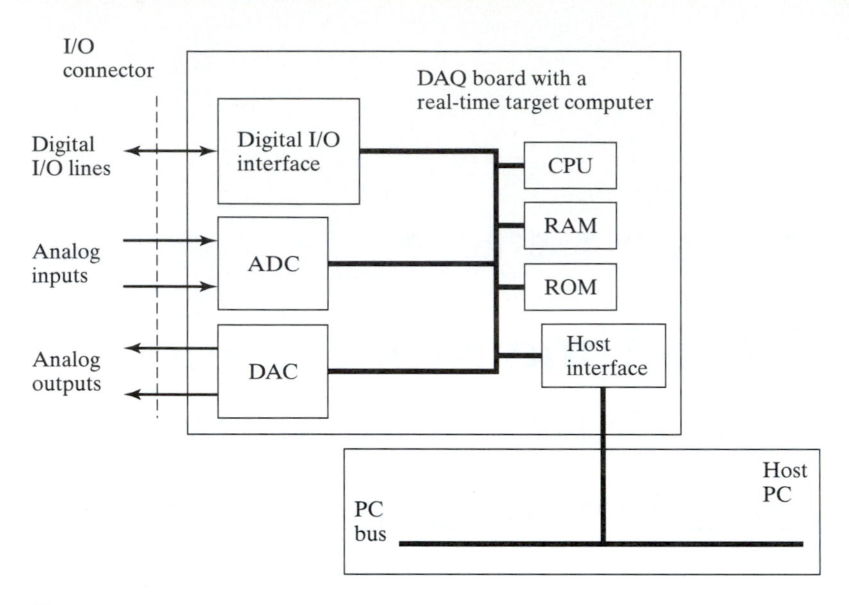

Figure 7.1

Block diagram of a DAQ board real-time target computer

PC monitor during the execution of the RT code on the target computer or for changing the parameters of the RT code. For executing the RT code on the RT target computer, an RT kernel is required [93]. The RT kernel allows the RT target computer to carry out RT task-management operations.

Listed in Section 1.2.1 were DSP-in-the-loop and embedded computer configurations for real-time applications with deterministic features. In this chapter, real-time computer-based systems will be presented in more detail. Presented in Section 7.2 will be real-time data acquisition systems based on PC plug-in boards with a specialized real-time target computer, a DSP. Section 7.3 discusses LabVIEW real-time data acquisition systems based on PC plug-in boards with a dedicated real-time target computer based on a generic microprocessor. MATLAB toolboxes for real-time applications on various real-time target computers (target PC, DSP, microcontrollers, etc.) are examined in Section 7.4. In these configurations, the real-time kernels carry out deterministically all real-time tasks, while the host PC continues to process non-real-time operations.

Embedded computers, presented in Section 7.5, need no PC and are dedicated to real-time tasks with no direct input/output interface with human operators. An embedded computer, as shown in Fig. 7.2, contains in a single integrated circuit CPU, RAM and ROM memory and interface devices for digital and analog signals.

Often, microcontrollers are used embedded in the system for monitoring and control [93]. Due to the lack of any built-in user interface, embedded computers can be linked by a serial port connection with a PC. The development system for embedded computer applications use normally a PC. For downloading the code developed on a PC and for operator access to the embedded computer during the real-time operation, serial connection to the PC is normally used.

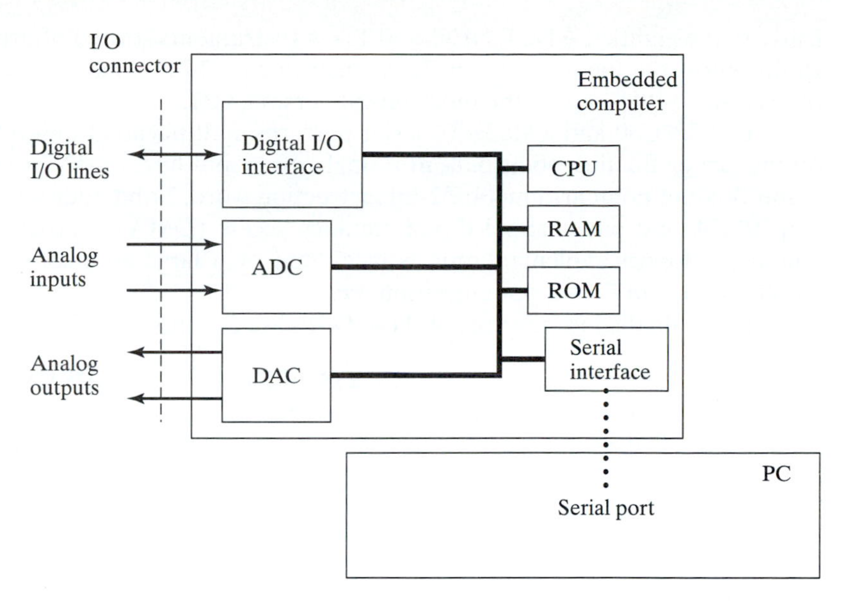

Figure 7.2

Block diagram of an embedded computer

7.2 DIGITAL SIGNAL PROCESSORS FOR REAL-TIME APPLICATIONS: dSPACE DEVELOPMENT SYSTEM

7.2.1 Digital Signal Processors

Digital signal processors are specialized processors, which have hardware multipliers for speeding up the computation of weighted averages. Hardware multipliers are particularly useful for the computations requires for linear digital filters and spectrum analyzers.

For example, consider the following:

(a) An IIR filter, presented in Section 6.4.6, is computed with the linear difference recursive equation

$$y_i = \frac{1}{a_0}\sum_{j=0}^{N-1} b_j x_{i-j} - \sum_{k=1}^{M-1} a_k y_{i-k},$$

which calculates the output y_i (for $i = 1, 2, \ldots$) as a weighted sum of the inputs x_{i-j} (for $j = 0$ to $N - 1$) and the previous outputs y_{i-k} (for $k = 1$ to $M - 1$) [87].

(b) The discrete Fourier transform (DFT), presented in Section 6.4.3, is given by a linear difference equation,

$$X_k = \sum_{i=0}^{n-1} x_i [\cos\left(\frac{2\pi k}{n} i\right) - j\sin\left(\frac{2\pi k}{n} i\right)],$$

which calculates the output X_k (for $k = 1, 2, \ldots, n - 1$) as a weighted sum of the sampled input signal x_i (for $i = 0, 1, \ldots, n - 1$).

Early in the eighties, AT&T, NEC, and Texas Instruments started offering DSPs [107]. In this book, the focus will be on Texas Instrument 32-bit, floating-point TMS320C30, offered since 1988 and on the more recent version, C31.

TMS320C30 and TMS320C31 can perform multiplication and ALU operations on integer or floating-point data in a single cycle and have16–32-bit integer and 32–40-bit floating-point arithmetic, 32-bit instruction word, 24-bit addresses, 2K 32-bit on-chip RAM, and one-channel direct memory access (DMA) coprocessor. High-level language support implementation is facilitated by a large address space, flexible instruction set, and floating-point arithmetic.

The main distinct features of these DSPs are as follows [107]:

	TMS320C30-33	**TMS320C31-60**
instruction cycle time	60-ns	33.3-ns
MFLOPS	33.3	60
serial ports	two	one
32-bit timers	two	two
package	208-pin or 181-pin	132-pin

DSP-based packages for real-time applications were offered by dSPACE and ISI, lately part of Wind River [109, 110]. In this chapter, the real-time development system offered for university users by dSPACE will be presented.

7.2.2 dSPACE Development System

In the early nineties, dSPACE started to offer implementation packages of real-time control applications based on hardware containing TMS320 DSPs and software accepting MATLAB and MATRIXx programs. dSPACE has a large variety of hardware using DSP, Alpha, and PowerPC processors and numerous I/O boards as well as software for automatic implementation of MATLAB/Simulink models, generation of the real-time code, and experiment control [109]. The presentation in this section focuses on the ACE kit for university users containing

DS 1102 DSP controller board with
  TMS 320 C31 DSP
  comprehensive on board I/O.
software for implementation of MATLAB/Simulink models on dSPACE hardware by generating the real-time code as well as for experiment control.

The DS 1102 DSP controller board, shown in Fig. 7.3, has the bottom access to plugging it in a half-length 16-bit ISA slot of a PC and a right-hand side access to digital-to-analog I/O signals. The board characteristics of the blocks shown in Fig. 7.3(b), are the following [109]:

(a) TMS320C31 processor described in Section 7.2.1
(b) 128K 32-bit on board RAM
(c) analog input, ±10-V input voltage range:

2 parallel 16-bit channels, 4-μsec conversion time
2 parallel 12-bit channels, 1.25-μsec conversion time

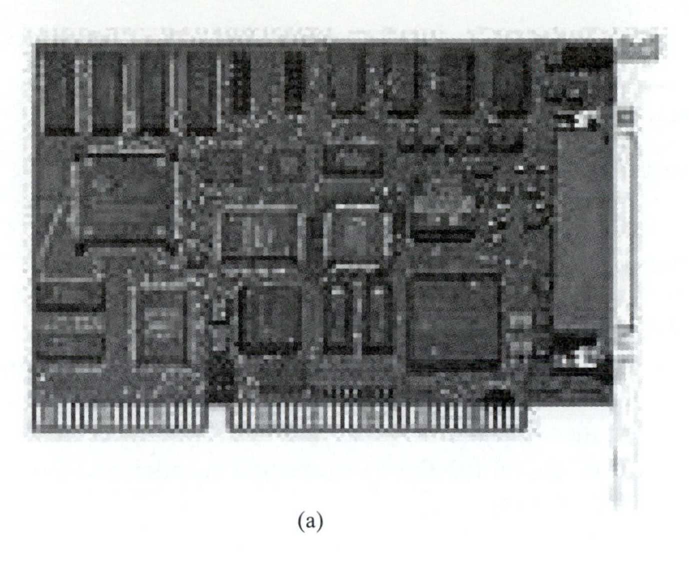

(a)

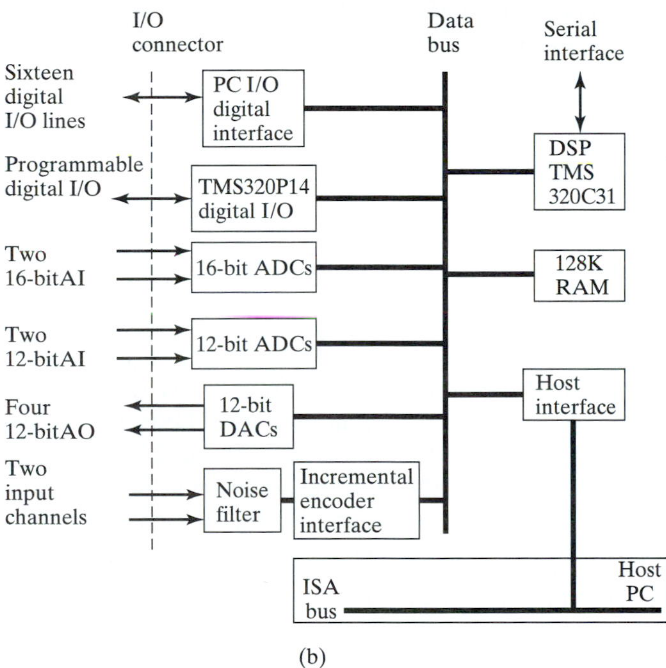

(b)

Figure 7.3

DS 1102 DSP controller board (a) and block diagram (b) (*Source*: dSpace [109])

(d) analog output, ±10-V input voltage range:

4 parallel 12-bit channels, 4-μsec typical settling time

(e) digital I/O:

programmable I/O subsystem based on a 25-Mhz TMS320P14 DSP
16 digital I/O lines
up to 6 channels PWM generation, 40-nsec resolution

(f) incremental encoder interface with noise filter, 8.3-MHz max count frequency, 24-bit counter:

2 parallel input channels

(g) frequency capture/generation

up to 8 channels

ACE kit software contains the following:

Real-time interface (RTI) for:

link between MATLAB/Simulink/Real-Time Workshop and dSPACE tools for real-time simulation, model implementation, compilation, and download to the DS 1102 DSP controller board:

configuration of ADC, DAC digital I/O lines, incremental encoder, and PWM generation with Simulink blocks.

MLIB/MTRACE for:

online experiment automation from MATLAB;

automatic parameter optimization;

data logging;

ControlDesk for:

experiment and data management;

online parameter tuning;

real-time data acquisition;

virtual instruments design.

Comparing the block diagrams of the DS 1102 DSP controller board from Fig. 7.3(b) and the block diagram of a DAQ board from Fig. 6.3, it can be observed that the differences are due to the addition of the DSP and its memory in Fig. 7.3(b). This addition permits us to dedicate a real-time target computer, the DSP, for actual monitoring and control task. The PC can be used for interface with the developer and the user and does not operate in real time. Another notable addition in Fig. 7.3 is the incremental encoder interface and a noise filter, which allow us to acquire signals from high-accuracy position sensors (incremental encoders or resolvers) used in advanced positioning devices.

7.3 LABVIEW REAL-TIME DATA ACQUISITION AND CONTROL

LabVIEW RT extends LabVIEW to the development of real-time applications. Programming LabVIEW RT applications is carried out on a host PC and the code is downloaded and executed on an independent RT target computer. The RT target computer is part of a RT DAQ board, plugged into the host PC. In this configuration, RT code runs on an RT operating system, an RT kernel, which gives deterministic control. For example, in case of a crush of the host PC under Windows operating system, the independent RT target computer continues to run the embedded executable RT code [96, 112]. As a result, this RT target can be seen as embedded in the process under monitoring and control. Such applications are common in machine and process monitoring and automation, engine control, or mine drainage control [111].

During the execution of RT code, the host PC and the RT hardware target can communicate through [113]:

RT DAQ hardware shared-memory access (Fig. 7.4);

Transmission Control Protocol/Internet Protocol (TCP/IP), a standard protocol for computer network communications

-VI server function calls.

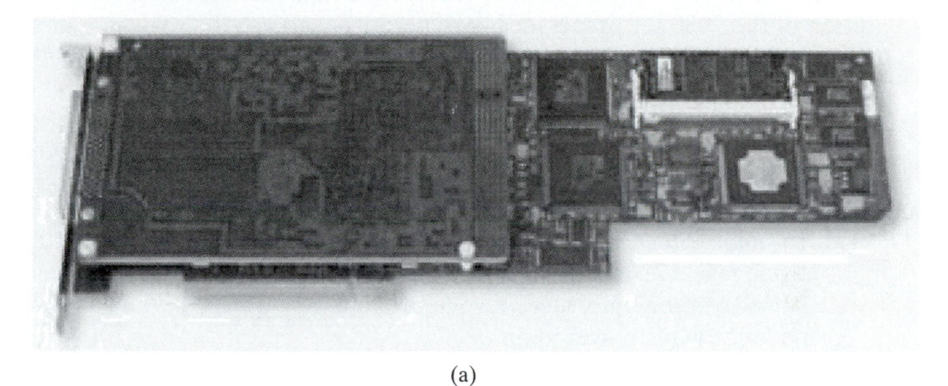

(a)

(b)

Figure 7.4

RT DAQ boards 7030/6040 (a) and block diagram (b)
(*Source*: National Instruments [109])

This presentation will focus on RT DAQ boards 7030/6040, documented as one of the National Instruments RT series RT DAQ boards [113]. RT DAQ 7030/6040 is shown in Fig. 7.4. The RT processor board contains an AMD 486DX5 133 MHz processor and 8 MB DRAM user-programmable memory.

The RT DAQ consists of two boards [113]:

7030 RT Series processor board;
6040E DAQ board PCI-MIO-16E-4.

The 6040E multifunction I/O board resembles to the DAQ board from Fig. 6.3. The addition of the 7030 RT processor board permits us to use the real-time target computer, the AMD 486 DX5, for actual monitoring and control tasks. Given that the processor board has no disk drive, keyboard, monitor, or mouse, the host PC can be used for interfacing with the developer or the user.

The 6040E PCI-MIO-16E-4 is a typical multifunction DAQ PCI board with the following characteristics [113]:

(a) 16 single-ended or 8 differential 12-bit analog input channels; ±0.05, ±0.1, ±0.25, ±0.5, ±1, ±2.5, ±5, and ±10-V input voltage ranges; 250-kS/sec sampling rate; 3 μsec settling time, successive approximation ADC; 100-GΩ amplifier input impedance when powered on
(b) two channels, 12-bit analog output, ±10-V input voltage range, 1-Ms/sec output rate, 0.1-Ω maximum output impedance, ±5 mA maximum current drive
(c) eight digital I/O lines (5V/TTL compatible for logic 1 and logic 0 [14])
(d) two up/down counters/timers, 24-bit resolution
(e) analog and digital triggers.

LabVIEW RT software consists of [96, 112, 113]

RT development system;
RT engine.

The RT development system is installed in the host PC under Windows and can be used for LabVIEW graphical programming, debugging, and downloading the RT code to the RT series hardware target.

The RT engine is installed in the RT series processor board, which executes, in real time, the downloaded and embedded LabVIEW RT programs. This system, shown in Fig. 7.4, enables deterministic real-time control of embedded executables uninterrupted by host PC operation under Windows. Downloading the RT code from the host PC to the target RT series processor board using the RT development system is carried out by pull-down menu selection. The execution of the RT code, on the target RT series processor board, continues after exiting the RT development system, running on the host PC.

Shown in Fig. 7.5 is the real-time LabVIEW RT Control.vi Diagram (Control Example Screen Shot.vi) [96].

This diagram shows a simple example of a PID control, which compares the desired set point value with the actual value of the controlled process variable, and computes the PID control output. PID controller parameters or gains and the desired set

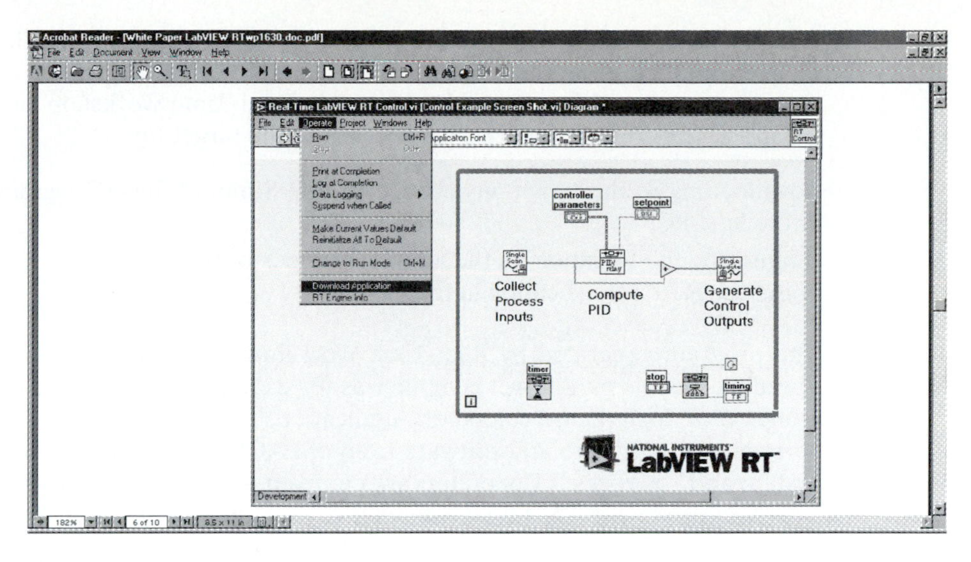

Figure 7.5

Real-Time LabVIEW RT Control.vi diagram
(*Source*: National Instruments [96])

point value can be changed, during the execution of the RT code on the target RT series processor board, from the front panel of the LabVIEW RT Control.vi, shown in Fig. 7.5, on the monitor of the host PC. Moreover, the indicators on this front panel permit dynamic viewing of chosen process variables.

The RT performance of the LabVIEW RT is due to the following factors [83, 96]:

The RT engine runs on an RT operating system of the target RT series processor board and ensures run-time performance for higher priority threads of the multithread operation;

The target RT series processor board, running the RT engine, has no disk drivers, network drivers, or other device drivers or virtual memory that could be source of nondeterministic operation.

7.4 MATHWORKS TOOLS FOR REAL-TIME DATA ACQUISITION AND CONTROL

MATHWORKS offers tools for real-time applications on a variety of hardware platforms, while dSPACE and National Instruments software packages for real-time applications, presented in Sections 7.2 and 7.3, were developed for their own hardware. Three tools from MATHWORKS will be presented in this section:

- **(a)** Real-Time Workshop®, for RT code generation from Simulink models on the host PC [103];
- **(b)** Real-Time Windows target, for running Simulink models in real-time on the host PC [104];
- **(c)** xPC target, for running Simulink models in real-time on a target PC [102].

(a) Real-Time Workshop allows the user to generate, debug, and compile the code from a Simulink program in order to create an RT executable using various operating systems and for various DAQ boards. Real-Time Workshop has an external mode feature that connects the host PC to the target for

downloading to the target any changes in the Simulink block diagram made in the host PC;

monitoring the variation on the Simulink scopes of the variables of interest during the RT code execution in the target.

The program generated by Real-Time Workshop can run in real-time existing Simulink models on a target computer as for example a PC, a DSP, a microcontroller, or a single-board computer. Applications for real-time embedded processors can use hardware and software from dSPACE [109], Tornado development system and VxWorks RT operating system from Wind River [114], or other real-time tool producers. The target computer can be under a real-time operating system or under DOS or Windows operating systems.

Real-Time Workshop allows us to select among various code formats:

embedded C-code format, which produces reduced size codes, but not for continuous time systems;

Real-Time code format, which allows changing parameters and display in real time in Simulink, supports both continuous and discrete time systems and allocates memory by static declaration at compilation time;

Real-Time Malloc code format, which is similar to Real-Time code format, but allocates memory by dynamic declaration.

dSPACE offers a Real-Time interface, which permits the code generated by Real-Time Workshop to run on dSPACE hardware.

(b) Real-Time Windows target permits the use of a PC under Windows to run, in real time, the code generated by Real-Time Workshop from Simulink. In this case, the same PC under Windows runs Simulink models and the real-time code. Simulink scopes can be used for monitoring the variation of the variables of interest during the RT code execution. Changes of parameters in the Simulink model are automatically transferred to the real-time model during its execution. Real-time monitoring and control with the Real-Time Windows target uses real-time inputs and outputs from a DAQ board plugged into the PC. Sensors and actuators can be connected to the DAQ board and interfaced directly with a Simulink model.

The sequence of operations for signal processing and control applications is as follows:

program graphically in Simulink the algorithm for signal processing or control;

generate automatically C-code with Real-Time Workshop and build an RT executable;

run theRT executable on the Real-Time Windows target.

Windows can be used in this case for RT applications because the real-time kernel addition runs at the highest priority in the Windows environment.

The gateway from the DAQ board to the real-time model is provided by RT in and RT out blocks.

(c) xPC target provides a real-time kernel for applications on an RT target computer, different from the host used for Simulink model development. xPC Target permits us to add input/output blocks to Simulink block diagrams.

Figure 7.6 shows the block diagram of host PC -RT target PC using xPC.

The sequence of operations for real-time applications is:

program graphically in Simulink,
generate code with Real-Time Workshop,
download the code to a second (target) PC under xPC target real-time kernel, and
run the real-time code.

During the operation of the target PC under xPC, it runs as a RT computer, while after closing xPC and restarting under Windows, the same PC can execute any Windows program stored in the hard drive. This permits us to use any 386 or 486 PC as an RT target PC. Target scope can have update rates for display comparable to those of digital oscilloscopes.

The communication between the host PC and the target PC is shown in Fig. 7.6 using serial ports. A faster communication option, up to 10 Mbit/sec, is achieved with TCP/IP standard protocol for PCs equipped with Ethernet cards.

Target PC can achieve standalone, embedded operation using the xPC target embedded option without communication with the host PC.

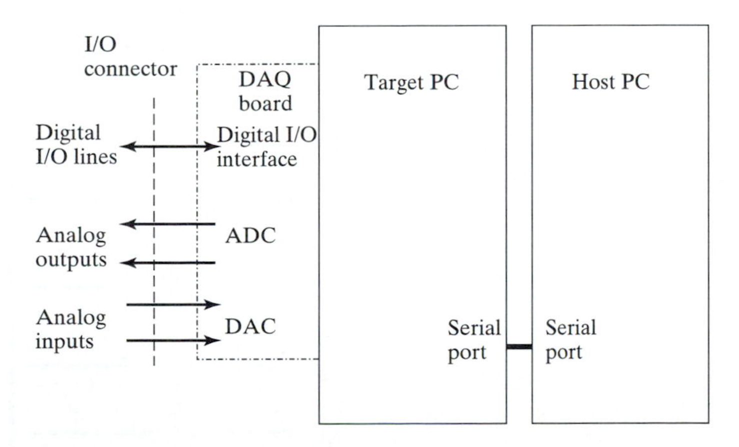

Figure 7.6

Block diagram of host PC–RT target PC using xPC

7.5 EMBEDDED SINGLE-CHIP COMPUTERS FOR SYSTEM INTEGRATION

7.5.1 General Issues

The distinction between microcontrollers and microprocessors is the result of difference of purpose, generation of outputs for process control in the case of microcontrollers, and data processing for the case of microprocessors.

A microprocessor is a single chip integrated circuit containing a central processing unit (CPU). The components of the CPU are the arithmetic logic unit (ALU) for digital calculations, registers for temporary storage, and a controller for program execution.

A microcontroller is a single-chip microcomputer for embedded control applications (i.e., an embedded computer). Besides the CPU, the simple microcontroller (shown in Fig. 7.7) has on-chip memory (ROM and RAM) and input/output interface.

Most CPUs are used in embedded applications as microcontrollers. Microcontrollers are available as 4-, 8-, 16-, or 32-bit chips. Complex applications require microcontrollers with data acquisition capabilities and take the form of embedded computers with the typical block diagram shown in Fig. 7.8.

In Fig. 7.8, compared to Fig. 7.7, new components are added in the single-chip microcontroller:

ADC for acquiring analog input (AI) signals from sensors;

DAC for sending analog output commands to actuators;

pulse width modulation (PMW) module for efficient control of DC motors using variable duty cycle in accordance with the required magnitude of the output;

interrupt controller for handling interrupt request priorities.

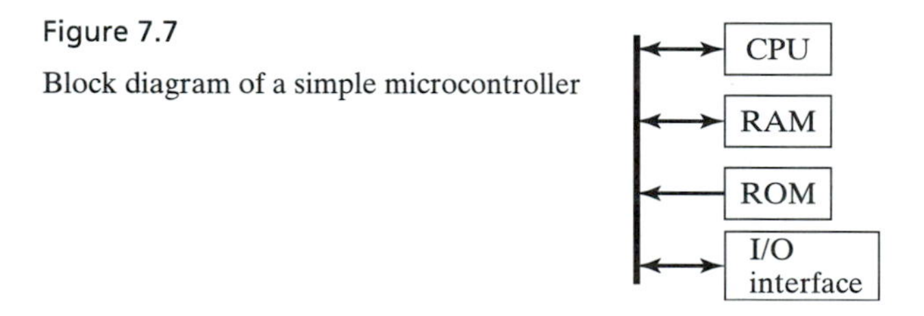

Figure 7.7

Block diagram of a simple microcontroller

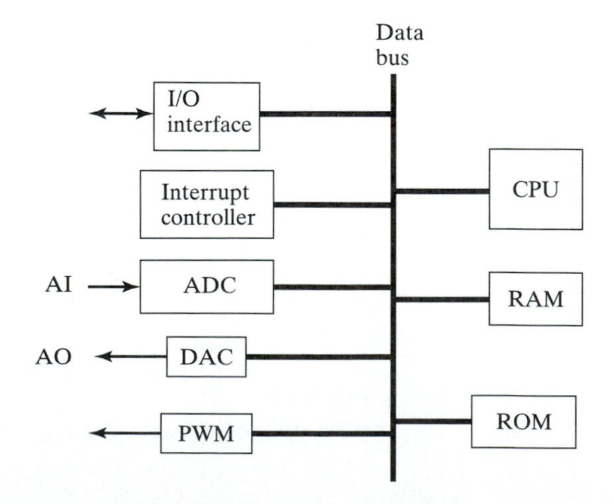

Figure 7.8

Block diagram of a typical embedded computer

Besides microcontrollers, programmable controllers with specific functions for discrete event sequencing control are used frequently in industry. Lately, programmable controllers and microcontrollers share the same hardware structure of a microcomputer and are distinguishable mostly in software [92].

Microcontrollers are produced by Intel, Motorola, Hitachi, Texas Instruments, Thomson CSF, etc. In this chapter, the presentation will focus on Intel embedded microcontroller 80C196KB and Parallax Basic Stamp II integrated circuit computer. Simple applications of these embedded computers are also presented.

7.5.2 Intel Embedded Microcontroller 80C196KB

The 80C196KB microcontroller is a member of MCS-96 family of Intel microcontrollers with a 16-bit CPU and minimum of 230 bytes of on-chip RAM [97]. These microcontrollers are used for real-time implementation of controllers and digital signal processing for modems, electric motors, engines, antilock brakes, air conditioning systems, medical instruments, computer peripherals, etc. These microcontrollers perform arithmetic operations, including multiplication and division.

Figure 7.9 shows the block diagram of the 80C196KB microcontroller. The blocks of interest for the application, presented in this chapter, are as follows:

- CPU with 232-byte register file
- interrupt controller
- optional 8-KByte on-chip ROM
- two 16-bit timers
- serial port
- pulse width modulator (PWM)
- 10-bit A/D converter for eight analog inputs, sample and hold (S/H), and 8-to-1 multiplexer (MUX)
- high-speed I/O
- clock generator

Port 0 transfers up to eight analog input lines to the A/D converter. Ports 2, 3, and 4 are bi-directional and share their pins with the address/data bus. Timer2 increments are produced by external positive and negative transition. PWM provides output waveform with variable duty cycle that can drive various types of motors.

Figure 7.10 shows the 68-pin PLCC package of the 80C196KB microcontroller. The pins of interest for the application presented in Fig. 7.11 are

1	Vcc = 5 V	main supply voltage
4-11	P0.0 to P0.7	port 0 8-bit high impedance inputs to be used for analog inputs to A/D converter (pin 6 ACH0/P0.0 is used for analog input from current sensing circuit)
12	ANGND	reference ground for the A/D converter
13	Vref = 5 V	reference voltage for the A/D converter
14 and 68	Vss = 0 V	digital circuit ground
33	P2.6/T2UP/DN	direction control for the Timer2 clock as an up and down counter
38	P2.7	quasi-bidirectional line

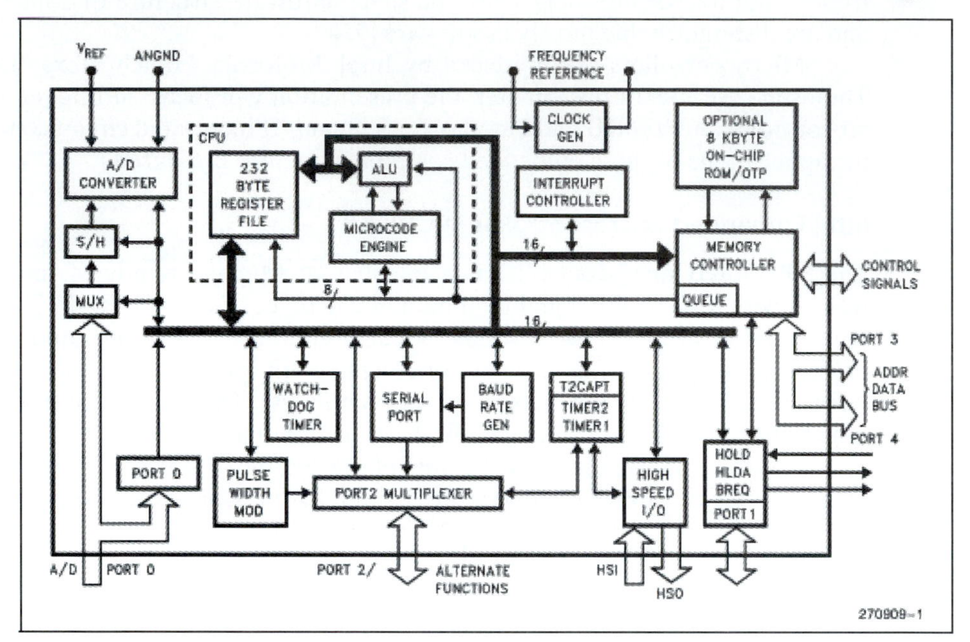

Figure 7.9

Block diagram of 80C196KB microcontroller

(*Source*: Photo courtesy of Intel Corporation. Copyright Intel [97])

39	PMW	PWM/P2.5
44	T2CLK/P2.5	Timer2 input or serial port input

Figure 7.11 shows the wiring diagram for position control of a DC servomotor with a 80C196KB microcontroller [117]. The main components are as follows:

80C196KB microcontroller

DC motor

2-track incremental encoder for relative angular displacement measurement and direction of motion sensing,

motor driver using a bidirectional scheme with power transistors

current sensing circuit for limiting the current to safe values.

The program for real-time embedded computer 80C196KB can be written in assembly language in order to avoid software overloading.

7.5.3 Parallax Basic Stamp II Integrated Circuit Computer

Basic Stamp II BS2-IC integrated circuit computer has features that make it particularly suitable for educational use. Programming BS2-IC is facilitated by the PBASIC

8XC196KB/8XC196KB16

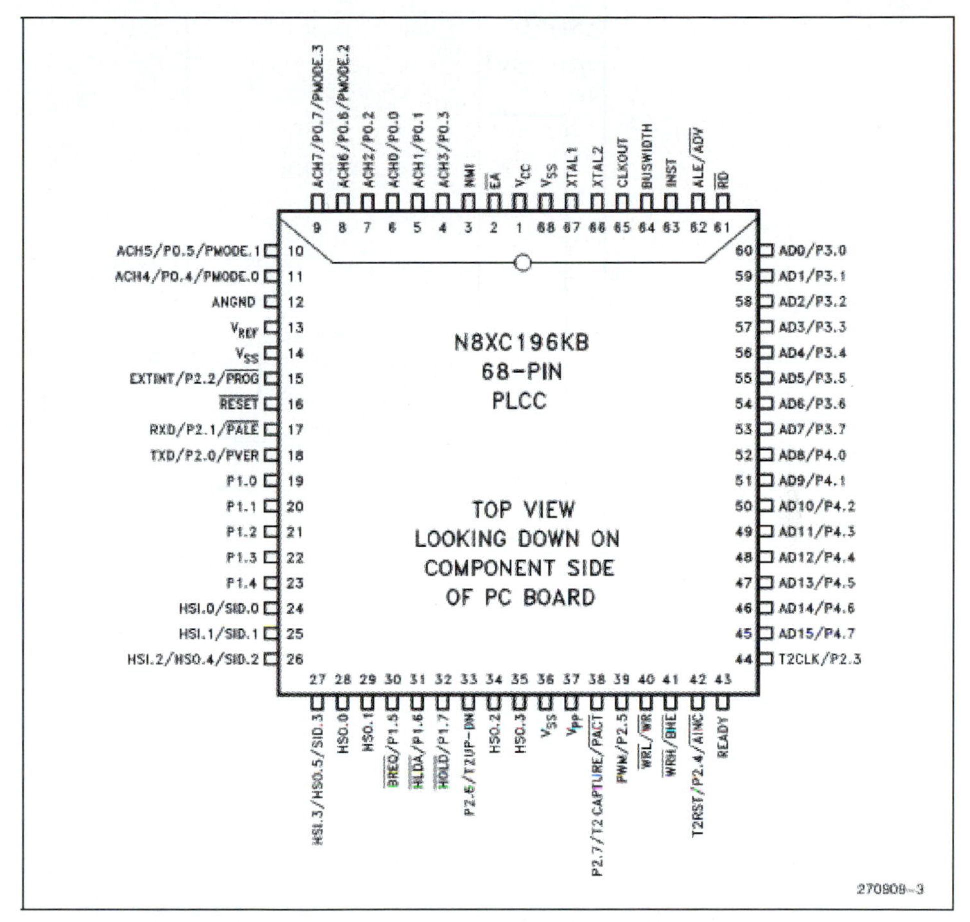

Figure 7.10

68-pin PLCC package of 80C196KB microcontroller

(*Source*: Photo courtesy of Intel Corporation. Copyright Intel 2001 [97])

interpreter stored permanently on the chip ROM. PBASIC is a form of BASIC language adapted for embedded computers. BS2-IC stores and executes only one program at a given time and the program can be stored in the EEPROM (electrically erasable programmable read-only memory).

Figure 7.12 shows the BASIC Stamp II BS2-IC module and its block diagram. The main components of the BS2-IC module are [116] as follows:

PIC 16C57 microcontroller with 20-MHz resonator executing 4000 instructions/sec:

2 Kbytes of EEPROM for nonvolatile storage of the program (up to 500 lines of PBASIC) and data;

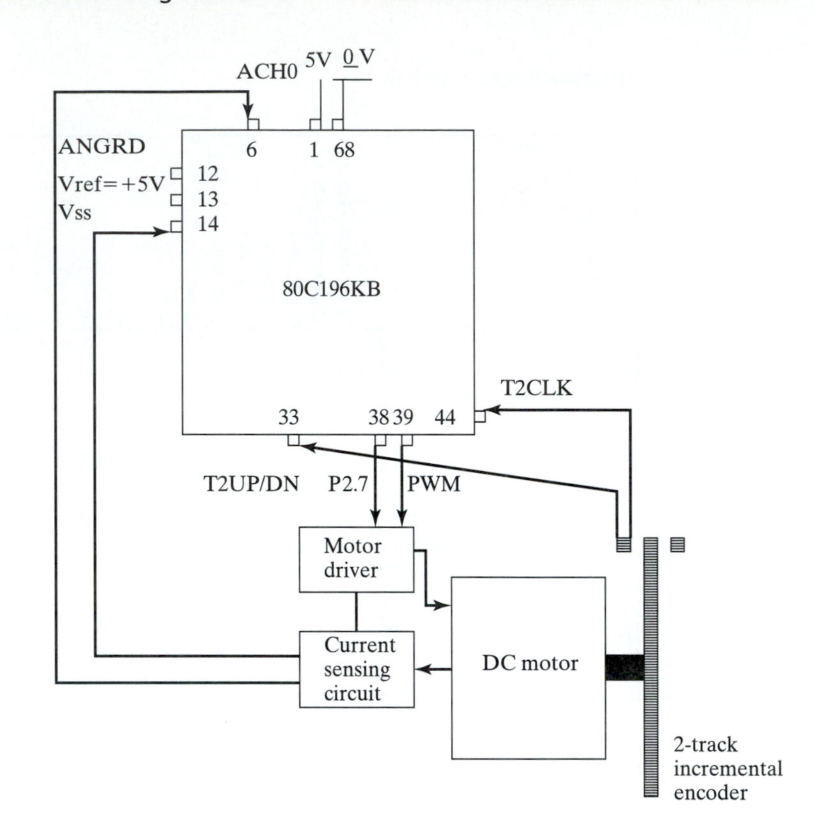

Figure 7.11

Position control of a DC servomotor with 80C196KB microcontroller

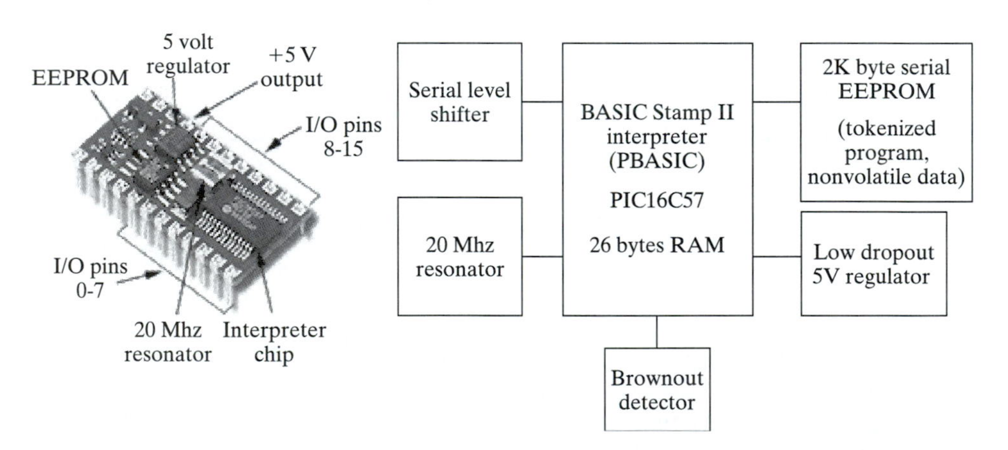

Figure 7.12

BASIC Stamp II BS2-IC module
(*Source*: Parallax [116])

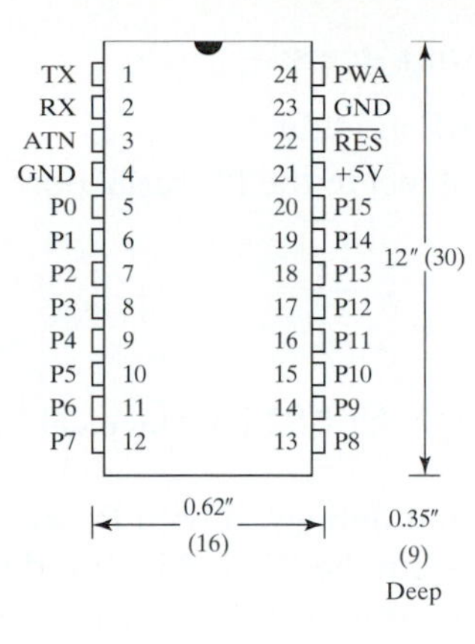

Figure 7.13

BASIC Stamp II BS2-IC 24-pin DIP module (*Source*: Parallax [116])

32 bytes of RAM;

16 digital input/output; 50-kbaud serial I/O;

low-dropout 5-V regulator.

The BS2-IC package is a 24-pin DIP (dual in-line package) module, as shown in Fig. 7.13. Pin allocation is

1 to 4	TX, RX, ATN and GND, renamed in Fig. 7.14 as SOUT, SIN, ATN, and VSS, are used for serial connection
5 to 20	P0 to P15 for digital I/O with TTL compatible devices
21 + 5 V	VDD
22 /RES	reset
23 GND	VSS = 0 V, ground
24 PWR	VIN = 5 V for power supply (with power regulator, the supply voltage can be between 5 V and 15 V DC).

Programming BS2-IC for applications requires the following equipment [106]:

BS-2 IC module

BS2-IC carrier board

BS2 interface cable for serial connection of BS2-IC and the PC serial port Parallax developing system software, to be installed in the PC, for PBASIC program editing and debugging.

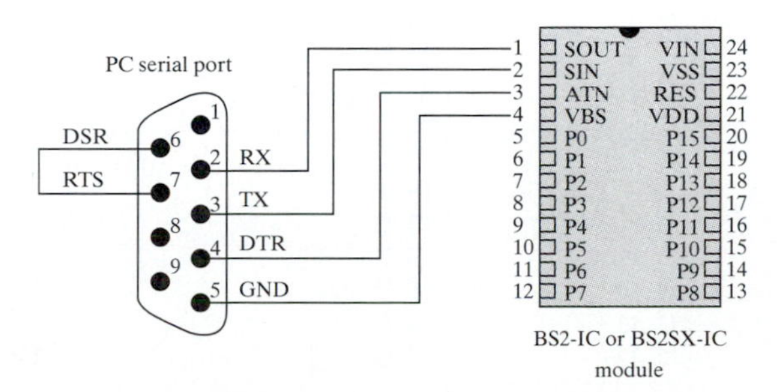

Figure 7.14

BS2-IC connection to PC serial port (*Source*: Parallax [116])

The carrier board contains [106] the following items:

a socket for plugging in the 24-pin DIP module
a serial connection for the cable to connect to the PC serial port, shown in Fig. 7.14
9-V battery clips
I/O header
reset button.

The serial connection of BS2-IC module, for the BS2 interface cable to access the PC serial port, is shown in Fig. 7.14.

The PBASIC source code is edited and debugged with Parallax developing system software. The program is downloaded from the PC and is stored in the 2 KByte on-chip EEPROM. EEPROM not used for program storage can be used for nonvolatile data storage [106].

The BS2-IC has a digital I/O communication port with only TTL-compatible devices, for example,

sensors (logic 1 for inputs of 2 V to 5.5 V and logic 0 for inputs of 0 V to 0.8 V)

and

actuators (logic 1 for outputs of 2.4 V to 5.5 V and logic 0 for outputs of 0 V to 0.6 V).

For example, TTL-compatible buttons, LED, and potentiometers can be directly connected to the BS2-IC I/O lines. Non-TTL-compatible devices with on/off operation can be connected to the BS2-IC I/O lines using a signal conditioning of a driver interface.

An example of open-loop control of a DC motor is shown in Fig. 7.15. In this example, one of the I/O lines, in this case pin 20, P15, of the BS2-IC is connected to a DC motor drive circuit. The components are [116] as follows:

a NPN transistor
a 100 ohm resistance between P15 and the base of the NPN transistor
a 10 ohm resistance between the emitter of the NPN transistor and one of the motor leads
a DC motor with one lead grounded.

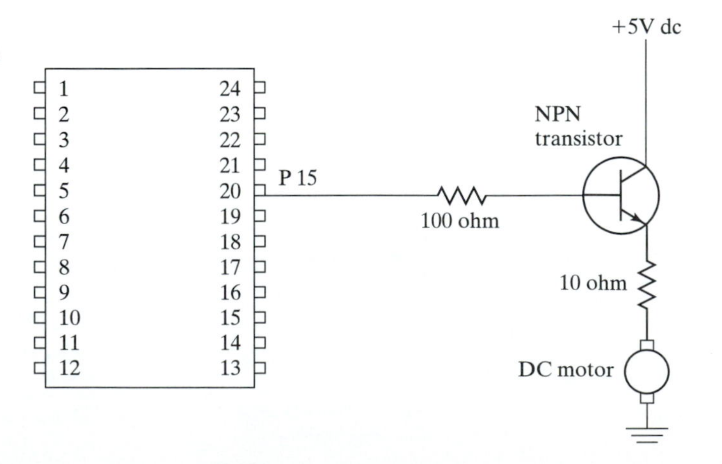

Figure 7.15
On/Off control of a DC motor using I/O pin of BS2-IC

The PBASIC program for achieving a repeated cycle with a period of 2 sec with a duty cycle of 25% (i.e., the motor voltage supply on for 0.5 sec and off for 1.5 sec) is the following [106, 116]:

start:

low 15	changes pin 15 output setting to Low to have the motor off
pause 500	pause for 500 msec to have the motor voltage supply off for 0.5 sec
high pin 15	changes pin 15 to output set to High to turn on the motor
pause 1500	pause for 1500 msec to have the motor voltage supply off for 1.5 sec
goto start	to repeat the cycle as long as the power supply of BS2-IC is on.

For acquiring analog input signals, the LTC 1298 ADC, shown in Fig. 7.16, can be used [116].

LTC 1998 has:

two channels ADC IN^+ and IN^-
a real-time clock with maximum 200 kHz rate
22 mV resolution.

A specialized interface for velocity or position control as well as direction of motion control of DC motors with BS2-IC is Motor Mind B driver module, shown in Fig. 7.17.

Motor Mind B has the following characteristics:

voltage supply up to 30 V DC
peak current up to 3.5 A
continuous current up to 2 A
serial communication interface
input for tachometers and optical encoders up to 65 kHz frequency
PWM with frequency of 64 Hz and variable duty cycle in 254 discrete steps
9-pin SIP (single in line package).

In Fig. 7.17, the pin allocation of Motor Mind B is [116]

VMOTOR	pin for voltage supply for motor
MOTOR+	DC motor lead connection pin

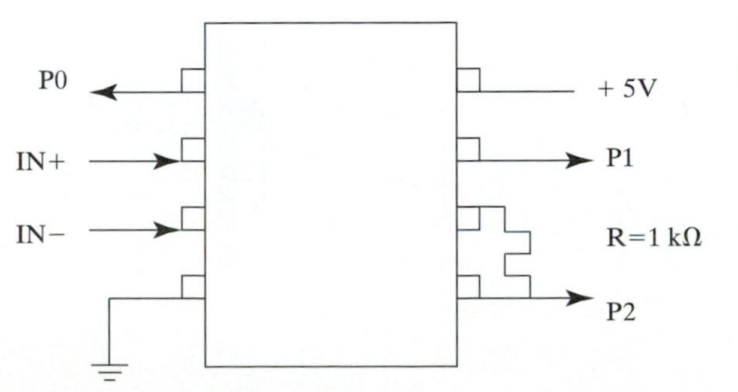

Figure 7.16
LTC 1298 ADC Diagram

MOTOR−	DC motor lead connection pin
TACH_IN	input pin for tachometers and encoders
/BRAKE	pin for motor stop signal when low and with no resetting of the module
FM	pin for data reception from master (BS2-IC)
TM	pin for data sending to master (BS2-IC)
GND	common ground potential pin
Vcc	supply voltage input pin for control circuits.

An example of wiring for angular position and direction of motion control of a DC motor and incremental encoder data acquisition using a BS2-IC and a Motor Mind B is shown in Fig. 7.17. The components in Fig. 7.17 are

BS2-IC
Motor Mind B
DC motor
incremental encoder.

The wiring is the following:

(a) from BS2-IC to Motor Mind B

pin 23 GND with pin GND
pin 21 VDD = +5 V with Vcc

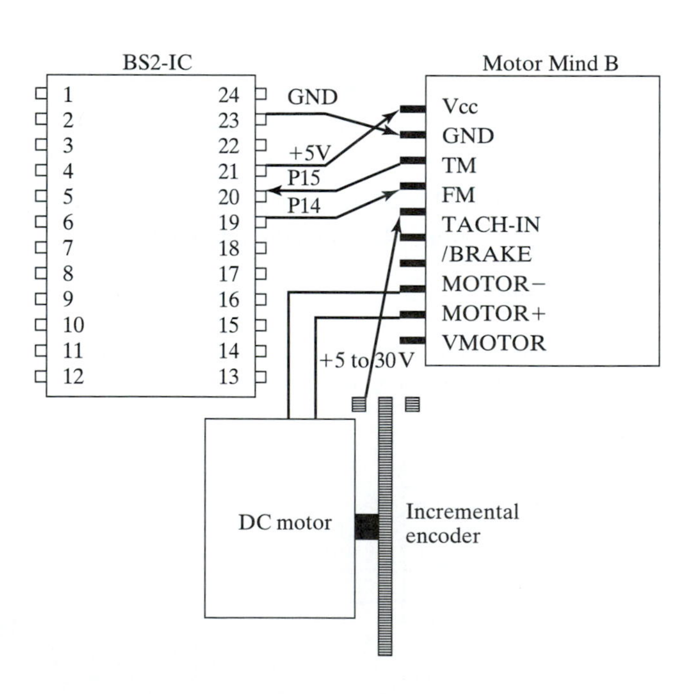

Figure 7.17
Position control of a DC servomotor using BS2-IC and Motor Mind B

pin 20 P15 with pin TM
pin 19 P14 with pin FM

(b) from Motor Mind 2 to the DC motor

pins MOTOR+ and MOTOR− with the leads from the motor

(c) from Motor Mind B to the incremental encoder

pin TACH-IN with output from incremental encoder.

Asynchronous serial communication between BS2-IC and Motor Mind B is carried out on TM (from BS2-IC to Motor Mind B) and FM (from Motor Mind B to BS2-IC) wires.

The serial communication takes place as follows: one start bit, eight data bits, and one stop bit. The communication can only be initiated by the BS2-IC and starts with '55'h (or binary 0101 0101) sync byte sent to FM pin. Motor Mind B can send data to BS2-IC only in response to a request from BS2-IC. Motor Mind B understands and executes commands sent from BS2-IC, as for example [116],

REV command '01'h, or binary 0000 0001, which changes the direction of motion of the motor;

TACH command '02'h , or binary 0000 0010, which causes the Motor Mind B to send two bytes value of the frequency value.

These commands sent by BS2-IC are the ones obtained from a PBASIC program executed by BS2-IC, which sends the binary command to the Motor Mind B using pin FM.

PROBLEMS

7.1. Compare the block diagram of a standard DAQ board with the block diagram of a DAQ board containing a real-time target computer. Identify the blocks specific to the DAQ board containing a real-time target computer and describe their functions. Analyze the role of the real-time target computer with regard to the response time.

7.2. Does an embedded computer need a permanent link to a PC for performing real-time tasks?

7.3. What is the difference between a digital signal processor and a standard processor?

7.4. Compare the block diagram of the DS 1102 DSP controller board with the block diagram of a embedded computer. Identify the differences.

7.5. Search the characteristics of an advanced microcontroller—for example Intel [97]—and evaluate if it can be interfaced with an incremental encoder, which generates 1 million pulses per second?

7.6. Determine the domain of real-time applications of single-chip computers as opposed to real-time DAQ boards for PC. Can an embedded computer perform as complex data analysis as a real-time DAQ board plugged into a PC equipped with signal analysis software?

7.7. Obtain the block diagram of the position control of a two degrees of freedom robot using two DC motors and a microcontroller.

7.8. Obtain the block diagram of the position control of a DC motor with both optical encoder and tachometer feedback using BS2-IC and Motor Mind B.

7.9. Obtain the block diagram of the position control of a two degrees of freedom robot using two DC motors with both optical encoder and tachometer feedback using BS2-IC and Motor Mind B. Can the control be implemented with a single Motor Mind B?

CHAPTER 8

Laboratory Experiments for Mechatronics

8.1 OVERVIEW

Mechatronics is a field of study that cannot be complete without hands-on experiments. Mechatronics experimental study has to cover computer integration of electromechanical systems. Experimental set-ups for this purpose should contain a variety of mechanical and electrical components, interfaces with computers, and programmable digital components for data acquisition and system automation.

Several commercial packages are available for mechatronics experimental study. Many packages were initially developed for automatic control applications but, as long as they contain motors, mechanical loads, digital-to-analog and analog-to-digital converters, and a computer for monitoring and control, they can be used for experimental study in mechatronics as well.

An example is Servo Fundamentals Trainer 33-001 shown on the web site of Feedback Instruments [119]. The components of this kit are

- a mechanical unit with input and output potentiometers, motor, tachometer, power supply, etc.;
- an analog unit with amplifiers, PID controller, etc.;
- a digital unit, A/D & D/A converters, linear, or PWM motor drives;
- discovery software for PC-based instrumentation, which includes a chart recorder.

MATLAB is used for digital signal processing and control design.

Electromechanical Linear Positioning System EMPS300 from dSPACE is another example of a system developed for automatic control applications, but which can be used for experimental study in mechatronics as well [109]. This system, designed for educational use, contains a DC motor, a DC tachometer, a PWM servo amplifier, a ball-screw transmission for transforming rotational and translational motion, and an incremental encoder. The load shaft of the DC motor is equipped with a spring element to simulate flexible transmission effects. All analog and digital signals from EMPS300 are available through a cable for connection to the DAQ system.

Mechatronics experimental set-ups are offered by Feedback Instruments for studying 16 common industrial transducers, AC and DC instrumentation, and some signal processing methods (Kits TK2942, TK2941M [119]).

Commercial packages for hardware-in-the-loop experimentation are also suitable for mechatronics experimental study. The dSPACE simulator can be used for testing controllers, electronic control units, and other components in a hardware-in-the-loop environment [109]. ETAS offers hardware-in-the-loop test systems for electronic control units (ECU) intended for automobile industry applications. The main components are LabCar models, developer, operator automation, hardware, and ECU [123].

University laboratories also developed experimental facilities relevant to mechatronics study. For example, the Reconfigurable Hardware Testbed for Elastically Coupled Systems was developed in graduate thesis work at University of Washington, particularly for PC-based experimental test of control systems of elastically coupled rectilinear mass-spring-damper systems [122]. The main components of the system are a DC motor, motor power supply, servo amplifier, AC100 real-time control system from ISI [110], analog amplifier, sensors, and a Sun workstation plus Xmath/System Build software [110]. Servo Fundamentals Trainer 33-001 from Feedback Instruments [119], EMPS300 from dSPACE [109], and the Reconfigurable Hardware Testbed for Elastically Coupled Systems from the University of Washington [122] have in common the goal of educational use in testing control strategies on mixed electromechanical systems and can also serve mechatronics study of electromechanical system integration using a PC.

Mechatronics books also present design examples:

timed switch using a programmable logic controller (PLC);
windscreen wiper motion control using a microcontroller;
bathroom scale using a microcontroller and LED display;
a pick and place robot;
car park barrier under PLC control;
automatic camera using a microcontroller;
car engine management using a microcontroller;
bar code reader.

In addition, mechatronics books present various mechatronics applications and case studies [30]:

on-line quality monitoring;
integration of heterogeneous manufacturing system;
real-time robotic interface application;

testing of transportation bridge surface materials;
transducer calibration;
strain gauge weighing system;
rotary optical encoder, etc.

Given the specificity of this book, in which LabVIEW, MATLAB, and dSPACE were used to illustrate PC-based monitoring and control of mechatronic systems, in the next three sections, experiments using these computer packages will be presented:

experimental set-ups for interfacing sensors and actuators using LabVIEW (Section 8.2),
MATLAB sound data acquisition and analysis (Section 8.3), and
robot control using dSPACE (Section 8.4).

The experimental set-ups presented in Sections 8.2 and 8.3 were developed using readily available components (mostly from Radio Shack) and are presented with sufficient details such that they can be easily reproduced.

8.2 INTERFACING SENSORS AND ACTUATORS USING LABVIEW

8.2.1 Thermistor

Shown in Fig. 8.1 is the experimental set-up for a thermistor (shown as $R + \Delta R$) interfaced to the analog input of V_o of PC DAQ board.

The components of this experimental set-up are

INTERTAN model 271-110A thermistor, 27 kΩ ±1% at 0°C [118];
three R = 27-kΩ resistors;
9-V battery, connected to the bottom wires;
GSP board model GB2-354 circuit board

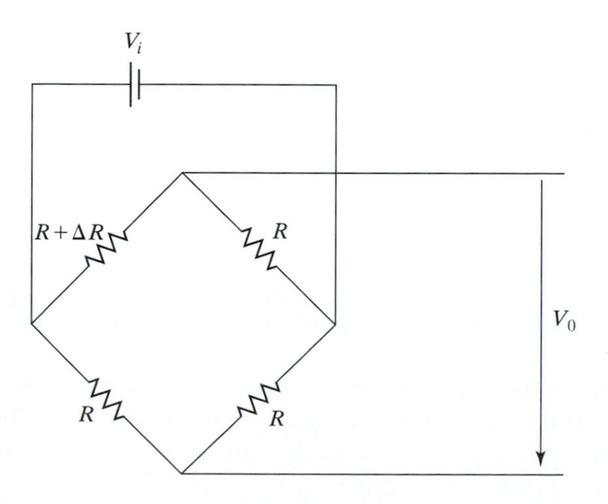

Figure 8.1

Wheatstone bridge for resistance variation ΔR measurement with deflection method

The temperature-versus-resistance look-up table, given by the manufacturer, is the following:

Temperature [°C]	−50	−45	−40	−35	−30	−25	−20	−15	−10	−5
Resistance [kΩ]	329.2	247.5	188.4	144.0	111.3	86.39	67.74	53.39	42.45	33.89

Temperature [°C]	0	5	10	15	20	25	30	35	40	45	50
Resistance [kΩ]	27.28	22.05	17.96	14.68	12.09	10.00	8.31	6.941	5.83	4.91	4.16

Temperature [°C]	55	60	65	70	75	80	85	90	95	100	105	110
Resistance [kΩ]	3.54	3.02	2.59	2.23	1.92	1.67	1.45	1.37	1.11	0.97	0.86	0.76

The measurement of the resistance variation due to temperature measurement is achieved using a Wheatstone bridge, shown in Fig. 8.1 and the deflection method, presented in Section 2.3.3. The resistance of the thermistor of $R = 27.28$ [kΩ] at 0°C and changes to $R + \Delta R$ at nonzero temperatures.

The relationship between voltage output V_0 and ΔR obtained in Section 2.3.3 follows:

$$V_0 = V_i \frac{R\Delta R}{4R^2 + 2R\Delta R}.$$

For selected values of the temperature versus resistance from the previous look-up table, the output voltage V_0 values calculated with the foregoing equation are

Temperature [°C]	$R+\Delta R$ **[kΩ]**	**Voltage [V_0]**
−50	329.2	3.52
0	27.28	0.02
10	17.96	−0.83
15	14.68	−1.27
25	10.0	−1.91
35	6.941	−2.46
100	0.9735	−3.86
110	0.7599	−3.92

The results show that V_0 varies between 3.52 V and −3.92 V for temperatures between -50 and 110°C and does not exceed the maximum value accepted by the DAQ board for the analog input of 10 [V]. The temperature versus resistance look-up table is not convenient in PC-based instrumentation, and a polynomial $T = p(V)$ is considered more convenient in this case.

The power equation $T = -V_0^{\,a}$, where a is a constant to be determined, can be fitted to the look-up table data. The value of the constant a can be obtained using the minimum mean square error method. The value $a = 3.9$ can be used for this illustrative example. The equation $T = -V_0^{\,3.9}$ is used in the LabVIEW program to convert the measured V_0 into temperature T.

The experimental set-up contains the thermistor with resistance $R + R$, three $R = 27$-kΩ resistors and the battery for $V_i = 9$ V, connected in accordance to Fig. 8.1. The voltage output V_0 is connected to channel 0 analog input of a National Instruments DAQ board plugged into a PC, as shown in Fig. 6.3. The DAQ board is configured as device 1 of the PC.

The measurement is carried out with the thermistor1.vi LabVIEW program shown in the block diagram form in Fig. 8.2. An AI Acquire Waveform.vi is used to acquire the analog signal V_0 from channel 0 of the device 1, with a sampling rate of 500 [samples/sec] for 0.4 sec. (i.e., for a total number of samples 200 [samples]). The acquired voltage V_0 is converted into temperature T using the equation $T = -V_0^{3.9}$, programmed in the block diagram by Multiply.vi with -1 and Power Of X.vi with the constant $a = 3.9$. The results are displayed in the front panel waveform graph labeled Temperature [degree Celsius], shown in Fig. 8.3.

The results of temperature measurement show a random variation in a range of about ±0.2°C, about an average room temperature of 26.4°C. The random variation can be explained by the effects of the electric noise in the thermistor and in the analog signal transmission.

The measurements can take place for the following cases, for which the temperature is known such that the temperature measurement error can be calculated:

	Temperature *T* [°C]
boiling water	100
body temperature	36
melting ice	0

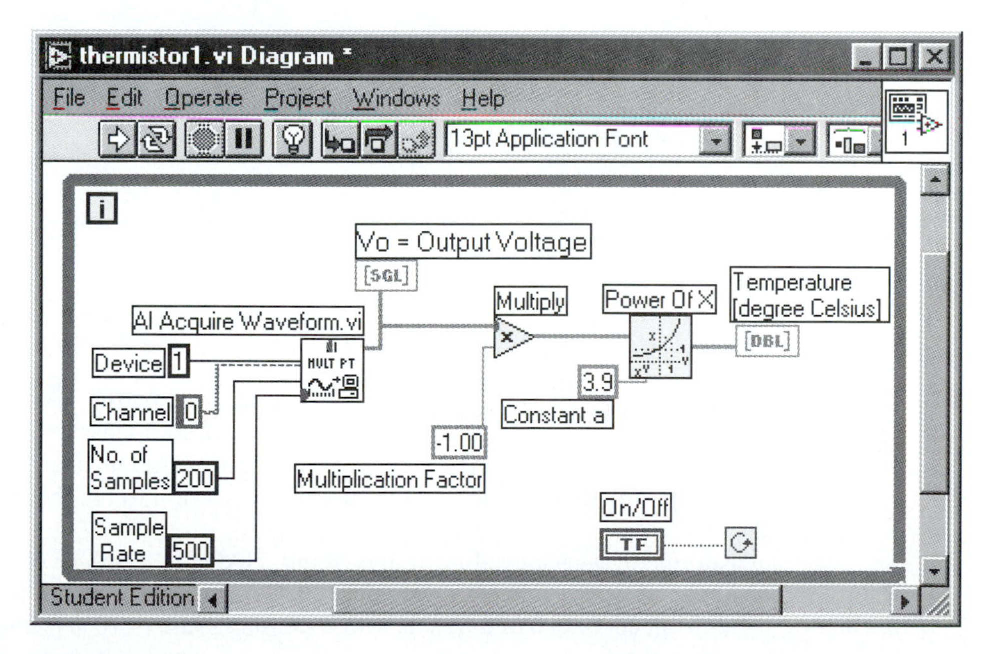

Figure 8.2

Thermistor1.vi Block diagram

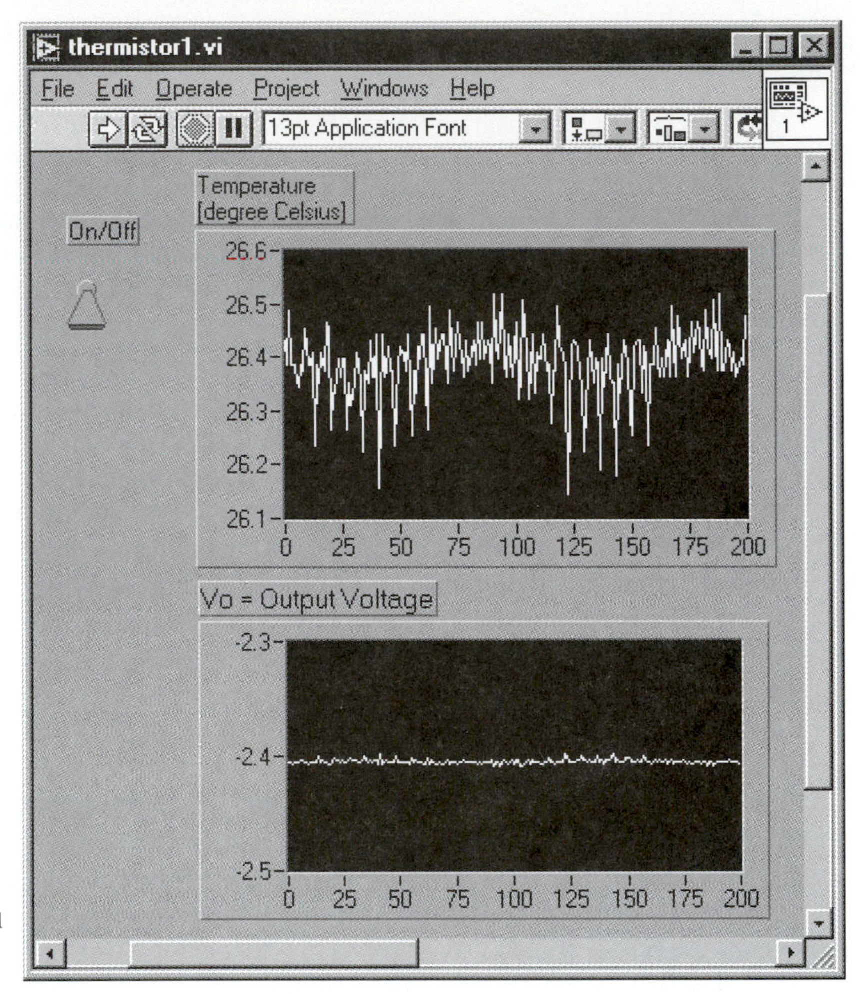

Figure 8.3

Thermistor1.vi front panel for room temperature measurement

8.2.2 Thermocouple

The thermocouple type K used in this experiment, shown in Fig. 8.4, was presented in Section 2.2.

The voltage output from the thermocouple is conditioned by the TAC-386 thermocouple-to-analog converter [16]. TAC-386 linearizes and compensates the output voltage from the thermocouple to 1 mV/°C $=1/10^3$ [V/°C] for DAQ measurement [16].

Figure 8.4 shows the diagram of the DAQ board connection to the thermocouple K and the TAC-386 box.

The measurement is achieved with the Temp_DAQ1.vi LabVIEW program shown in block diagram form in Fig. 8.5. An AI acquire Waveform.vi is used to acquire the analog signal from channel 0 of device 1, with a sampling rate of 1000 [samples/sec] for 0.2 sec. (i.e., for a total number of 200 [samples]). The acquired voltage is converted in temperature T using multiplication with a scaling factor of 1000.00 [°C/ V]. The results are displayed in the front panel waveform graph labeled Noisy Temperature,

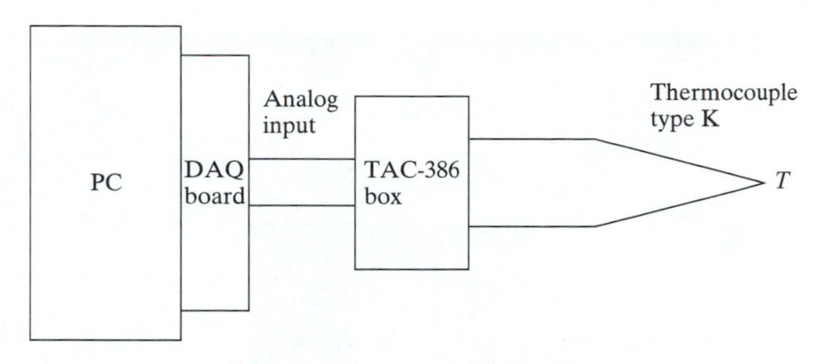

Figure 8.4

DAQ board-based temperature measurement with a thermocouple

shown in Fig. 8.6. The noise is filtered out with a digital filter Mean.vi and the results are displayed as a waveform graph labeled Temperature and as a thermometer.

The front panel contains two waveform graphs for displaying the noisy temperature [degree Celsius] and temperature, respectively, as well as a thermometer with a vertical 1-D display and a numerical display. The on/off switch acts on the while loop for acquisition in Fig. 8.2.

As in the case of the thermistor, presented in Section 8.2.1, the measurements can take place for cases for which the temperature is known such that the temperature measurement error can be calculated:

	Temperature *T* [°C]
boiling water	100
body temperature	36
melting ice	0

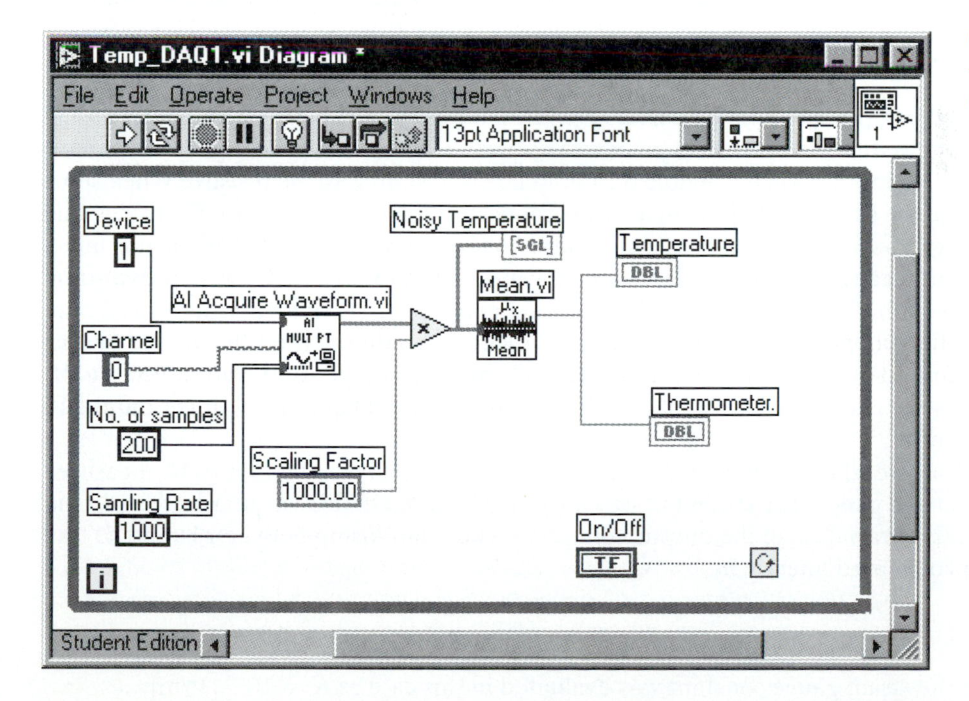

Figure 8.5

Temp_DAQ1.vi block diagram

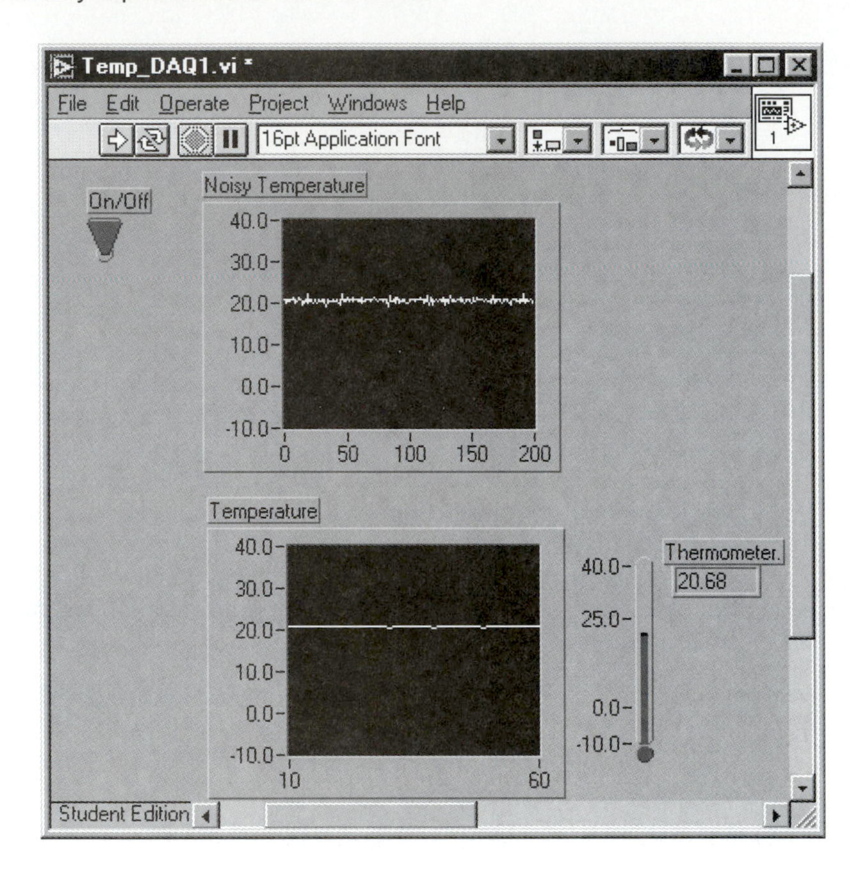

Figure 8.6
Temp_DAQ1.vi front panel

8.2.3 Strain Gauge

Strain gauges, presented in Section 2.3, are used for strain, stress, acceleration, force, and torque measurement. The strain gauge experimental set-up shown in Fig. 8.8 contains a strain gauge, mounted in the middle on a short beam, a board with the resistive Wheatstone bridge (in the bottom left of the figure), the connecters for a 9 V battery for the bridge, and the connectors for the wires to transmit bridge voltage output V_0 to the DAQ analog input.

Strain gauges are available in a large variety of shapes [16, 128]. A resistive strain gauge diagram is shown in Fig. 8.7. This strain gauge consists of a very fine wire bonded to a flexible plaque and has two terminals for connection in the Wheatstone bridge.

Figure 8.8 shows the diagram of the Wheatstone bridge used for converting the strain gauge resistance variation ΔR, due to the strain applied, into voltage output V_0 variation.

The lateral deflection dY of the free end beam, results in a strain to be measured by the strain gauge. The strain induces a strain gauge resistance variation ΔR and, consequently, a variation of the output voltage V_0. The relationship between V_0 and dY can be approximated linearly by

$$dY = KV_0,$$

where the strain gauge constant was evaluated in this case as $K = 10^{-4}$ [V/m].

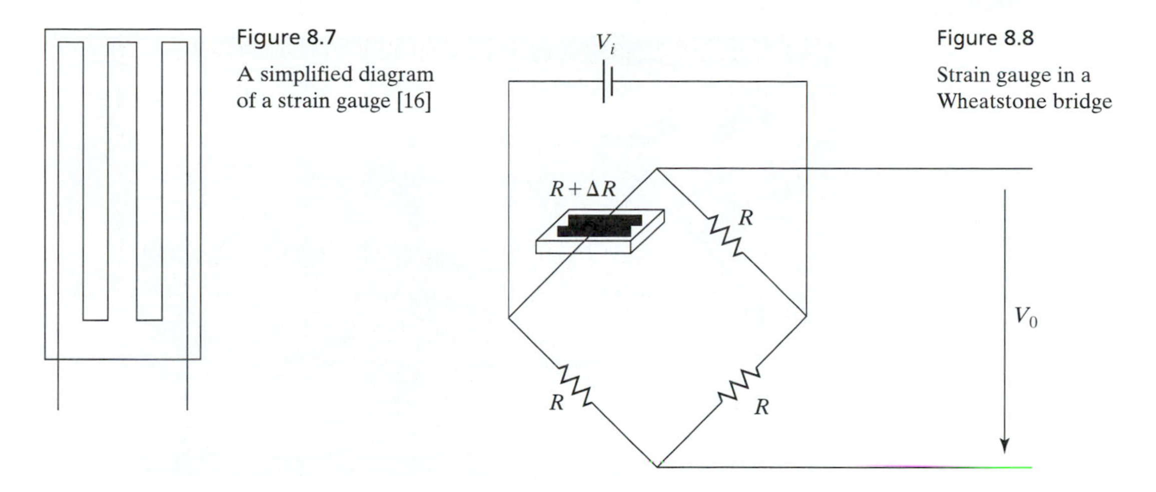

Figure 8.7
A simplified diagram of a strain gauge [16]

Figure 8.8
Strain gauge in a Wheatstone bridge

The LabVIEW program StrainGage.vi block diagram is shown in Fig. 8.9. The AI wave icon corresponds to the VI, which acquires device 1 channel 0 analog voltage input samples at a sampling rate of 500 [samples/sec] for 0.5 [sec] (i.e., 100 [samples]). The Wheatstone bridge requires us to determine by trial and error a voltage bias correction

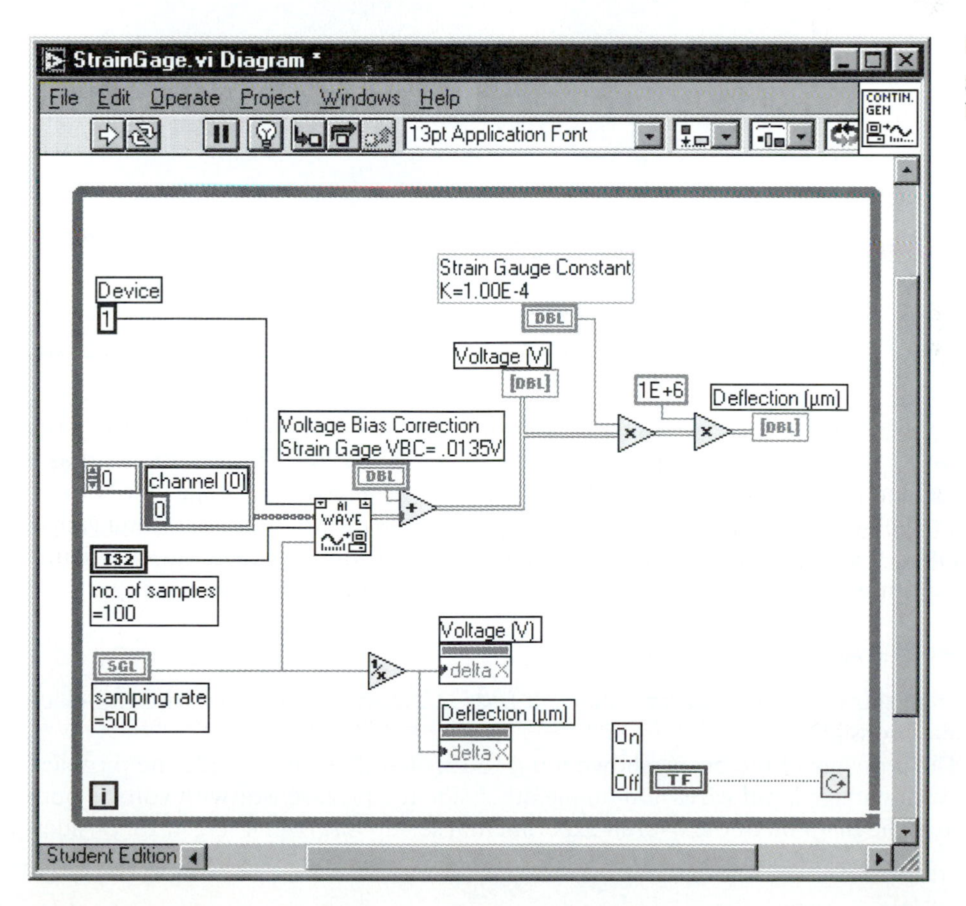

Figure 8.9
StrainGage.vi block diagram

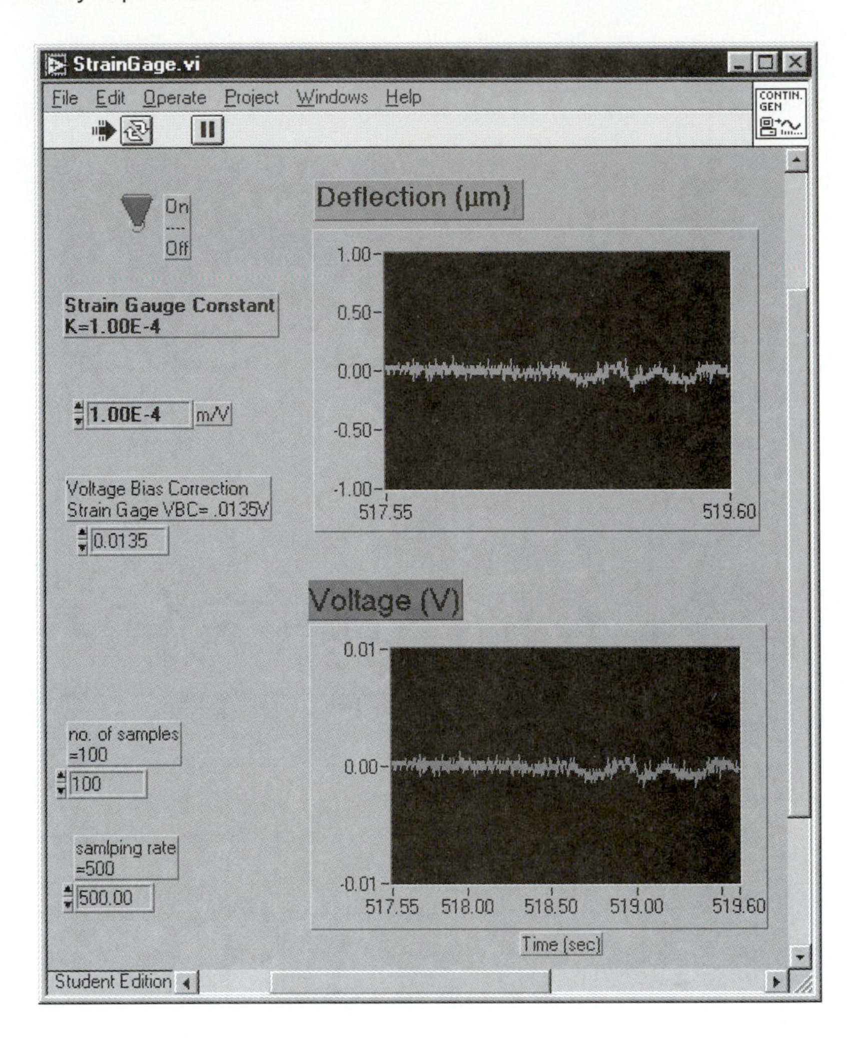

Figure 8.10
StrainGage.vi front panel

to achieve $V_0 = 0$ for $dY = 0$. In the case shown in Fig. 8.10, the voltage bias correction strain gauge VBC was found to be approximately +0.0135 [V]. The resulting voltage measurement is shown in Fig. 8.10 in the waveform graph labeled "Voltage (V)."

The voltage measurement is converted in deflection dY [m] using the strain gauge constant $K = 10^{-4}$ [V/m]. After multiplication by 10^6 [µm/m], the resulting deflection measurement is shown in Fig. 8.10 in the waveform graph labeled "Deflection (µm)."

8.2.4 Piezoelectric Actuator and Sensor

Figure 8.11 shows the piezoceramoc element BM532 Type V with the two wires for electric connections [22].

The top view of the beam, shown in Fig. 8.12, identifies on one side the piezoactuator with voltage input wires and on the other side the piezosensor with voltage output wires. The diagram of the overall experimental set-up is shown in Fig. 8.13. Besides

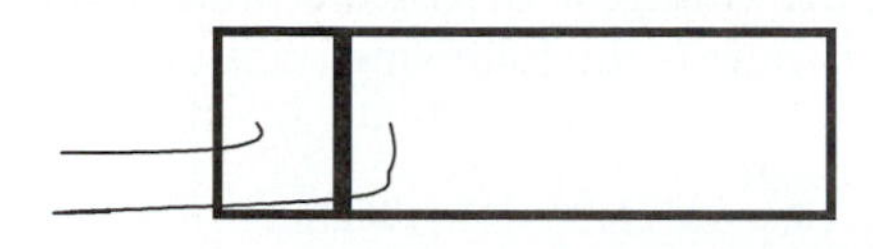

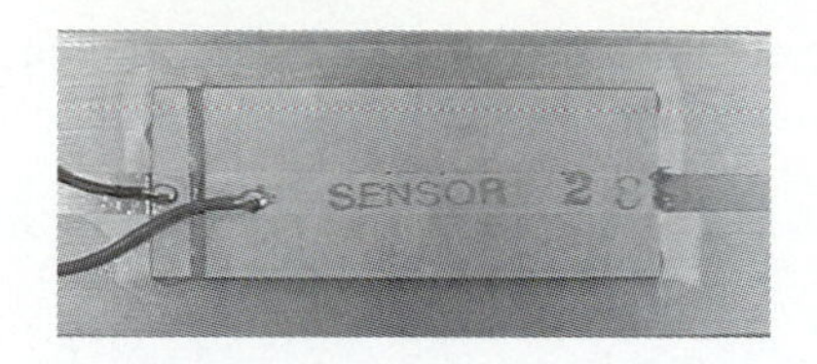

Figure 8.11

BM532 Type V piezoelement (*Source: Sensor Technology [22]*)

the beam equipped with the piezoactuator and the piezosensor, Fig. 8.13 shows a voltage follower and a DAQ board plugged in a PC. The beam has length L = 810 mm, width W = 31.2 mm, and thickness T = 1.58 mm, and the piezoelements are glued at 80 mm from the fixed end of the beam. The voltage follower is required to protect the DAQ board from excessive voltages—more than 10 V—which can damage the DAQ board. The voltage follower can be designed to have a gain equal to 1 (i.e., to have no effect on the measured voltage).

The piezoelements BM532 TypeV have length l = 37.7 mm, width w = 15.7 mm, thickness t = 0.5 mm, dielectric constant $c = 2.877 \cdot 10^{-8}$ [F/m], piezoelectric voltage coefficient $g_{31} = -7.5 \cdot 10^{-3}$ [Vm/N] [22], and modulus of elasticity $E = 7.14 \cdot 10^{10}$ [N/m^2].

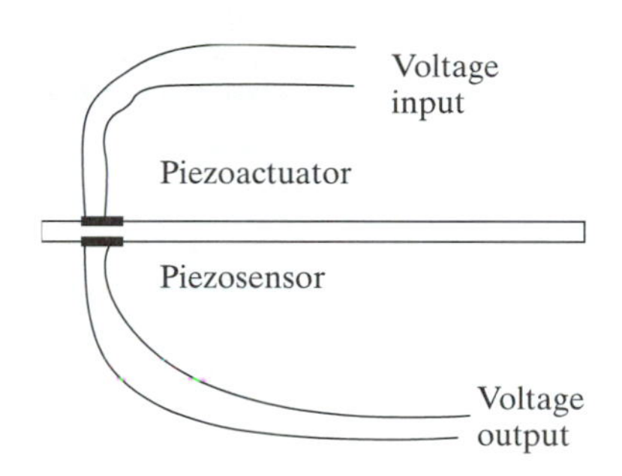

Figure 8.12

Top view of the beam with piezoactuator and piezosensor

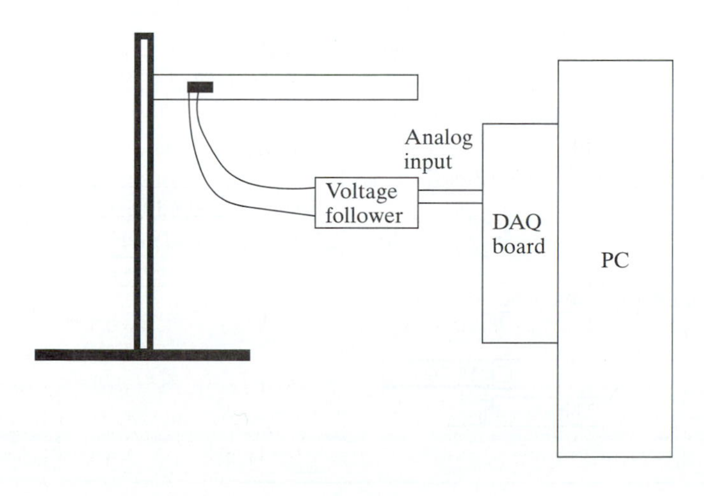

Figure 8.13

Diagram of the overall experiment set-up

In Section 2.4 it was shown that the between the voltage V across the thickness t and the strain $\Delta l/l$ along the piezoelement of length l, there is the following relationship:

$$V = g_{31} Et\Delta l/l,$$

such that

$$\Delta l/l = V/(g_{31} Et).$$

The block diagram of the PZTexp1.vi, shown in Fig. 8.14, contains on the top half the analog output from the function generator for the AO Generate Waveform.vi to supply voltage input to the piezoactuator and in the bottom half, the analog input data acquisition from the analog output of the piezosensor, using the AI Waveform Scan.vi.

The function generator has the signal source set for Sine Wave.vi and has other inputs, received from the terminals of the controls of the front panel, shown in Fig. 8.15. These inputs are the sampling rate, the frequency and the amplitude of the gener-

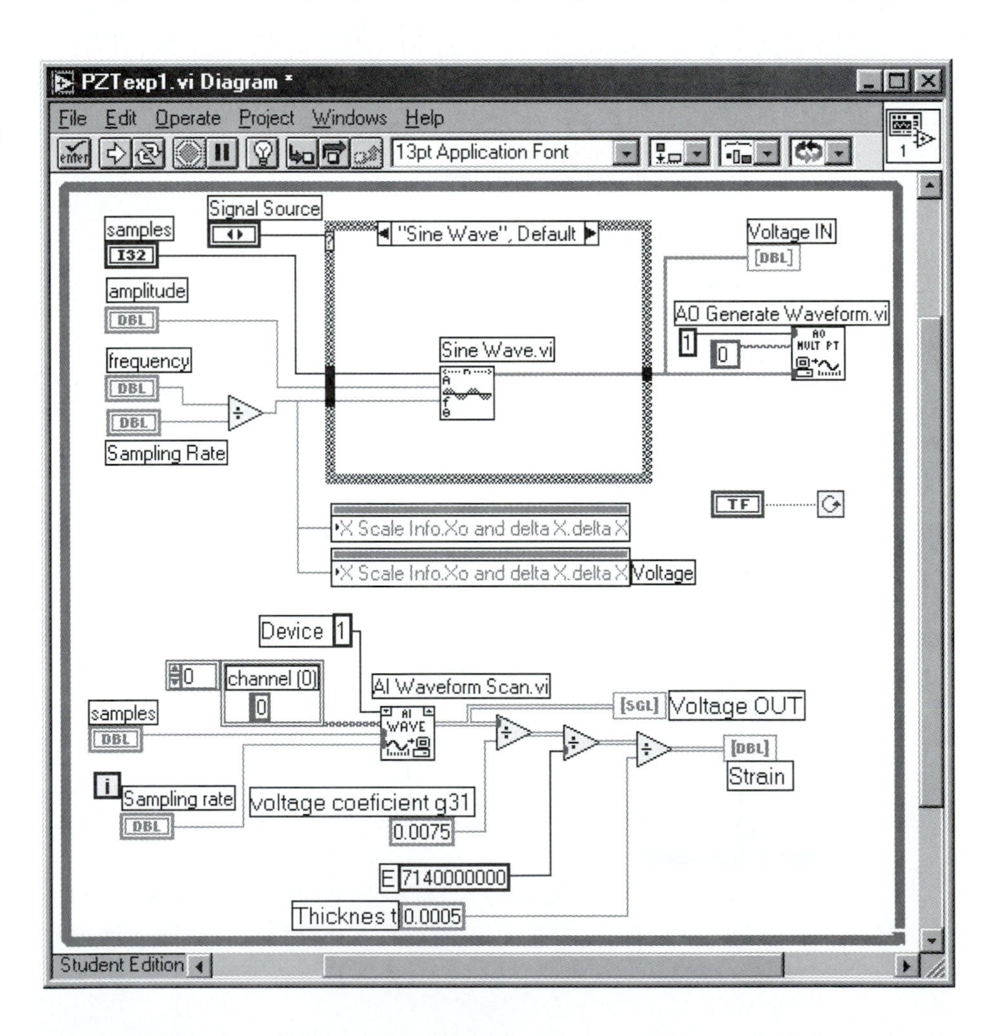

Figure 8.14
PZTexp1.vi block diagram

ated wave, and the number of samples. The generated wave is displayed on a waveform graph on the front panel, labeled "Voltage IN." For the time scale X of all waveform graphs on the front panel shown in Fig. 8.15, time spacing is calculated with X Scale info Xo and delta X delta X.vi. Voltage IN is the analog input to AO Generate Waveform.vi. This VI sends an analog output as channel 0 to the DAQ installed as device 1. This analog output is transmitted to the piezoactuator as voltage input.

The data acquisition part of the PZTexp1.vi uses AI Waveform Scan.vi to acquire the voltage output from the piezosensor as channel 0 as analog input to the DAQ board installed as device 1. This VI requires inputs regarding the sampling rate and the number of samples. The output of this VI is the measurement of the voltage V from the piezosensor and is displayed on the waveform graph on the front panel labeled "Voltage OUT", as shown in Fig. 8.15. This voltage V can be converted into strain measurement using the equation $\Delta l/l = V/(g_{31}Et)$ for the values of g_{31}, E, and thickness t given by the piezoelement manufacturer and listed earlier in this chapter [22]. The results of the experiment are shown in Fig. 8.15 for a sine wave voltage IN input with 4.3 V am-

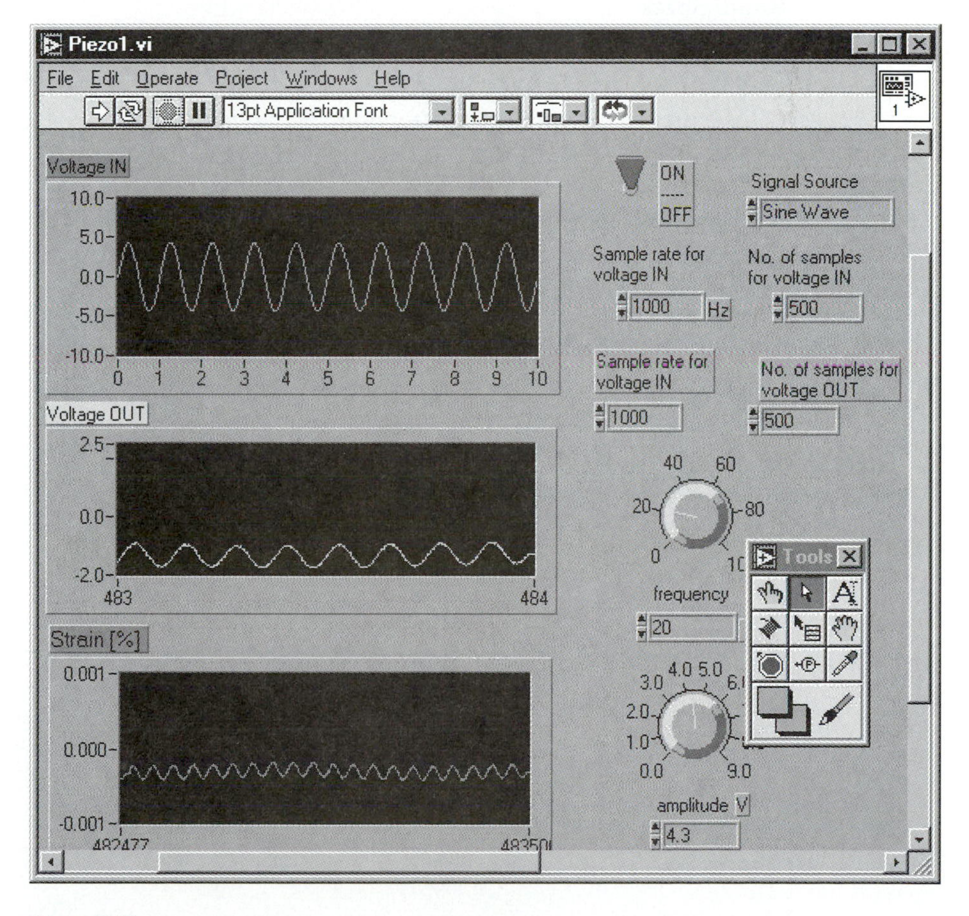

Figure 8.15

PZTexp1.vi front panel

plitude, 20 Hz frequency, sampled at 1000 [samples/sec] for 500 [samples]. Voltage OUT is also sampled at 1000 [samples/sec] for 500 [samples]. Strain is also displayed in Fig. 8.15 after its computation from voltage OUT measurement. Both voltage OUT and strain have, as expected, the same frequency as the voltage IN signal.

8.2.5 Light Emitting Diode and Photosensor

Light emitting diodes (LED) are often used for simple binary display [98].

Infrared photoresistors are used for sensing infrared emission from LED. Light emitting diodes and photosensors are available as an assembly consisting of an infrared light emitting diode and a phototransistor detector as, for example, Intertan 276-142 from Radio Shack [118].

The circuit diagram for the experimental set-up is shown in Fig. 8.16. The components shown in Fig. 8.16 are as follows:

assembly of infrared LED and phototransistor detector, Intertan 276-142, Radio Shack

operational amplifier 741, 276-007 Radio Shack [118]

1.5 V battery for LED power supply

two 9 V batteries for operational amplifier 741 power supply

three resistances, $R_o = 8.15$ KΩ, $R_1 = 1$ KΩ÷ and $R_F = 100$ KΩ.

The characteristics of the infrared-emitting diode are the following:

reverse voltage 2 V

continuous forward current 40 mA

wavelength at peak emission 915 mm.

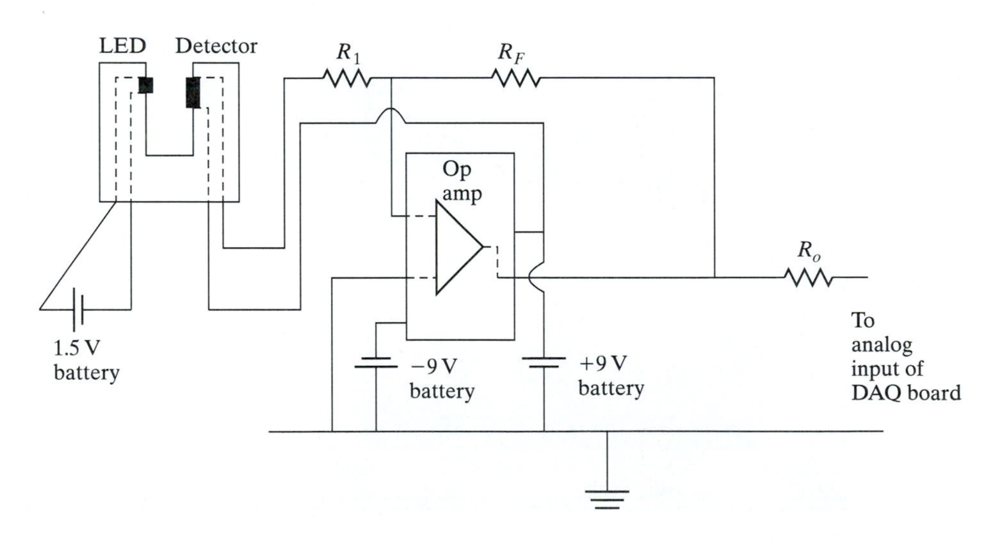

Figure 8.16

Diagram of experimental set-up for LED-detector

To avoid overloading the LED, a 1.5-V battery was used as power supply.

The characteristics of the phototransistor detector are as follows:

collector-emitter voltage 20 V
collector current 25 mA.

The DAQ board analog input requirements are maximum 10 V and maximum 5 mA.

To condition the signal from phototransistor detector in order to match the analog input requirement of the DAQ board, the operational amplifier 741 and the three resistances—$R_o = 8.15$ K, $R_1 = 1$ KΩ, and $R_F = 100$ KΩ—were used for the circuit interfacing the LED-detector pair and the DAQ board. The results of measurements with a digital voltmeter gave, in this case, approximately 0.0007 V for the phototransistor detector output and approximately 4 V for the voltage sent by the signal conditioning circuit to the analog input of the DAQ board.

The output voltage from the signal conditioning circuit, denoted as "To analog input of DAQ board" in Fig. 8.16, is acquired and displayed by the LabVIEW program LED-detector.vi shown in the block diagram of Fig. 8.17.

The program uses the AI Acquire Waveform.vi to acquire the analog input from device 1 channel 0 at a sampling rate of 500 [samples/sec] for 200 [samples]. On the front panel from Fig. 8.18, the results are displayed as a time-varying voltage tending towards 4.48 V with a range of variation of about ±0.01 V on the waveform graph. These results are in agreement with the digital voltmeter measurement of the voltage sent by the signal conditioning circuit to the analog input of the DAQ board. The front

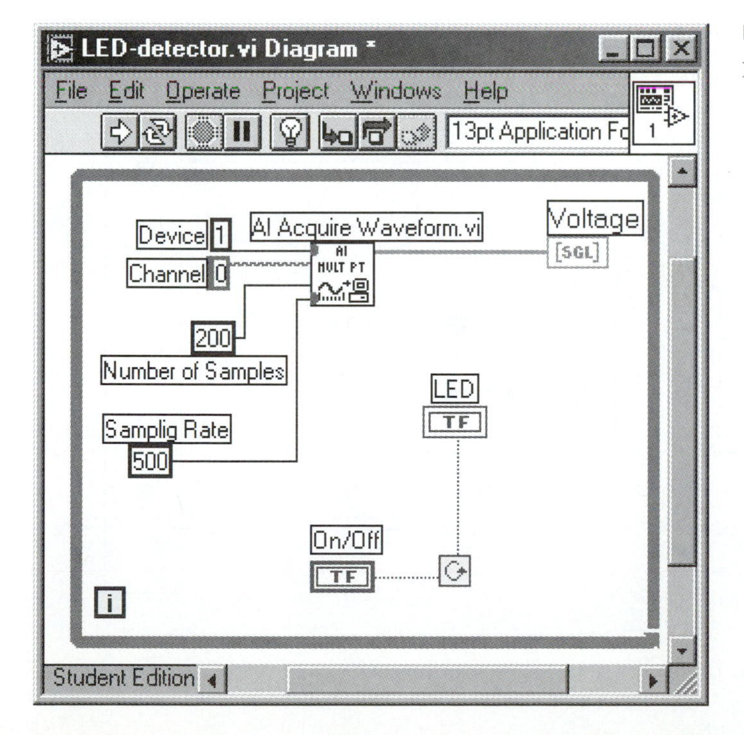

Figure 8.17

Block diagram of the LED-detector.vi

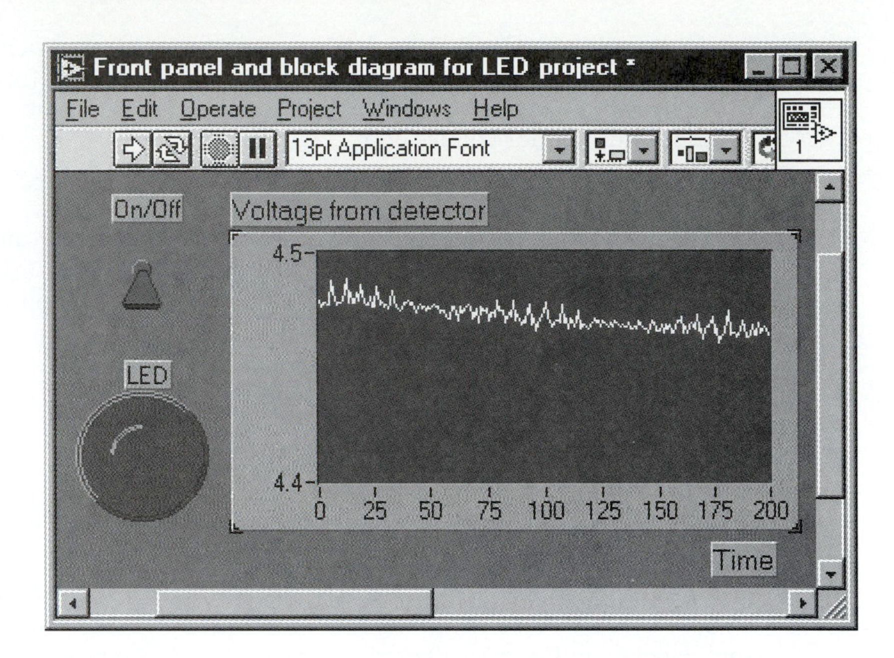

Figure 8.18

Front panel of the LED project vi

panel displays on a LED indicator the state of the on/off switch used in the while loop of the block diagram from Fig. 8.17.

8.2.6 DC Motor Velocity Control and Hall Sensor Measurement

DC motor and Hall sensor experimental set-up and the circuit diagram are shown in Fig. 8.19.

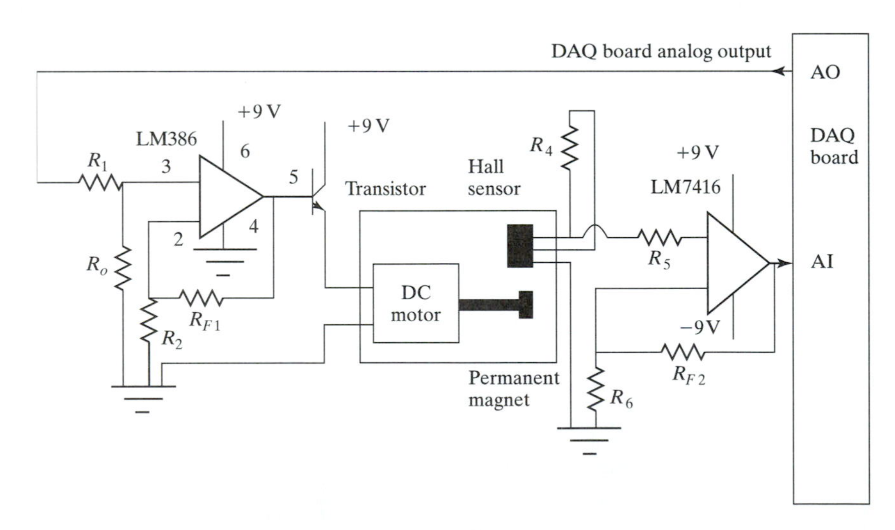

Figure 8.19

Circuit diagram of the DC motor and Hall sensor experimental set-up

The components of the DC motor and Hall sensor experimental set-up shown in Fig. 8.19 are the following:

DC motor, 273-223 Radio Shack [118]
permanent magnet with N and S polarities at two opposed ends of the disk
Hall effect sensor HE3135, Cherry Corporation [130]
power amplifier LM386, National Semiconductor [131]
operational amplifier LM741, National Semiconductor [131]
transistor 2N4401, 276-2058 Radio Shack [118]
eight resistances, $R_o = 1\ \text{K}\Omega$, $R_1 = 10\ \text{K}\Omega$, $R_2 = 1\ \text{K}\Omega$, $R_3 = 1\ \text{K}\Omega$, $R_{F1} = 2.2\ \text{K}$, $R_4 = 2.2\ \text{K}\Omega$, $R_5 = 3\ \text{K}\Omega$, $R_6 = 22\ \text{K}\Omega$, and $R_{F2} = 18\ \text{K}\Omega$.

Analog output from the DAQ board is connected to the input of the amplifier LM386, which provides the input voltage to the power transistor driving the DC motor. The output from the operational amplifier LM741 is the analog input to the DAQ board and represents the measurement of the signal from the Hall sensor output signal in order to evaluate the angular velocity of the DC motor.

The characterisics of the components of the experimental set-ups are as follows:

The DC motor, 273-223, Radio Shack [118], has voltage supply from 1.5 to 3 [V] and at 1.5 [V] voltage supply, no load angular velocity of 5700 [rpm], no load current of 0.2 [A], and resistance of 0.25 [Ω];
the Hall Effect Latch Assembly, Cherry Corp [130], has the operating voltage from 4.5 to 24 [V], maximum current supply of 9 mA, and maximum output voltage of 0.4 [V];
operational amplifier LM741, National Semiconductor [131], shown in Fig. 8.20;
power amplifier LM386, National Semiconductor [131], with the same pin numbers in Fig. 8.21.

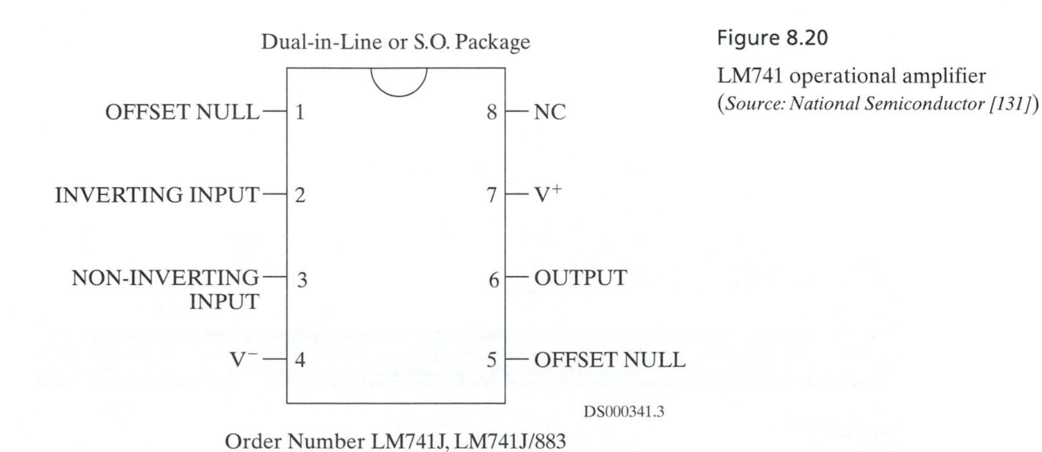

Figure 8.20
LM741 operational amplifier (*Source: National Semiconductor [131]*)

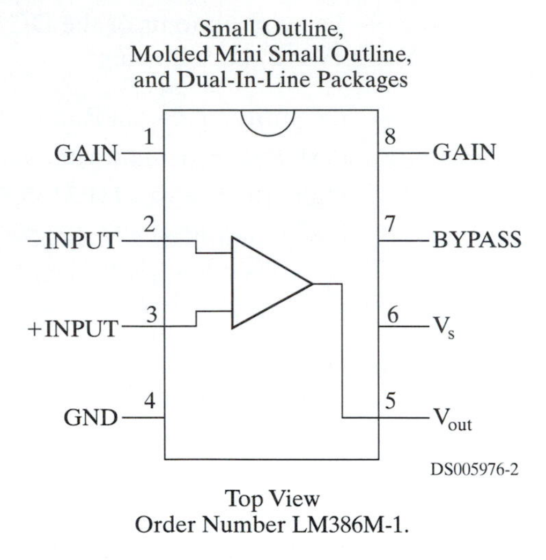

Figure 8.21

LM386 power amplifier (*Source: National Semiconductor [131]*)

As shown in Fig. 8.19, the output voltage from LM741 in is acquired and displayed by the LabVIEW program dcmotor.vi. The block diagram of this program is presented in Fig. 8.22.

The program uses the AO Update Channel.vi to provide the analog output (AO) voltage at device 1 channel 0 of the DAQ board. The voltage value sent to AO

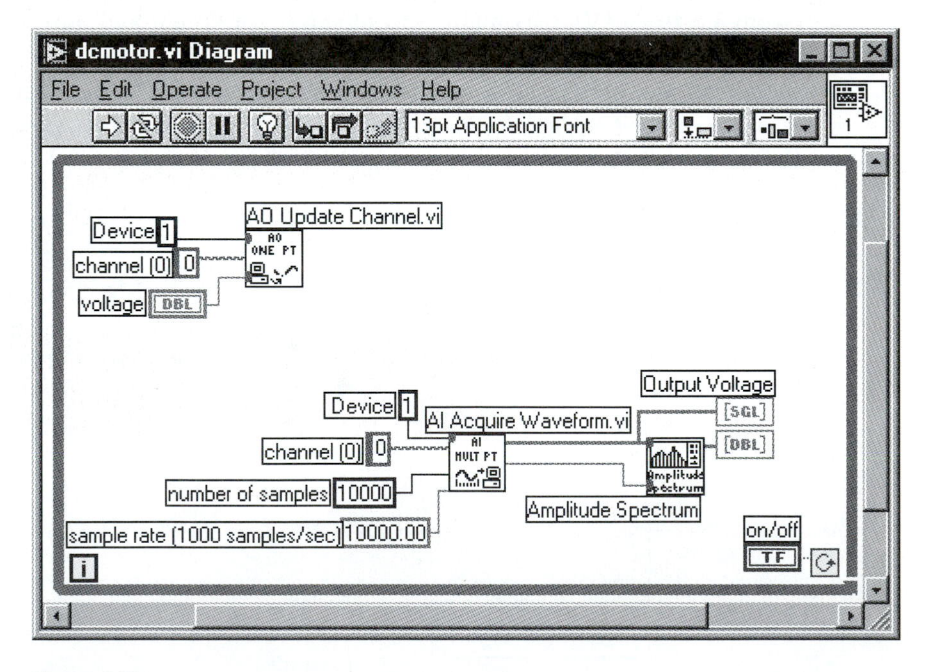

Figure 8.22

Block diagram of dcmotor.vi

is set by the front panel knob labeled "Voltage." In Fig. 8.23, the voltage value for AO is set at 5 V.

Also, the program uses the AI Acquire Waveform.vi to acquire the analog input (AI) signal from device 1 channel 0 at a sampling rate of 10000 [samples/sec] for 10000 [samples]. The results are displayed on the waveform graph of the front panel from Fig. 8.23 as a voltage, of constant frequency, decreasing in time. The signal from the AI Acquire Waveform.vi is send to Amplitude Spectrum.vi and the frequency-dependent results are displayed in a waveform graph on the front panel of Fig. 8.23 labeled "Amplitude Spectrum.".

For the AO value of 5 V supplied to the operational amplifier LM from Fig. 8.19, the maximum amplitude, in the amplitude spectrum waveform graph from Fig. 8.23, corresponds to a frequency of approximately 68 Hz. This frequency, obtained from the spectrum analysis of the Hall sensor voltage output measurement, indicates a motor angular velocity of 68 [revolutions/sec] or 4080 [rpm].

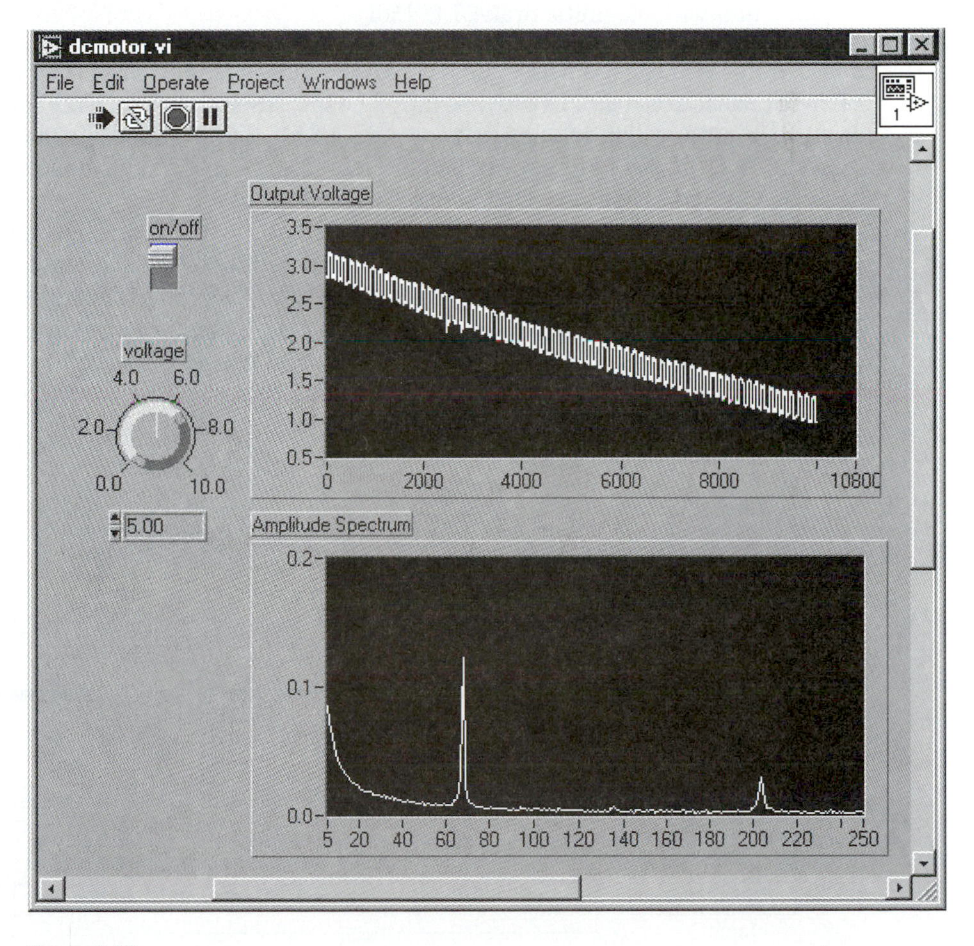

Figure 8.23

Front panel of dcmotor.vi

8.2.7 DC Position Servomotor Control

Figure 8.24 shows the diagram of the MS 150 servomotor set-up.

The components of the MS 150 Mk II Modular Servo System are the following, from Feedback Instruments [120]:

input potentiometer, IP 150H
output potentiometer, OP 150K
operational amplifier unit, OA 150A
attenuator unit, AU 150B
preamplifier unit, PA 150C
drive amplifier, SA 150D
power supply, PS 150E
DC motor unit and tachometer, MT 150F
load unit with magnetic brake, LU 150L.

The MS-150 was designed with analog electronics implementation of the position controller, for educational use. In order to interface the MS-150 with a DAQ board, it is required to condition the voltage signals from the MS150, available in the range ±15 across the 10 KΩ input and output potentiometers (see Fig. 8.25, left-hand side), to less than ±10 V, acceptable to the DAQ board.

The current in the potentiometer is 30V/10 KΩ = 3 mA. For this purpose, two 5.65-KΩ resistors are connected in series with the input potentiometer (and, similarly with the output potentiometer), as shown in the right-hand side of Fig. 8.25. The current becomes 30 V/23 KΩ = 1.4 mA such that the voltage drop on the 10-KΩ potentiometer is reduced to (1.4 mA)(10 K) = 14 V or ±7 V. This voltage divider permits us to collect a maximum of ±7 V from the terminals of the 10-KΩ potentiometer as signal to be sent to the analog

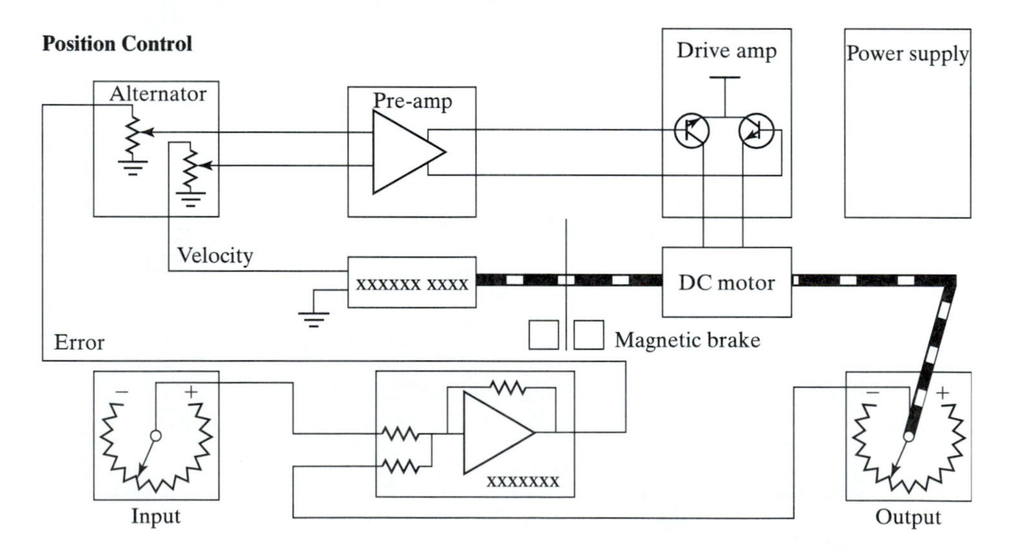

Figure 8.24
Diagram of the MS 150 position servomotor, (*Source: Feedback Instruments [120]*)

10 KΩ
−15V +15V
$i = 3\text{mA}$

5.65 KΩ 10 KΩ 5.65 KΩ
−15V −7V +7V +15V
$i = 1.4\text{ mA}$

Figure 8.25

Signal conditioning for interfacing MS 150 with DAQ board

input channel 0 for measurement and to be received as a command from the analog output channel 0 of the device 1 of the DAQ board plugged into the PC.

A LabVIEW program, DC-ServoMotor-DAQ.vi, shown in the block diagram form in the bottom part of the Fig. 8.26, uses the AO update Channel.vi to produce the analog output voltage at device 1 channel 0 of the DAQ board. The desired angular position is set by the front panel knob, labeled "Desired Angular Position" in Fig. 8.27, and is displayed versus time in the waveform graph, labeled "Desired Angular Posi-

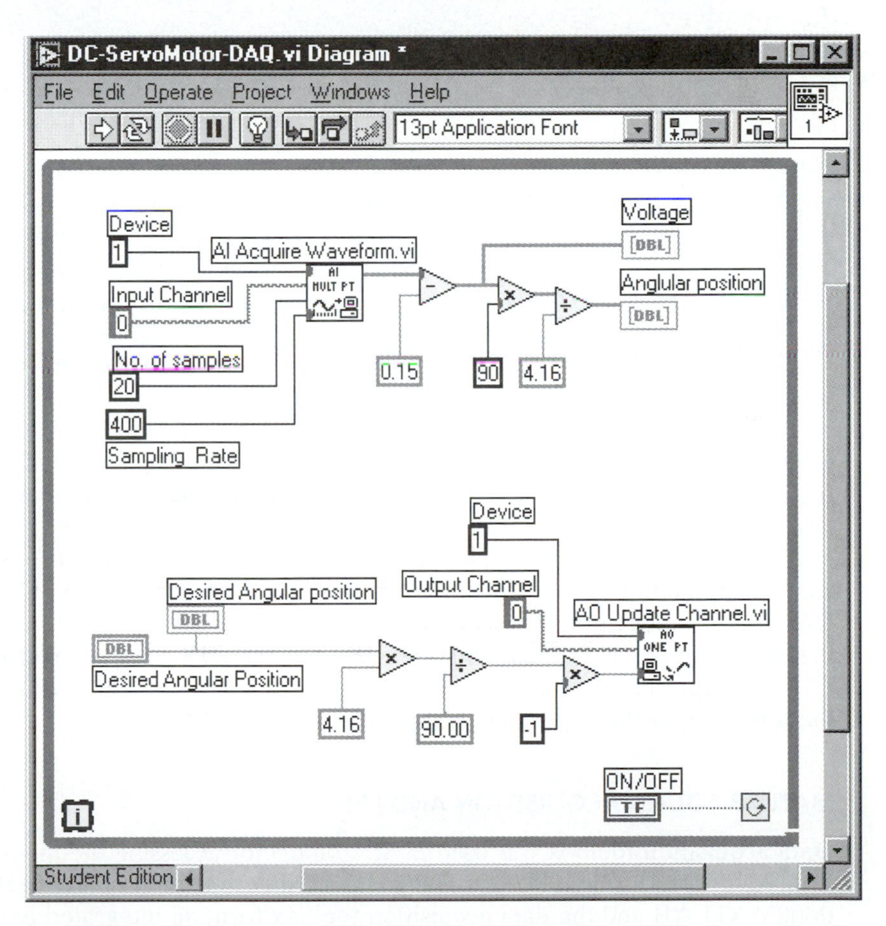

Figure 8.26

Block diagram of DC-servoMotor-DAQ.vi

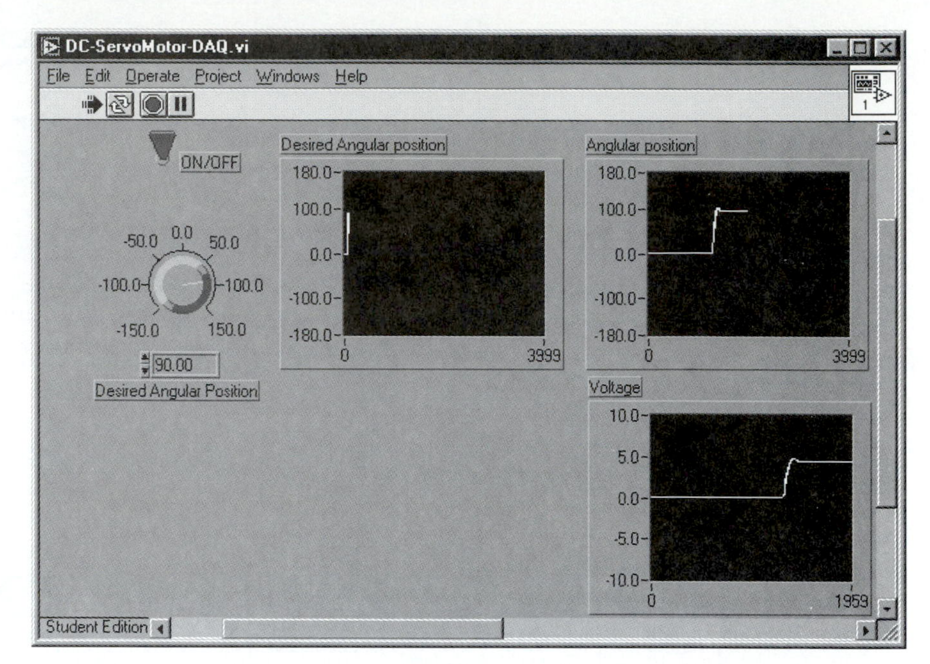

Figure 8.27

Front panel of DC-ServoMotor-DAQ.vi

tion." As shown in block diagram form in Fig. 8.26, the angle value is converted from angle to voltage using the conversion factor 4.16/90 [V/degree] and the sign is changed. Then, the signal is sent to the AO Update Channel.vi to produce the analog output of the DAQ board connected to the input potentiometer, shown in Fig. 8.24.

The voltage from the output potentiometer, after signal conditioning, is acquired by the LabVIEW program DC-ServoMotor-DAQ.vi, shown in the top part of the block diagram from Fig. 8.26.

The program DC-ServoMotor-DAQ.vi uses the AI Acquire Waveform.vi to acquire the analog signal from device 1 channel 0 at a sampling rate of 400 [samples/sec] for 20 [samples]. The measured voltage is displayed versus time in the waveform graph labeled "Voltage" from the front panel of Fig. 8.27.

The measured voltage is converted in angular position of the motor shaft using the conversion factor 90/4.16 [degree /V], as shown in the top part of the block diagram of DC-ServoMotor-DAQ.vi from Fig. 8.26. The angular position versus time is displayed in the waveform graph, labeled "Angular Position," from the front panel of Fig. 8.27.

As expected, after a transient regime, the angular position reaches 90 [degrees], the set value for the desired angular position.

8.3 MATLAB SOUND ACQUISITION AND FFT

Data acquisition toolbox was used in Section 6.3 for accessing the measurement data from MATLAB, using a PC equipped with a data acquisition board. In this configuration, MATLAB and the data acquisition toolbox form an integrated environment for data acquisition and data analysis, for example for Fast Fourier Transform (FFT).

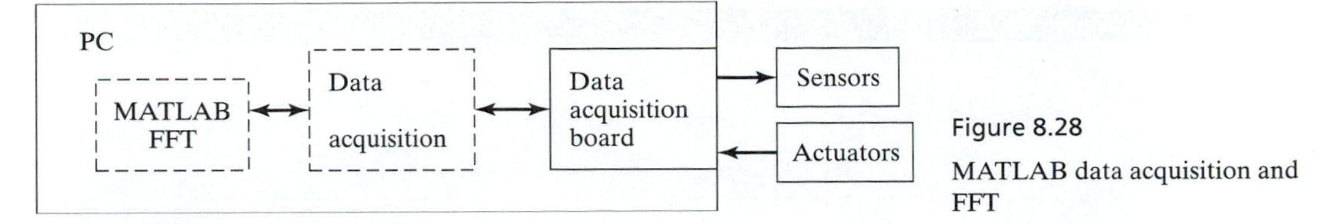

Figure 8.28
MATLAB data acquisition and FFT

Figure 8.28 shows the same data acquisition configuration as shown in Fig. 6.24, with the addition of the MATLAB FFT tool. In this figure, the PC has plugged-in data acquisition board and has installed MATLAB and the data acquisition toolbox. The data acquisition board permits analog and digital inputs from sensors and analog and digital outputs to actuators.

Figure 8.29 shows the experimental set-up for the same example of data acquisition with the MATLAB data acquisition toolbox for acquiring a musical sound as shown in Fig. 6.25, again with the addition of the FFT tool.

Figure 8.30 shows the listing of the Getsound2.m program [100]. This textual program uses the commands available in MATLAB and the data acquisition toolbox. Beside the commands are the explanations, which start with % and, obviously, are not executable.

The results for the acquired data for a high Do sound for an acquisition duration of *ad*=1 [sec] are plotted in Fig. 8.31.

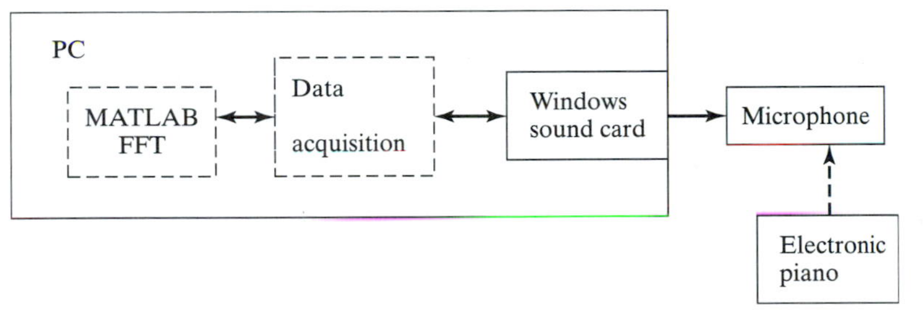

Figure 8.29
Experimental set-up for acquiring and FFT of a musical sound

```
%Getsound2.m
AI = analoginput ('winsound');          %creation of AI object
addchannel (AI, 1);                     %configuration of channel 1
sr = 20000;                             %sampling sate [samples /sec]
set (AI, 'SampleRate', sr)
ad = 1;                                 %acquisition duration [sec]
set(AI, 'SamplesPerTrigger', ad*sr);
start (AI)                              %starts data acquisition
data = getdata (AI);                    %measurement data retrieval
delete (AI)                             %deletion of AI object
xfft = abs (fft (data));      %calculation of absolute values of FFT of data
mag = 20*log10 (xfft);                  %conversion of amplitude to dB
mag = mag (1 : end/2);
    plot (mag)                %plotting command for mag [dB] versus

                              frequency [Hz]
```

Figure 8.30
Getsound2.m program for data acquisition

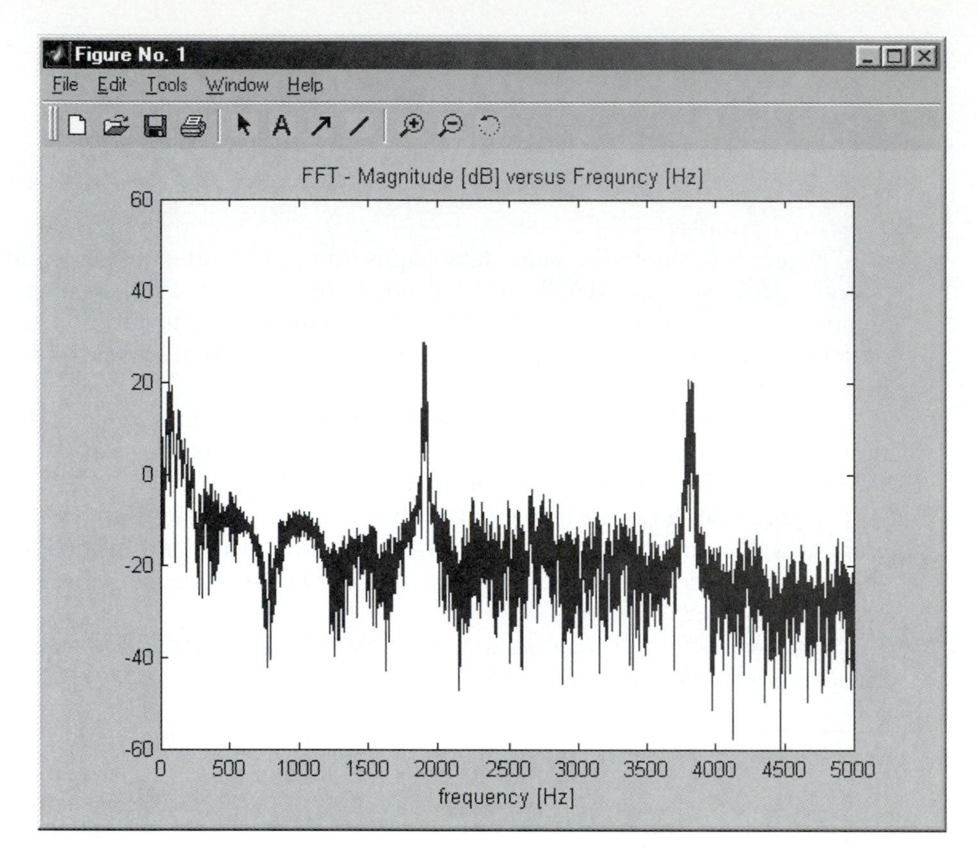

Figure 8.31

Plot of the results of Getsound2.m program

In order to eliminate the effect of finite time sampling, a Hanning window is included in the GetsoundHanningWindow.m program, listed in Fig. 8.32.

The results are displayed in Fig. 8.33. These results show that Hanning window leads to an amplitude spectrum, which contains the same main frequencies, but has a simpler spectrum due too the elimination of the effect of the finite time signal acquisition.

8.4 ADVANCED MONITORING AND CONTROL EXPERIMENTS

8.4.1 Robotics Experiments

Presented in this section will be two illustrative examples of advanced monitoring and control applications: a dual arm robot in Section 8.4.2 and a mobile robot in Section 8.4.3.

These two experimental set-ups share the same real-time computer—the dSPACE digital signal processor based system—presented in Section 7.2. Figure 8.34 shows the dSPACE system configuration used for these experiments.

```
%GetsoundHanningWindow.m
AI = analoginput ('winsound');                     %creation of AI
object
addchannel(AI, 1);                                 %configuration of
channel 1
sr=10000;                                          %sampling sate
[samples/sec]
set (AI, 'SampleRate', sr)
ad = 0.01;                                         %acquisition
duration [sec]
set (AI, 'SamplesPerTrigger', ad*sr);
start (AI)                                         %starts data
acquisition
data = getdata (AI);                               %measurement data
retrieval
delete (AI)      %deletion of AI object
data = data (1 : end);
dt=1/sr;
t=0:dt:(ad-dt);
subplot (2, 2, 1); plot (t, data); grid; hold on;  %plotting data (sampled ampl)
vs.time [sample]
title ('Signal') ;
ylabel ('Amplitude'); xlabel ('Time [sec]');
xfft = abs (fft (data));                     %calculation of absolute values of FFT
of data
mag = 20*log10 (xfft);                             %conversion of amplitude
to dB
mag =mag (1:end/2);
N=length (mag)-1;
f=(0:N);
nn=sr*ad;           %number of samples [samples]
df=sr/nn;           %sampling spacing
fHz=f*df;
subplot (2, 2, 3); plot (fHz, mag); grid;    %plotting  mag [dB] versus
frequency [Hz]
title('AMPL SPECTRUM-Original Signal');
ylabel ('Amplitude'); xlabel ('frequency [Hz]');
Th=ad;                     %period of hanning window=duration of acquisition
duration[sec]
Fh=1/ Th;
omegah=2*pi*Fh;
yh=0.5*(1-cos (omegah*t));               %hanning window
y1=data'.*yh;                            %windowed signal
subplot (2, 2, 2); plot (t, y1); grid;
title ('Windowed Signal');
ylabel ('Amplitude'); xlabel ('Time [sec]');
yy1=fft (y1);                            %FFT of windowed sinus wave
m1=abs (yy1);                            %amplitude of FFT
m2=20*log10 (m1);   %conversion of amplitude to dB
m2=m2(1:end/2);
N1=length (m2)-1;
f1= (0:N1);
fHz1=f1*df;
subplot (2, 2, 4); plot (fHz1,m2); grid;
title ('AMPL SPECTRUM-Windowed Signal');
ylabel ('Amplitude'); xlabel ('frequency [Hz]');
```

Figure 8.32

GetsoundHanningWindow.m program listing

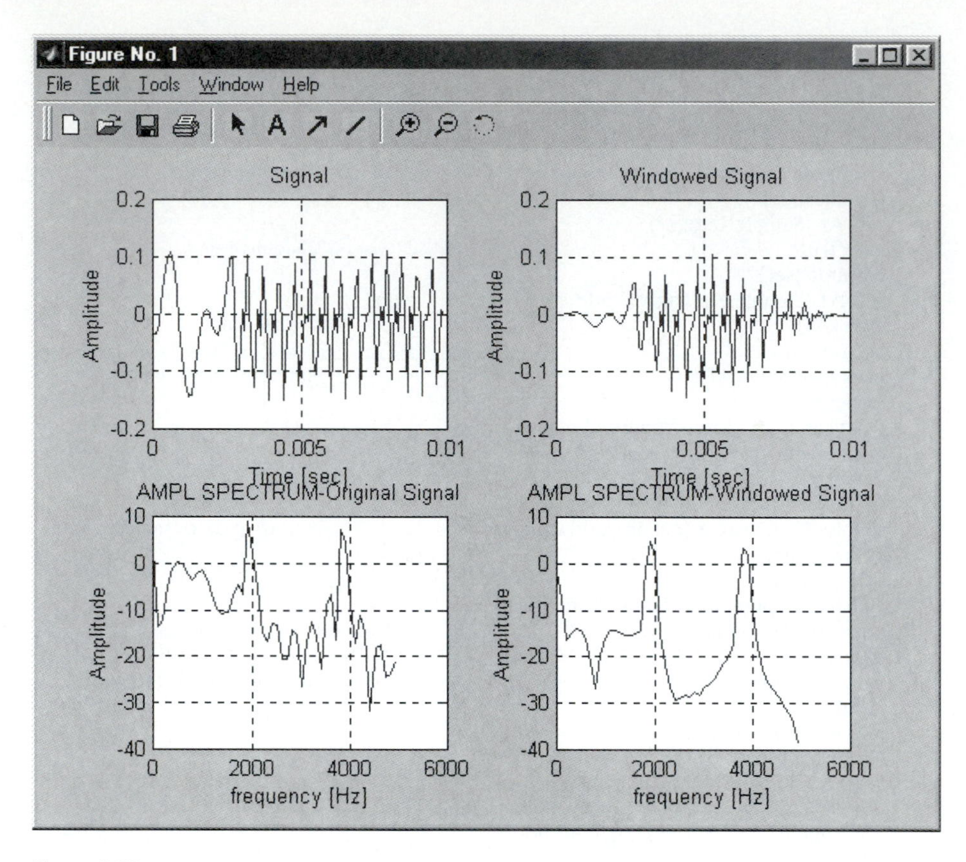

Figure 8.33

GetsoundHanningWindow.m results

Figure 8.35 shows the floating-point processor board contains a floating-point DSP TMS 320C30 from Texas Instruments, memory (e.g., 128 K words on-board RAM) and modules for interfacing with PHS bus and with the host PC.

The TMS 320C30 has a 33-MHz clock rate, 2 K words of 32-bit on-chip RAM, and 40-bit floating-point multiply. The PHS bus interface is a 32-bit I/O bus with 25 MB/sec transfer rate. The host interface for AT bus has five 8-bit I/O ports. The input/output boards connected to the PHS bus are as follows:

A DS2101 DAC board with five parallel DAC 12-bit channels, 5-μsec settling time, and programmable ±5-V, ±10-V, and 0-to-10-V analog output ranges;

DS2002, a multichannel ADC board with two 16-bit ADC converters, 5-μsec conversion time, and programmable ±5-V and ±10-V analog input ranges;

DS3001, an incremental encoder board with five parallel encoder interface channels, a 24-bit position counter, and a digital noise pulse filter.

These boards are plugged into a PX20 expansion box, which expands the PC-AT host bus.

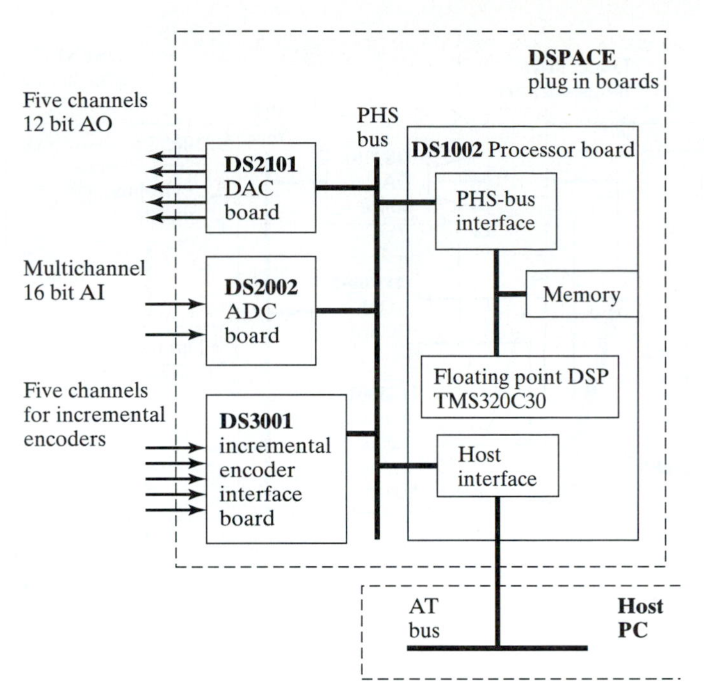

Figure 8.34
dSPACE controller for robotics experiments

The main components of the dSPACE software for these applications are as follows:

TRACE, for tracing variables of the object code executed on the DS 1002 processor board. The traced parameters and signals can be displayed graphically;

IMPEX, an application expert that supports the implementation of high speed linear invariant controllers. This software can be interfaced with MATLAB.

MATLAB can be used to design linear controllers and IMPEX can then generate the C-code for the controller. Nonlinear parts can be added eventually to this C-code. The C-code is compiled with a compiler for the DSP.

8.4.2 Dual-Arm Robot Manipulation Control

Shown in Fig. 8.35 is the dual-arm robot. This robot is monitored and controlled by a dSPACE-based controller, developed during the PhD research work of R. Jassemi-Zargani [124, 125].

The dimensions of the arms are the link between shoulder and elbow motors, 0.221 [m] long, and the second link, 0.16 [m] long. The center of the shoulder motors is at 0.9 [m] apart [125]. The maximum torques of the NSK direct drive motor are 88.2 [Nm] and 9.8 [Nm], respectively. The maximum velocities are 3 [rps] and 4.5 [rps], respectively. The angular positions of the motors shafts are measured with brushless resolvers. The resolutions of the resolvers are 614,400 [pulses/revolution] and 409,600

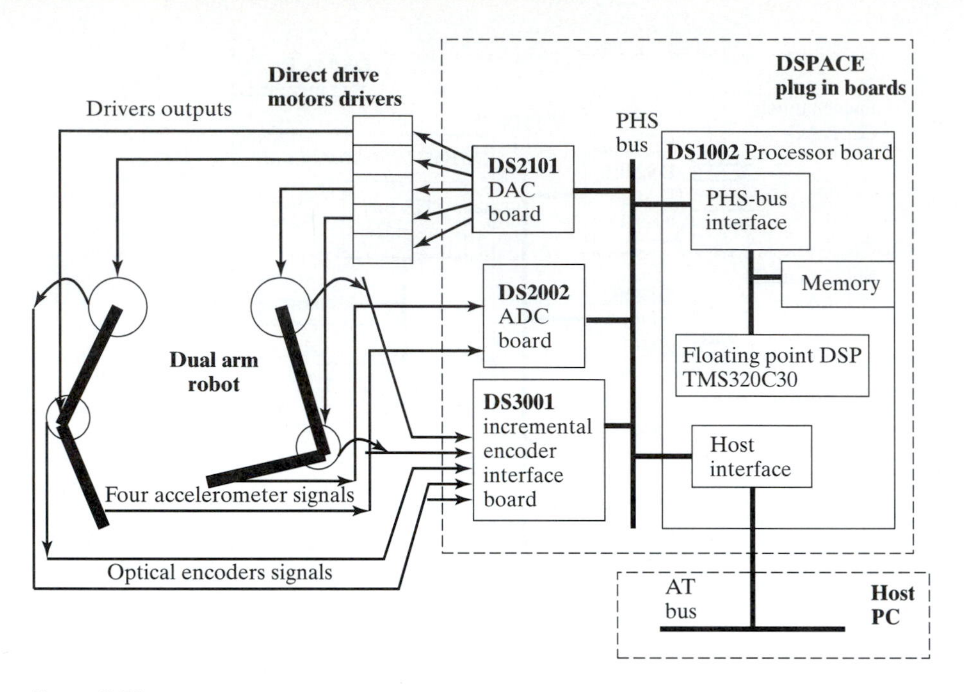

Figure 8.35

Diagram of the dSPACE controller and the dual arm robot

[pulses/revolution], respectively. The maximum pulse rate from the first link is (614,400 [pulses/revolution])·(3 [revolutions/sec]) = 1,843,200 [pulses/sec], a pulse rate that can be processed by the DS3001 incremental encoder interface board. Two orthogonal accelerometers were positioned at the upper link end-effector:

A NAS Series Accelerometer, with silicon cantilever sensor, from Nova Sensor, ±2g range, 1.25 [V/g] sensitivity, 0–5 [V] signal conditioned output, and 0–200 [Hz] frequency response [133]

A V-ACCESS capacitive accelerometer from ACCESS Sensors , ±1-g range, 2 [V/g] sensitivity, 0–2.5 [V] signal conditioned output, and 0–500 [Hz] bandwidth [134].

Figure 8.35 shows the diagram of the dSPACE controller and the wiring with the dual-arm robot. The dual-arm robot is under impedance control. This control approach uses a linarization scheme, which requires good estimation of motor shafts angular accelerations. Acceleration estimation was carried out using sensor fusion of resolvers and accelerometers signals [124, 125].

As shown in Fig. 8.36, the results of the point-to-point control of the arm-end point are close to ideal straight-line trajectory.

The graph represents points sampled with a constant sampling rate, and consequently, the more distant points at the beginning of the motion correspond to a higher robot velocity.

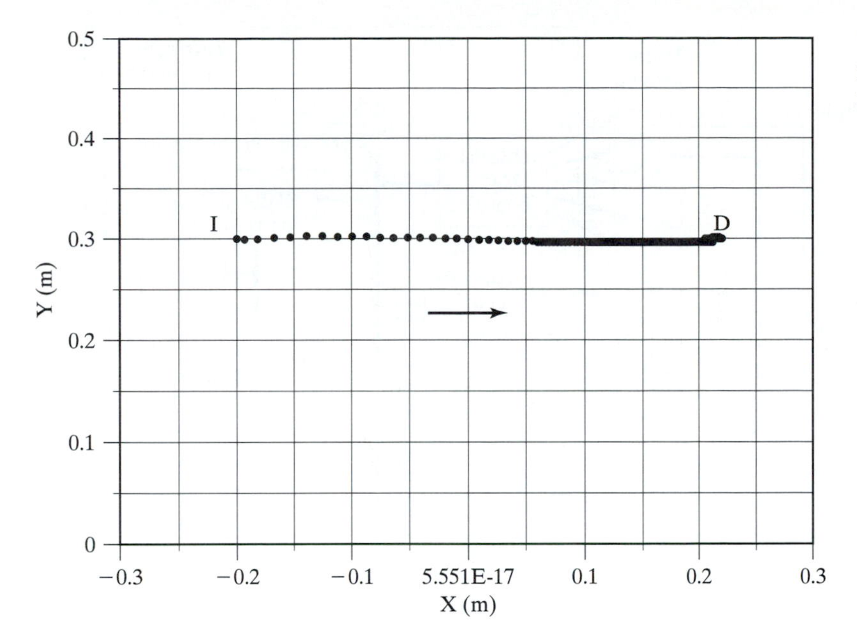

Figure 8.36

End-point motion from I to D under impedance control

8.4.3 Mobile Robot Motion Control

Shown in Fig. 8.37 is the mobile robot, which is monitored and controlled by a dSPACE controller. The mobile robot was designed by V. Lonmo, during his Master's thesis work [127].

The weight with the two batteries installed is 24.5 Kg.

The dimensions of the mobile robot are

length 0.51 [m]
width 0.51 [m]
height 0.53 [m].

Shown in Fig. 8.38 is the diagram of the dSPACE controller for mobile robot experiments [126, 127].

The main components of the mobile robot are as follows:

two Pitmo 14202 series DC motors with 75.1-to-1 planetary gear trains, 0.75 [Nm] peak (stall) torque, 0.0411 Nm/(W)$^{0.5}$ motor constant, 333 [W] power at peak torque, 400 [rad/sec] no-load speed, 1.47 [msec] electrical time constant, and 8.5 [msec] mechanical time constant [135]. Each motor has an incremental optical encoder with 1000 [pulses/revolution]; given the gear train ratio of 75.1, the gear train output shaft angular position is measured with a resolution equivalent to 75100 [pulses/revolution]

two PA 228 amplifier motor drivers, 160 [W] continuous output power, 7.5 [A] continuous output current, and 19–32 [V] power supply DC voltage [135]

two 12 [V], 10 [Ah] batteries.

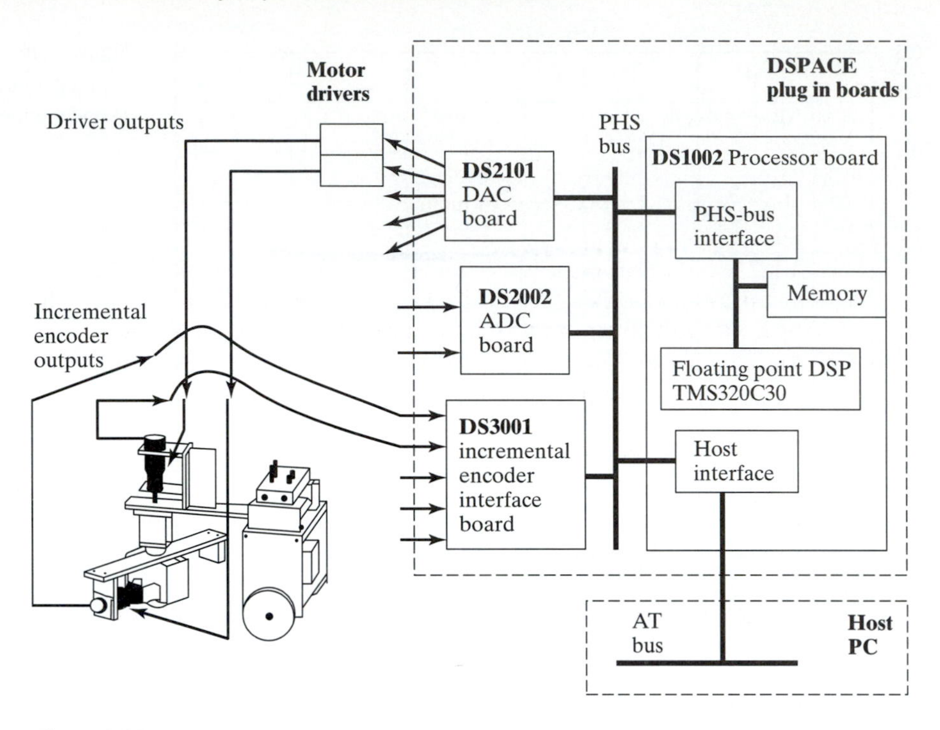

Figure 8.37

dSPACE controller for mobile robot experiments [126, 127]

Using a simple kinematics-based controller, the mobile robot trajectory (shown as a plain line) can track a desired sinusoidal path (shown as a dotted line in Fig. 8.38) [126, 127].

The results show that when the path has sharp curvature, the trajectory has small tracking errors.

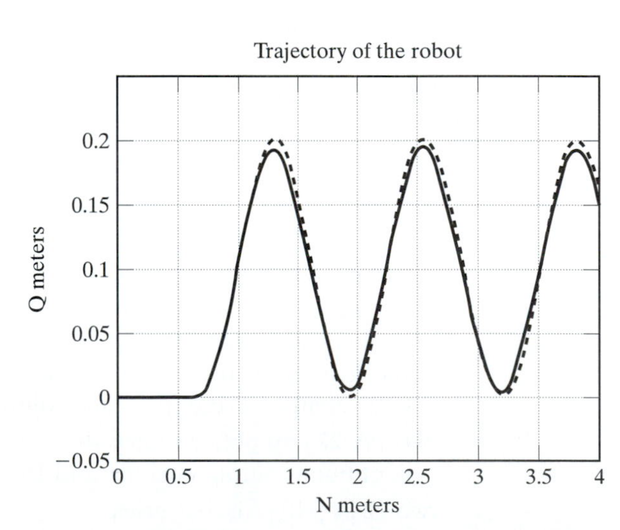

Figure 8.38

Mobile robot experimental results of sinusoidal path tracking [126, 127]

PROBLEMS

8.1. Modify the virtual instrument Thermistor1.vi to include an option to compute and display the results in either degrees Celsius or degrees Fahrenheit.

8.2. The results of temperature measurement with a thermocouple are noisy. Modify the virtual instrument Thermistor1.vi to include a Mean.vi filter and to display both the noisy and the filtered results.

8.3. The virtual instrument used for temperature measurement with a thermocouple, Temp.DAQ1.vi, uses a Mean.vi to filter the noisy measurement values. Replace the filter by Butterworth filter of order 5 with a cut-off frequency of 10 Hz.

8.4. The virtual instrument used for deflection measurement with a strain gauge shows a noisy signal. Assume that it is used to measure the deflection of a beam due to vibrations with significant frequencies lower than 1000 Hz. Determine the required sampling rate and include in the StrainGage.vi a median filter of Mean.vi filter.

8.5. The experimental set-up from Section 8.2.4 uses using a piezoelectric actuator and a piezolectric sensor. Could the experimental set-up be modified to use a single piezoelectric element for both sensing and actuating?

8.6. Include in the PZTexp1.vi the calculation and the display of the difference between the sine wave command sent to the actuator, voltage IN, and the voltage output of the sensor, voltage OUT.

8.7. Modify the virtual instrument for the light emitting diode and photosensor such that the LED is supplied from an analog output of maximum 2 V voltage from the DAQ board, using AO Generate Waveform.vi with a simple on/off switch on the front panel. Include in the front panel the display of the analog output voltage. Is it required to condition the analog output signal for supplying the LED?

8.8. The DC motor velocity control and Hall sensor measurement experiment uses a virtual instrument with an analog output voltage for motor velocity command and a voltage output from the Hall sensor. Would the calculation results of the difference of these voltages have any practical meaning?

8.9. Modify the DC ServoMoto-DAQ.vi to include the calculation and the display of the position error.

8.10. Modify the MATLAB program GetsoundHanningWindow.m to calculate and display the difference between the actual signal and the windowed signal as well as the calculation of the maximum amplitude of the amplitude spectrums.

References

[1] C.W. De Silva, *Control Sensors and Actuators*. Prentice Hall, 1989.

[2] D.M. Auslander, C.J. Kempf, *Mechatronics: Mechanical System Interfacing*. Prentice Hall, 1996.

[3] J.W. Dally, W.F. Riley, K. G. McConnell, *Instrumentation for Engineering Measurement*. John Wiley, 1984.

[4] R. Isermann, *Mechatronic Systems, A Challenge for Control Engineering*. AACC Conference, 1997.

[5] K. Ohnishi, M. Shibata, and T. Muramaki, *Motion Control for Advanced Mechatronics, IEEE/ASME Trans. on Mechatronics*. Vol. 1, No. 1, March 1996, 56–67.

[6] W. Stadler, *Analytical Robotics and Mechatronics*. McGraw-Hill, 1995.

[7] M. Jufer, *Traducteurs Electromécaniques*. Editions Georgi, 1979.

[8] I. Cochin, H. J. Plass, *Analysis and Design of Dynamic Systems*. Harper & Row, 1990.

[9] V. Del Toro, *Electrical Engineering Fundamentals*. Prentice Hall, 1986

[10] J.K. Eaton, L. Eaton, LabTutor, *A Friendly Guide to Computer Interfacing and LabVIEW Programming*. Oxford University Press, 1995.

[11] *The McGraw-Hill Computer Handbook*. Edited by H. Helms, McGraw-Hill Book Company, 1983.

[12] *DAQPad™ – MIO – 16XE-50, User Manual, 16-Bit Data Acquisition and Control for Parallel Port*. National Instruments Corp., 1995.

[13] Y, Dote, *Servo Motor and Motion Control Using Digital Signal Processors*. Texas Instruments, 1990.

[14] OMEGA, *Universal Guide to Data Acquisition and Computer Interfaces™*. Omega Engineering Inc., http://www.omega.com/pdf/.

[15] R. Helsel, *Visual Programming with HP VEE*. Prentice Hall PTR, 1996.

[16] OMEGA, *The Temperature and The Pressure, Strain and Force Handbooks*. Omega Engineering Inc., http://www.omega.com/pdf/.

[17] R. Figliola, D. Bealey, *Theory and Design for Mechanical Measurements*. J. Wiley, 1995.

[18] *The Encyclopedia of Inventions*. Galahad Books, 1977.

[19] E.O. Doebelin, *Measurement Systems*. Fourth edition, McGraw-Hill, 1990.

[20] *Process Instruments and Controls Handbook*. Edited by D. M. Considine, McGraw-Hill, 1985.

[21] H. Bühler, *Mecatronique*. EPFL, 1992.

[22] Sensor Technology Ltd., *Piezoelectric Ceramics Catalogue*. Callingwood, Canada www.sensortech.ca.

[23] T. Sokira, W. Jaffe, *Brushless DC Motors. Electronic Commutation and Control*. Tab Books, 1990.

[24] W. Snyder, *Industrial Robots. Computer Interfacing and Control*. Prentice Hall, 1985.

[25] Apex. *Power Integrated Circuits*. Vol. 7, Apex Microtechnology, 1996.

[26] K. Ogata, *Modern Control Engineering*. Prentice Hall, 1997.

[27] G. Olson, G. Piani, *Computer Systems for Automation and Control*. Prentice Hall, 1992.

[28] V. Del Toro, *Basic Electric Machines*. Prentice Hall, 1990.

[29] W. Bolton, *Mechatronics*. Addison Wesley Longman, 1999.

[30] D. Shetty, R. Kolk, *Mechatronics System Design*. PWS, 1997.

[31] E. Doebelin, *System Dynamics*. Marcel Dekker, 1998.

[32] J. Bollinger, N. Duffie, *Computer Control of Machines and Processes*. Addison Wesley, 1989.

[33] S. Schwarz, W. Oldham, *Electrical Engineering*. HRW, 1984.

[34] D. Hanselman, B. Littlefield, *The Student Edition of MATLAB®, Version 5, User's Guide*. Prentice Hall, 1997,

[35] J. Dabney, T. Harman, *The Student Edition of SIMULINK®, User's Guide*. Prentice Hall, 1998.

[36] *LabVIEW Data Acquisition Basics Manual*. National Instruments, 1998.

[37] P. W. Ross, *The Handbook of Software for Engineers and Scientists*. CRC Press, 1996.

[38] J. Rumbach et al. *Object Oriented Modeling and Design*. Prentice Hall, 1991.

[39] R. Bishop, *Learning with LabVIEW*. Addison Wesley, 1998.

[40] P.A. Fishwick, *Simulation Model Design and Execution*. Prentice Hall, 1995.

[41] F. Cellier, *Continuous System Modeling*. Springer Verlag, 1996.

[42] E. Part-Enander et al., *The MATLAB Handbook*. Addison-Wesley, 1996.

[43] L. Ljung, T. Glad, *Modeling of Dynamic Systems*. Prentice Hall, 1994.

[44] D.J. Murray-Smith, *Continuous System Simulation*. Chapman & Hall, 1995.

[45] M.H. MacDougall, *Simulating Computer Systems*. MIT Press, 1987.

[46] W. Kreutzer, *System Simulation. Programming Styles and Languages*. Addison-Wesley, 1986.

[47] S.B. Gershwin, *Manufacturing Systems Engineering*. PTR Prentice Hall, 1994.

[48] F. Neelamkavil, *Computer Simulation and Modeling*. John Wiley, 1987.

[49] P. Ramalingam, *System Analysis for Managerial Decisions*. John Wiley, 1976.

[50] S.M. Lee, L. J. Moore, B.W. Taylor, *Management Science*. WCB, 1981.

[51] *Handbook of Industrial Engineering*. Edited by G. Salvendy, John Wiley, 1982.

[52] P.L. Meyer, *Introductory Probability and Statistical Applications*. Addison-Wesley, 1970.

[53] D.M. Etter, *Engineering Problem Solving with MATLAB*. Prentice Hall, 1993.

[54] I. Pohl, *Object-Oriented Programming Using C++* . Benjamin/Cummings Publ. Comp., 1993.

[55] *Introduction to C++*. Microsoft Corp., 1996.

[56] D. P. Sydow, *Jumping to Java. Fast Track for C and C++ Programmers*. IDG Books, 1996.

[57] S.Gupta, J. Gupta, *PC Interfacing and Process Control*. Instrument Society of America, 1994.

[58] E. Mitchell, J. Gauthier, *ACSL: Advanced Continuous Simulation Language-User Guide and Reference Manual*. Massachusetts: Mitchell&Gauthier Assoc., 1986.

[59] H. Elmqvist, *SIMNON-An Interactive Simulation Program for Nonlinear Systems*. Report, Department of Automatic Control, Lund Institute of Technology, 1975.

[60] G. Korn, *Interactive Dynamic System Simulation*. McGraw-Hill, 1989.

[61] L. Wells, J. Travis, *LabVIEW for Everyone*. Prentice Hall, 1997.

[62] Hewlett Packart, *Exploring HP VEE*. 1995.

[63] *SYSTEM-BUILD User's Manual*. Integrated Systems Inc., 1984.

[64] *PSpice User's Manual*. MicroSim, 1987.

[65] F. Cellier, "Simulation Software: Today and Tomorrow" *in Simulation in Engineering Sciences*. Edited by J. Burger and Y. Jarny, Elsevier Science Publishers, 1983, 3–19.

[66] J. E. Hopcroft, *Finite Automata and Regular Expressions*. Ch. 2 in Introduction to Automata Theory, Languages and Computation, Addison-Wesley, 1979, pp. 13-24.

[67] F. Cellier, H. Elmqvist, *Automated Formula Manipulation Supports Object-Oriented Continuous-System Modeling, IEEE Control Systems*. 1993, 28–38.

[68] H. Elmqvist, F. Cellier, M Otter, *Object-Oriented Modeling of Hybrid Systems*. ESS'93 European Simulation Symposium, Delft, The Netherland, Oct. 25–28, 1993, 1–11.

[69] H. Elmqvist, M. Otter, F. Cellier, *Inline Integration: A New Mixed Symbolic/Numeric Approach for Solving Differential-Algebraic Equation Systems*. Proc. 1995 European Simulation Multiconference, Prague, 5–7 June, 1995, pp. XXIII–XXXIV.

[70] V. Rao, H. Rao, *C++ Neural Networks and Fuzzy Logic*. MIS Press, 1995.

[71] *Working Model, Demonstration Guide and Tutorial*. Knowledge Revolution. 1995.

[72] *LabVIEW, An In-depth Look at Graphical Programming*. NI Part Number 350150D-01, National Instruments, Dec. 1997.

[73] S. Mattson, M. Andersson, K. Astrom, "Object-Oriented Modeling and Simulation" *in CAD for Control Systems*. edited by D. Linkens, Marcel Dekker, 1993, 31–68.

[74] B. Simeon, C. Fuhrer, P. Rentrop, *Differential-Algebraic Equations in Vehicle System Dynamics*. Surveys on Mathematics for Industry, I, 1–37.

[75] H.M. Paynter, *Analysis and Design of Engineering Systems*. MIT Press, 1960, 31–69.

[76] F. Cellier, H. Elmqvist, M. Otter, "Modeling from Physical Principles" in *The Control Handbook*. Edited by W. S. Levine, CRC Press, 1996, 99–108.

[77] M. Otter, F. Cellier, "Software for Modeling and Simulating Control Systems" in *The Control Handbook*. Edited by W. S. Levine , CRC Press, 1996, 415–428.

[78] D.S. Necsulescu, J.M. Skowronski, H. Shaban-Zanjani, "Low Speed Motion Control of a Hamiltonian System" in *Mechanics and Control*. Edited by R.S. Gutalu, Plenum Press, 1994, 201–212.

[79] B.P. Zeigler, *Object-Oriented Simulation with Hierarchical, Modular Models*. Academic Press, 1990.

[80] J. TenEyck, "Object-Oriented Programming" in *The Handbook of Software for Engineers and Scientists*. Edited by P. Ross, CRC Press, 1996, 315–339.

[81] E. Petriu, ed., *Instrumentation and Measurement Technology and Applications*. IEEE Press, 1998.

[82] R. Jamal, H. Pichlik, *LabVIEW Applications and Solutions*. Prentice Hall, 1999.

[83] J. Essick, *Advanced LabVIEW Labs*. Prentice Hall, 1999.

[84] M. Chugani et al., *LabVIEW Signal Processing*. Prentice Hall, 1998.

[85] *LabVIEW User Manual*. National Instruments, 1998.

[86] *LabVIEW Function and VI Reference Manual*. National Instruments, 1998.

[87] *G Programming Reference Manual*. National Instruments, 1998.

[88] A. Biran, M. Breiner, *MATLAB for Engineers*. Addison Wesley, 1995.

[89] Signal Processing Toolbox, The MATHWORKS, 1998.

[90] L. Cristaldi, A Ferrero, V. Piuri, *Programmable Instruments, Virtual Instruments and Distributed Measurement Systems*. IEEE Instrumentation & Measurement Magazine, September 1999, 20–27.

[91] T. O. Boucher, *Computer Automation in Manufacturing*. Chapman&Hall, 1996.

[92] P.A. Laplante, *Real-Time Systems Design and Analysis*. IEEE Press, 1993.

[93] P.H. Garrett, *Advanced Instrumentation and Computer I/O Design*. IEEE Press, 1993.

[94] R.D. Hersch, *Informatique Industrielle*. Presses polytechnique et universitaires romandes, 1997.

[95] National Instruments, *LabVIEW RT and the RT Series Real-Time DAQ Boards*. http://sine.ni.com/apps/, May 99.

[96] Intel, *Embedded Microcontrollers*. http://developer.intel.com/design/mcs96.

[97] J. Axelson, *The Microcontroller Idea Book*. LakeView Research, 1994.

[98] R. Baican, D. Necsulescu, *Applied Virtual Instrumentation*. WIT Press, 2000.

[99] Data Acquisition Toolbox for use with MATLAB, 9701v00 5/99, Mathworks, 1999.

[100] *LabVIEW Evaluation Guide*. National Instruments, 1998.

[101] xPC Target, *Real-time Rapid Prototyping and Deployment on PC*. 9798v00 11/99, Mathworks, 1999 http://www.mathworks.com/products/xpctarget/.

[102] *Real-Time Workshop 3.0*. 9400v01 06/99, Mathworks, 1999. http://www.mathworks.com/products/rtw/

[103] *Real-Time Windows Target*. 9677v00 2/99, Mathworks, 1999 http://www.mathworks.com/products/rtwt/.

[104] M. Tooley, *PC-based Instrumentation and Control*. Newnes, 1995.

[105] C. Kuhnel, K. Zahnert, *BASIC Stamp*. Newnes, 1997.

[106] E.A, Lee, *Programmable DSPs: A Brief Overview*. IEEE Micro, Oct, 1990, pp. 14-16.

[107] Texas Instruments, *Floating Point DSP Family TMS320C3x*. http://www-s.ti.com/cgi-bin/sc/family3.cgi?family=TMS320C3X+FLOATING+POINT+DSP.

[108] dSPACE, *Solutions for Control*. Catalog 2000, www.dspaceinc.com.

[109] *RealSim PC AC-104*. www.isi.com/products.

[110] A. Burns, A. Welling, *Real-Time Systems and Programming Languages*. Addison-Wesley, 1997.

[111] R. Jamal, H. Illig, *Die Leichtigkeit des Echtzeitprogrammierens*. Design & Electronik, Febr. 1999, 24–26.

[112] National Instruments, *The Measurement and Automation Catalog 2000*. http://www.ni.com/catalog/.

[113] *Tornado Development system*. Wind River, http://www.wrs.com/products/html/tornado2.html.

[114] J. Peatman, *Design with Microcontrollers*. McGrawHill, 1988.

[115] *Parallax Product Catalog 2000*. http://www.parallaxinc.com.

[116] T. Schafer, M. Chevalier, *Distributed Motor Control Using the 80C196KB, Application Note AP-428*. Intel, Dec. 1993.

[117] Radio Shack, *Answers Catalogue*. www.radioshack.ca.

[118] Feedback Instruments, *Catalog*. www.fbk.com.

[119] Feedback Instruments, *Modular Instructional Servosytem*. MS 150, www.fbk.com/products/Control and Instrumentation/MS150.

[120] A. Tzes et al., *Recent Developments in a Mechatronics-Oriented Design Project Laboratory*. IEEE Control Systems, February 1997, pp. 72-79.

[121] NS. Hardman, *A Reconfigurable Testbed for Elastically-Coupled Systems*. M.A. Sc Thesis, University of Washington, 1997.

[122] LabCar White Paper, *Hardware-in-the-Loop Test Systems for Electronic Control Units*. ETAS Inc., 1999.

[123] D.S. Necsulescu, R. Jassemi-Zargani, S. Kalaycioglu, "Dual Arm Operational Space Control Using Efficient Impedance Modulation," 8th CASI Conference, Nov. 1994, Ottawa, pp. 483-491.

[124] R. Jassemi-Zargani, *Impedance Control of a Dual-Arm Robot*. PhD. Thesis, Dept. of Mech. Eng., University of Ottawa, 1998.

[125] V. Lonmo, D. Necsulescu, G. Vukovich, *Experimental Verification of Operational Space Control Strategies for Autonomous Mobile Robots*. Canadian Conference on Electrical and Computer Engineering, Sept. 1995.

[126] V. Lonmo, *Dynamics Based Control of a Mobile Robot with Non-holonomic Constraints*. M.A. Sc. Thesis, University of Ottawa, 1996.

[127] *EA-Series Strain Gauges*. Measurement Group Inc.

[128] *Digi-Key Catalogue*. www.digikey.com.

[129] *Cherry Corporation Catalogue*. www.cherrycorp.com.

[130] *National Semiconductor Catalogue*. www.national.com.

[131] *Metatorque Motor system, User's Manual*. Motion and Control Technology NSK NIPPON SEIKO K.K, Japan, 1989.

[132] *NAS Series Accelerometer Notes*. Nova Sensor, CA 1989,

[133] *V-ACCESS Capacitive Accelerometer*. ACCESS Sensors SA, Switzerland, 1992.

[134] *Pitmo 14202 Series Servomotors*. The Pittman Corporation, PENN Eng & Mfg Corp., Harleysville, PA 19438.

[135] *PA228-001 Amplifier*. Cleveland Machine Controls.

[136] S.A. Nasar, *Electric Machines and Electromechanics*. Schaum's Outline Series, McGraw-Hill, 1981.

[137] *Pittman Catalogue*. http://www.pittmannet.com/quick_index.html.

[138] *Parker Hannifin Catalogue*. http://www.parker.com/parkersql.

[139] K. Ogata, *System Dynamics*, Prentice Hall, 1992.

[140] *Using Simulink and Stateflow in Automotive Applications*. Simulink-Stateflow Technical Examples, Mathworks, 9521v00, 2/98, http://www.mathworks.com/.

[141] *Nanomotion Catalogue*. Nanomotion Ltd, http://www.nanomotion.net.

Index

T

U

V

W

X